AF621005

Birkhäuser

Operator Theory: Advances and Applications

Volume 247

Founded in 1979 by Israel Gohberg

Editors:
Joseph A. Ball (Blacksburg, VA, USA)
Harry Dym (Rehovot, Israel)
Marinus A. Kaashoek (Amsterdam, The Netherlands)
Heinz Langer (Wien, Austria)
Christiane Tretter (Bern, Switzerland)

Associate Editors:
Vadim Adamyan (Odessa, Ukraine)
Wolfgang Arendt (Ulm, Germany)
Albrecht Böttcher (Chemnitz, Germany)
B. Malcolm Brown (Cardiff, UK)
Raul Curto (Iowa, IA, USA)
Fritz Gesztesy (Columbia, MO, USA)
Pavel Kurasov (Stockholm, Sweden)
Vern Paulsen (Houston, TX, USA)
Mihai Putinar (Santa Barbara, CA, USA)
Ilya M. Spitkovsky (Williamsburg, VA, USA)

Honorary and Advisory Editorial Board:
Lewis A. Coburn (Buffalo, NY, USA)
Ciprian Foias (College Station, TX, USA)
J.William Helton (San Diego, CA, USA)
Thomas Kailath (Stanford, CA, USA)
Peter Lancaster (Calgary, Canada)
Peter D. Lax (New York, NY, USA)
Donald Sarason (Berkeley, CA, USA)
Bernd Silbermann (Chemnitz, Germany)
Harold Widom (Santa Cruz, CA, USA)

Subseries
Linear Operators and Linear Systems

Subseries editors:
Daniel Alpay (Beer Sheva, Israel)
Birgit Jacob (Wuppertal, Germany)
André C.M. Ran (Amsterdam, The Netherlands)

Subseries
Advances in Partial Differential Equations

Subseries editors:
Bert-Wolfgang Schulze (Potsdam, Germany)
Michael Demuth (Clausthal, Germany)
Jerome A. Goldstein (Memphis, TN, USA)
Nobuyuki Tose (Yokohama, Japan)
Ingo Witt (Göttingen, Germany)

Tirthankar Bhattacharyya • Michael A. Dritschel
Editors

Operator Algebras and Mathematical Physics

24th International Workshop in Operator Theory and its Applications, Bangalore, December 2013

Birkhäuser

Editors
Tirthankar Bhattacharyya
Department of Mathematics
Indian Institute of Science
Bangalore, India

Michael A. Dritschel
School of Mathematics and Statistics
Newcastle University
Newcastle upon Tyne, UK

ISSN 0255-0156 ISSN 2296-4878 (electronic)
Operator Theory: Advances and Applications
ISBN 978-3-319-35743-0 ISBN 978-3-319-18182-0 (eBook)
DOI 10.1007/978-3-319-18182-0

Mathematics Subject Classification (2010): 47-06, 46-06

Springer Cham Heidelberg New York Dordrecht London
© Springer International Publishing Switzerland 2015
Softcover re-print of the Hardcover 1st edition 2015
This work is subject to copyright. All rights are reserved by the Publisher, whether the whole or part of the material is concerned, specifically the rights of translation, reprinting, reuse of illustrations, recitation, broadcasting, reproduction on microfilms or in any other physical way, and transmission or information storage and retrieval, electronic adaptation, computer software, or by similar or dissimilar methodology now known or hereafter developed.
The use of general descriptive names, registered names, trademarks, service marks, etc. in this publication does not imply, even in the absence of a specific statement, that such names are exempt from the relevant protective laws and regulations and therefore free for general use.
The publisher, the authors and the editors are safe to assume that the advice and information in this book are believed to be true and accurate at the date of publication. Neither the publisher nor the authors or the editors give a warranty, express or implied, with respect to the material contained herein or for any errors or omissions that may have been made.

Printed on acid-free paper

Springer International Publishing AG Switzerland is part of Springer Science+Business Media (www.birkhauser-science.com)

Contents

Appendix

Preface

In this volume, we are glad to present a collection of research papers and expository articles that mainly follow the lectures given at the XXIVth International Workshop on Operator Theory and its Applications (IWOTA) held at the Indian Institute of Science, Bangalore in December, 2013. The proceedings of the XXIVth IWOTA contains papers mainly in two areas: operator algebras and mathematical physics. The papers on operator algebras include works on Čebyšev subspaces of C^*-algebras, operator spaces and quantum group actions on von Neumann algebras. The papers on mathematical physics are on Krein's trace formulae and coherent states as well as on Schrödinger operators on hypersurfaces.

Many referees worked very hard and contributed significantly to the quality of the papers and hence our special thanks go to them. The large gathering of colleagues from different parts of the world in diverse fields related to operator theory that happened at Bangalore from 16 to 20 December, 2013 would not have been possible without the financial support of the following organizations:

National Science Foundation, USA;
Indo-US Science and Technology Forum;
International Centre for Theoretical Physics;
National Board for Higher Mathematics, India;
and
Indian Statistical Institute.

The organizers are thankful for their generous financial support.

As is the tradition, the proceedings of this IWOTA appears as a volume in the series Operator Theory: Advances and Applications published by Springer Basel. We appreciate the smooth interaction we had. It was a pleasure to work together with Springer Basel for this volume. Finally, it was a great learning experience to organize the XXIVth IWOTA and we thank all participants for the opportunity.

Tirthankar Bhattacharyya
Department of Mathematics
Indian Institute of Science
Bangalore 560012, India

Michael A. Dritschel
School of Mathematics and Statistics
Newcastle University
NE1 7RU, United Kingdom

Operator Theory:
Advances and Applications, Vol. 247, 1–36
© Springer International Publishing Switzerland 2015

Trace Formulae in Operator Theory

Arup Chattopadhyay and Kalyan B. Sinha

Abstract. In this article, we survey several kinds of trace formulas that one encounters in the theory of single and multi-variable operators. We give some sketches of the proofs, often based on the principle of finite-dimensional approximations to the objects at hand in the formulas.

Mathematics Subject Classification (2010). 47A13, 47A55, 47A56.

Keywords. Trace formula, spectral shift function, perturbations of self-adjoint operators, spectral integral, Stokes formula, multiple spectral integral.

1. Introduction

In the context of operator theory, trace formulas have arisen on many occasions and in many forms, some have been driven by intuitions arising from physical sciences and some others by mathematical curiosity and their beauty. Here we try to look at a brief survey of some aspects of only two of them, namely, Krein's formula and the Helton–Howe formula. Furthermore, we present a short introduction to a possible trace formula in multi- (more specifically 2-) variable operator theory. Some of these formulae have natural geometric (or index-theoretic) connections, however, any discussion of these aspects is outside the scope of this article.

Notation: In the following, we shall use the notations given below: $\mathcal{H}$, $\mathcal{B}(\mathcal{H})$, $\mathcal{B}_1(\mathcal{H})$, $\mathcal{B}_2(\mathcal{H})$, $\mathcal{B}_p(\mathcal{H})$ denote a separable Hilbert space, a set of bounded linear operators, a set of trace class operators, and a set of Hilbert–Schmidt operators and Schatten-p class operators respectively with $\|.\|_p$ as the associated Schatten norm. Furthermore by $\sigma(A)$, $E_A(\lambda)$, $D(A)$, $\rho(A)$, $R_z \equiv (A-z)^{-1}$ we shall mean a spectrum, a spectral family, a domain, a resolvent set, and the resolvent of a self-adjoint operator A respectively; and by $C(X)$, the Banach space of continuous functions over a compact topological space X and $L^p(Y)$, the standard Lebesgue space. Finally, $\mathrm{Tr}(A)$ will denote the trace of a trace class operator A in $\mathcal{H}$, and $D^{(k)}\phi(A) \bullet (\cdot)$ is the kth-order Fréchet derivative of ϕ at the self-adjoint operator A as a k-linear form on $\otimes^k \mathcal{B}(\mathcal{H})$.

Based partly on a talk given by the second author in the IWOTA conference held in Bangalore, India in December 2013.

2. Krein's theorem

Krein's original proof ([26], [27]; see also [40]) uses properties of perturbation determinants and the integral representation of holomorphic functions on the upper half-plane with a bounded positive imaginary part. In 1985, Voiculescu [47] approached the trace formula from a different direction and gave an alternative proof without using function theory for the case of bounded self-adjoint operators. Later Sinha and Mohapatra [40] extended these ideas to the unbounded self-adjoint and unitary cases [41]. There is also the interesting approach of Birman and Solomyak ([4], [5], [8]) using the theory of double operator integrals. More recently there has been an article by Potapov, Sukochev and Zanin [34] giving yet another proof of Krein's theorem. Let us begin by stating the theorem and its corollary in a finite-dimensional Hilbert space, the proof of which uses the minimax principles for eigenvalues [20].

Theorem 2.1. *Let H and H_0 be two self-adjoint operators in a finite-dimensional Hilbert space $\mathcal{H}$ in which $E_H(\lambda)$ and $E_{H_0}(\lambda)$ are the spectral families of H and H_0 respectively. Then there exists a unique real-valued bounded function ξ such that*

(i) $\xi(\lambda) = \mathrm{Tr}\{E_{H_0}(\lambda) - E_H(\lambda)\},\ \lambda \in \mathbb{R};$

(ii) $\int_{\mathbb{R}} \xi(\lambda) d\lambda = \mathrm{Tr}\,(H - H_0)\,,$

(iii) *for $\phi \in C^1(\mathbb{R})$ (set of all once continuously differentiable functions on $\mathbb{R}$),*

$$\mathrm{Tr}\,[\phi(H) - \phi(H_0)] = \int_{\mathbb{R}} \phi'(\lambda)\xi(\lambda)d\lambda. \tag{2.1}$$

Furthermore, ξ has support in $[a, b]$, where $a = \min\{\inf\sigma(H)\ , \inf\sigma(H_0)\}$, $b = \max\{\sup\sigma(H),\ \sup\sigma(H_0)\}$.

(iv) *If $H - H_0 = \tau|g\rangle\langle g|$ with $\tau > 0$, $\|g\| = 1$ (we have used Dirac notation for rank one perturbations), then ξ is a $\{0, 1\}$-valued function. More precisely,*

$$\xi(\lambda) = \sum_{j=1}^{r} \chi_{\Delta_j}(\lambda) \text{ for } r \text{ disjoint intervals } \Delta_j \subset \mathbb{R},\ 1 \leq r \leq n.$$

Corollary 2.1. For $t \in \mathbb{R}$, $\mathrm{Tr}\left(e^{itH} - e^{itH_0}\right) = it\int e^{it\lambda}\,\xi(\lambda)\,d\lambda$.

In an infinite-dimensional Hilbert space, the relation $\xi(\lambda) = \mathrm{Tr}\{E_{H_0}(\lambda) - E_H(\lambda)\}$ will not make sense in general because $E_{H_0}(\lambda) - E_H(\lambda)$ may not be trace-class. Next we give a counter-example due to Krein [26] where $H - H_0$ is rank one and yet $E_H(\lambda) - E_{H_0}(\lambda)$ is not trace-class.

Example. Let $\mathcal{H} = L^2[0, \infty)$ and $L = -\frac{d^2}{dx^2}$ be the differential operator with $D(L) = C_0^\infty(0, \infty)$. It is known that L has several self-adjoint extensions depending upon the boundary conditions on the corresponding differential equation. Of them we choose two, namely h_0 and h as

$$D(h_0) = \left\{ \begin{array}{l} f \in \mathcal{H} : f \text{ and } f' \text{ are absolutely continuous,} \\ f'' \in \mathcal{H} \text{ and } f(0) = 0 \end{array} \right\}$$

and

$$D(h) = \left\{ \begin{array}{l} f \in \mathcal{H} : f \text{ and } f' \text{ are absolutely continuous,} \\ f'' \in \mathcal{H} \text{ and } f'(0) = 0. \end{array} \right\}.$$

Note that both h_0 and h are positive operators, and one computes the spectral families $F_{h_0}(\lambda)$ and $F_h(\lambda)$ of h_0 and h respectively by solving the associated ordinary differential equations and get, for $\lambda \geq 0$,

$$F_{h_0}(\lambda)(x,y) = \frac{2}{\pi}\int_0^{\sqrt{\lambda}} \sin tx\ \sin ty\ dt \ \text{ and } \ F_h(\lambda)(x,y) = \frac{2}{\pi}\int_0^{\sqrt{\lambda}} \cos tx\ \cos ty\ dt. \tag{2.2}$$

Let $H_0 = (h_0 + I)^{-1}$ and $H = (h + I)^{-1}$. Green's functions associated with $(h_0 + I)^{-1}$ and $(h + I)^{-1}$ are

$$G_0(x,y) = \begin{cases} e^{-y}\sinh\ x, & \text{if } x \leq y, \\ e^{-x}\sinh\ y, & \text{if } x \geq y, \end{cases} \ \text{ and } \ G(x,y) = \begin{cases} e^{-y}\cosh\ x, & \text{if } x \leq y, \\ e^{-x}\cosh\ y, & \text{if } x \geq y \end{cases}$$

respectively. Then $H - H_0 = \frac{1}{2}|\psi\rangle\langle\psi|$, where $\psi(x) = \sqrt{2}\ e^{-x}$ so that $||\psi|| = 1$. Let $\mu = \frac{1}{1+\lambda}$. Then $E_{H_0}(\mu) = I - F_{h_0}(\lambda)$ and $E_H(\mu) = I - F_h(\lambda)$ are the spectral families of H_0 and H respectively and

$$E_H(\mu)(x,y) - E_{H_0}(\mu)(x,y) = F_{h_0}(\lambda)(x,y) - F_h(\lambda)(x,y) = (-2/\pi)\frac{\sin\ \sqrt{\lambda}(x+y)}{(x+y)}. \tag{2.3}$$

Note that $E_H(\mu) - E_{H_0}(\mu)$ is not trace-class since the Hilbert–Schmidt norm

$$\int\int |E_\mu(x,y)\ -\ E^0_\mu(x,y)|^2 dx\, dy = (2/\pi)^2 \int_0^\infty \int_0^\infty \frac{\sin^2 \sqrt{\lambda}(x+y)}{(x+y)^2} dx dy$$

$= \infty$ if $\lambda \neq 0$, equivalently if $0 < \mu < 1$; $= 0$ if $\lambda = 0$, equivalently if $\mu = 1$. If $E_H(\mu) - E_{H_0}(\mu)$ were trace-class, then since its integral kernel (2.3) is continuous, we could evaluate the trace (see p. 523 of [24]) as:

$$\begin{aligned} \mathrm{Tr}\{E_{H_0}(\mu) - E_H(\mu)\} &= \int_0^\infty \{E_{H_0}(\mu) - E_H(\mu)\}(x,x)dx = \frac{2}{\pi}\int_0^\infty \frac{\sin\ 2\sqrt{\lambda}\ x}{2x} dx \\ &= \frac{1}{2} \text{ if } 0 < \mu < 1 \ \text{ and } = 0 \text{ if } \mu = 1. \end{aligned} \tag{2.4}$$

Our next theorem, due to Krein [27], states Krein's trace formula (2.1) in an infinite-dimensional Hilbert space $\mathcal{H}$ (see also [14], [39], [40]).

Theorem 2.2. *Let H and H_0 be two self-adjoint operators in an infinite-dimensional Hilbert space $\mathcal{H}$ such that $V = H - H_0 \in \mathcal{B}_1(\mathcal{H})$. Let $\Delta(z) = \det(I + VR^0_z)$, for $\mathrm{Im} z \neq 0$, be the perturbation determinant (see, e.g., appendix of [40]). Then there exists a unique real-valued $L^1(\mathbb{R})$-function ξ satisfying*

(i) $\xi(\lambda) = \frac{1}{\pi} \lim_{\varepsilon \to 0+} \mathrm{Im} \ln \Delta(\lambda + i\varepsilon)$,

(ii) $\int_{-\infty}^{\infty} |\xi(\lambda)|\, d\lambda \leq \|V\|_1, \quad \int_{-\infty}^{\infty} \xi(\lambda) d\lambda = \operatorname{Tr} V,$

(iii) $\ln \Delta(z) = \int_{-\infty}^{\infty} \frac{\xi(\lambda)}{\lambda - z} d\lambda \quad \text{for } \operatorname{Im} z \neq 0,$

(iv) $\operatorname{Tr}(R_z - R_z^0) = -\int_{-\infty}^{\infty} \frac{\xi(\lambda)}{(\lambda - z)^2} d\lambda$ for $\operatorname{Im} z \neq 0.$

(v) *Let $\phi \in \mathcal{K}$, where the Krein class $\mathcal{K}$ consists of functions ϕ of the form* $\phi(\lambda) = \int_{-\infty}^{\infty} \frac{e^{it\lambda} - 1}{it} \nu(dt) + C$, *for some constant C and complex measure ν on $\mathbb{R}$. Then $\phi(H) - \phi(H_0) \in \mathcal{B}_1(\mathcal{H})$ and* $\operatorname{Tr}\{\phi(H) - \phi(H_0)\} = \int_{-\infty}^{\infty} \phi'(\lambda)\xi(\lambda) d\lambda.$

Note that such a function is necessarily continuously differentiable and the derivative is the Fourier transform of the measure ν, i.e., $\phi'(\lambda) = \int e^{it\lambda}\, \nu(dt)$. It is also worth observing that $\phi(H)$ and $\phi(H_0)$ are not necessarily bounded operators though defined on $D(H) = D(H_0)$.

Krein's original proof used the representation of the Herglotz function $\Delta(z)$; however, here following the proof of Voiculescu [47], we adopt the strategy of reducing the computation of $\operatorname{Tr}\left(e^{itH} - e^{itH_0}\right)$ to that of $\operatorname{Tr}\left(e^{itH_m} - e^{itH_{0,m}}\right)$ for suitable finite-dimensional approximations H_m and $H_{0,m}$ of H and H_0 respectively, then apply Theorem 2.1 and finally go to the limit. The next theorem is an extension of Weyl–von Neumann's result (see Lemma 2.2 on p. 523 of [24]).

Theorem 2.3. *Let A be a (possibly unbounded) self-adjoint operator in $\mathcal{H}$, $f \in \mathcal{H}$ and $\epsilon > 0$, K a compact set in $\mathbb{R}$. Then there exists a finite rank projection P in $\mathcal{H}$ such that*

(i) $\|(I - P)AP\|_2 < \epsilon$ *and* $\left\|(I - P)e^{itA}P\right\|_2 < \epsilon$ *uniformly for $t \in K$.*

(ii) $\|(I - P)f\| < \epsilon.$

Proof. Let E_A be the spectral measure associated with the self-adjoint operator A, and choose $a > 0$ such that $\|[I - E_A((-a, a])]f\| < \epsilon$. For each positive integer n and $1 \leq k \leq n$, set $E_k = E_A\left(\left(\frac{2k-2-n}{n}a, \frac{2k-n}{n}a\right]\right)$ so that

$$E_k E_j = \delta_{kj} E_j \quad \text{and} \quad \sum_{k=1}^{n} E_k = E_A\left((-a, a]\right).$$

We also set for $1 \leq k \leq n$, $g_k = \frac{E_k f}{\|E_k f\|}$ if $E_k f \neq 0$, and $= 0$ if $E_k f = 0$. Clearly $\{g_k\}_{k=1}^n \in D(A)$, and $A g_k = E_k(A g_k) \in E_k\mathcal{H}$. Let P be the orthogonal projection onto the subspace generated by $\{g_1, g_2, g_3, \ldots, g_n\}$ so that $\dim P(\mathcal{H}) \leq n$. Set $\lambda_k = \frac{2k-n-1}{n} a$, then

$$\|\,(A - \lambda_k)\, g_k\|^2 = \int_{\frac{2k-n-2}{n}a}^{\frac{2k-n}{n}a} (\lambda - \lambda_k)^2 \|E_A(d\lambda) g_k\|^2 \leq \left(\frac{a}{n}\right)^2.$$

Thus,

$$PAg_k = \sum_{j=1}^{n} \langle PAg_k, g_j \rangle g_j = \sum_{j=1}^{n} \langle Ag_k, g_j \rangle g_j = \langle Ag_k, g_k \rangle g_k \in E_k \mathcal{H},$$

since $Ag_k \in E_k\mathcal{H}$ and $g_j \in E_j\mathcal{H}$. Therefore, $(I-P)Ag_k \in E_k\mathcal{H}$, and therefore for $u \in \mathcal{H}$,

$$\|(I-P)APu\|^2 = \sum_{j=1}^{\infty} |\langle u, g_j \rangle|^2 \|(I-P)(A-\lambda_j)g_j\|^2 \leq \left(\frac{a}{n}\right)^2 \|u\|^2.$$

Thus $\|(I-P)AP\| \leq \left(\frac{a}{n}\right)$ and hence $\|(I-P)AP\|_2 \leq \sqrt{n}\, \|(I-P)AP\| \leq \left(\frac{a}{\sqrt{n}}\right)$. This leads to a Gronwall-type inequality (see page 831 of [40] for details):

$$\alpha(t) \leq 2\,a \int_0^t \alpha(s)ds + T\,\frac{a}{\sqrt{n}} \quad \text{for} \quad |t| \leq T, \text{ where } \alpha(t) \equiv \left\|(I-P)\mathrm{e}^{\mathrm{i}tA}P\right\|_2,$$

and this implies that

$$\left\|(I-P)\mathrm{e}^{\mathrm{i}tA}P\right\|_2 \leq \frac{(T\,a\,\mathrm{e}^{2aT})}{\sqrt{n}}.$$

On the other hand $E_A((-a,a])f = (I-P)\sum_{k=1}^{n} E_k f = \sum_{k=1}^{n} \|E_k f\| (I-P)g_k = 0$, so that

$$\|(I-P)f\| = \|(I-P)\left[I - E_A\left((-a,a]\right)\right] f\| \; \leq \| \left[I - E_A\left((-a,a]\right)\right] f\| < \epsilon.$$

The result follows by choosing n sufficiently large. □

Lemma 2.4. *Let H and H_0 be two self-adjoint operators in $\mathcal{H}$ such that $H - H_0 = \tau|g\rangle\langle g|$; $\tau > 0$ and $\|g\| = 1$. Then there exists a sequence $\{P_n\}$ of finite rank projections in $\mathcal{H}$ such that $P_n g \longrightarrow g$ as $n \longrightarrow \infty$ and for any $T > 0$,*

$$\mathrm{Tr}\{\mathrm{e}^{\mathrm{i}tH} - \mathrm{e}^{\mathrm{i}tH_0}\} = \lim_{n\longrightarrow\infty} \mathrm{Tr}\{P_n\left[\mathrm{e}^{\mathrm{i}tP_nHP_n} - \mathrm{e}^{\mathrm{i}tP_nH_0P_n}\right]P_n\}, \tag{2.5}$$

uniformly for all t with $|t| \leq T$.

Proof. Applying Theorem 2.3 with $A = H_0$ and $f = g$, we get a sequence $\{P_n\}$ of finite rank projections in $\mathcal{H}$ such that $\left\|P_n^{\perp} H_0 P_n\right\|_2 \longrightarrow 0$ as $n \longrightarrow \infty$; $\left\|P_n^{\perp} \mathrm{e}^{\mathrm{i}tH_0} P_n\right\|_2 \longrightarrow 0$ as $n \longrightarrow \infty$ and $\|(I-P_n)g\| \longrightarrow 0$ as $n \longrightarrow \infty$. Hence

$$\begin{aligned}\left\|P_n^{\perp} H P_n\right\|_2 &\leq \left\|P_n^{\perp} H_0 P_n\right\|_2 + \tau \left\||P_n^{\perp} g\rangle\langle P_n g|\right\|_2 \\ &\leq \left\|P_n^{\perp} H_0 P_n\right\|_2 + \tau \|(I-P_n)g\|\,\|g\|,\end{aligned}$$

which converges to 0 as $n \longrightarrow \infty$. Thus

$$\begin{aligned}
&\mathrm{Tr}\{\mathrm{e}^{\mathrm{i}tH} - \mathrm{e}^{\mathrm{i}tH_0}\} - \mathrm{Tr}\{P_n\left[\mathrm{e}^{\mathrm{i}tP_nHP_n} - \mathrm{e}^{\mathrm{i}tP_nH_0P_n}\right]P_n\} \\
&= \mathrm{Tr}\left\{\int_0^1 d\alpha\, \frac{d}{d\alpha}\left(\mathrm{e}^{\mathrm{i}t\alpha H}.\mathrm{e}^{\mathrm{i}t(1-\alpha)H_0}\right)\right\} \\
&\quad - \mathrm{Tr}\left\{P_n \int_0^1 d\alpha\, \frac{d}{d\alpha}\left(\mathrm{e}^{\mathrm{i}t\alpha P_nHP_n}.\mathrm{e}^{\mathrm{i}t(1-\alpha)P_nH_0P_n}\right)P_n\right\} \\
&= \mathrm{Tr}\left\{\int_0^1 d\alpha\, \mathrm{e}^{\mathrm{i}t\alpha H}\mathrm{i}t\; \tau|g\rangle\langle g|\; \mathrm{e}^{\mathrm{i}t(1-\alpha)H_0}\right\} \\
&\quad - \mathrm{Tr}\left\{\int_0^1 d\alpha\, P_n\mathrm{e}^{\mathrm{i}t\alpha P_nHP_n}\mathrm{i}t\; P_n\tau|g\rangle\langle g|P_n\; \mathrm{e}^{\mathrm{i}t(1-\alpha)P_nH_0P_n}P_n\right\} \\
&= \tau\, \mathrm{i}t \int_0^1 d\alpha\, \mathrm{Tr}\Big\{\left[\mathrm{e}^{\mathrm{i}t\alpha H} - \mathrm{e}^{\mathrm{i}t\alpha P_nHP_n}\right]P_n|g\rangle\langle g|\; \mathrm{e}^{\mathrm{i}t(1-\alpha)H_0} \\
&\quad + \mathrm{e}^{\mathrm{i}t\alpha H}P_n^{\perp}\; |g\rangle\langle g|\; \mathrm{e}^{\mathrm{i}t(1-\alpha)H_0} + \mathrm{e}^{\mathrm{i}t\alpha P_nHP_n}\; P_n|g\rangle\langle g|P_n^{\perp}\; \mathrm{e}^{\mathrm{i}t(1-\alpha)H_0} \\
&\quad + \mathrm{e}^{\mathrm{i}t\alpha P_nHP_n}\; P_n|g\rangle\langle g|P_n\; \left[\mathrm{e}^{\mathrm{i}t(1-\alpha)H_0} - \mathrm{e}^{\mathrm{i}t(1-\alpha)P_nH_0P_n}\right]\Big\}. \qquad (2.6)
\end{aligned}$$

In the first term of the expression (2.6):

$$\begin{aligned}
&\left\|\left[\mathrm{e}^{\mathrm{i}t\alpha H} - \mathrm{e}^{\mathrm{i}t\alpha P_nHP_n}\right]P_n\right\|_2 \\
&= \left\|\mathrm{i}t\,\alpha \int_0^1 d\beta\, \mathrm{e}^{\mathrm{i}t\alpha\beta H}[H - P_nHP_n]\mathrm{e}^{\mathrm{i}t\alpha(1-\beta)P_nHP_n}P_n\right\|_2 \\
&= \left\|\mathrm{i}t\,\alpha \int_0^1 d\beta\, \mathrm{e}^{\mathrm{i}t\alpha\beta H}P_n^{\perp}HP_n\mathrm{e}^{\mathrm{i}t\alpha(1-\beta)P_nHP_n}P_n\right\|_2 \leq T\left\|P_n^{\perp}HP_n\right\|_2,
\end{aligned}$$

and hence

$$\left\|\left[\mathrm{e}^{\mathrm{i}t\alpha H} - \mathrm{e}^{\mathrm{i}t\alpha P_nHP_n}\right]P_n|g\rangle\langle g|\; \mathrm{e}^{\mathrm{i}t(1-\alpha)H_0}\right\|_1 \leq T\; \left\|P_n^{\perp}HP_n\right\|_2\; \|g\|^2,$$

which converges to 0 as $n \longrightarrow \infty$, uniformly for $|t| \leq T$. Similarly for the fourth term in (2.6), we note that

$$\begin{aligned}
&\left\|\mathrm{e}^{\mathrm{i}t\alpha P_nHP_n}\; P_n|g\rangle\langle g|P_n\; \left[\mathrm{e}^{\mathrm{i}t(1-\alpha)H_0} - \mathrm{e}^{\mathrm{i}t(1-\alpha)P_nH_0P_n}\right]\right\|_1 \\
&\leq \left\|\mathrm{e}^{\mathrm{i}t\alpha P_nHP_n}\; P_n|g\rangle\langle g|\right\|_2 \left\|P_n\; \left[\mathrm{e}^{\mathrm{i}t(1-\alpha)H_0} - \mathrm{e}^{\mathrm{i}t(1-\alpha)P_nH_0P_n}\right]\right\|_2 \\
&\leq \left\|\mathrm{e}^{\mathrm{i}t\alpha P_nHP_n}\; |P_ng\rangle\langle g|\right\|_2\; T\left\|P_n^{\perp}H_0P_n\right\|_2 \leq\; T\left\|P_n^{\perp}H_0P_n\right\|_2\|g\|^2,
\end{aligned}$$

which converges to 0 as $n \longrightarrow \infty$, uniformly for $|t| \leq T$. Similarly, the second and the third terms in (2.6) can be shown to converge to 0 as $n \longrightarrow \infty$, uniformly for $|t| \leq T$. Therefore

$$\mathrm{Tr}\{\mathrm{e}^{\mathrm{i}tH} - \mathrm{e}^{\mathrm{i}tH_0}\} = \lim_{n\longrightarrow\infty} \mathrm{Tr}\{P_n\left[\mathrm{e}^{\mathrm{i}tP_nHP_n} - \mathrm{e}^{\mathrm{i}tP_nH_0P_n}\right]P_n\},$$

uniformly for all t with $|t| \leq T$. □

Now we are in a position to prove Krein's theorem in an unbounded self-adjoint case.

Theorem 2.5. *Let H and H_0 be two self-adjoint operators in $\mathcal{H}$ such that $H - H_0 \equiv V \in \mathcal{B}_1(\mathcal{H})$. Then there exists a unique real-valued function $\xi \in L^1(\mathbb{R})$ such that*

(i) $\mathrm{Tr}\{\mathrm{e}^{\mathrm{i}tH} - \mathrm{e}^{\mathrm{i}tH_0}\} = (\mathrm{i}t)\int_{\mathbb{R}} \mathrm{e}^{\mathrm{i}t\lambda}\ \xi(\lambda)\ d\lambda$,

(ii) $\int_{\mathbb{R}} \xi(\lambda)\ d\lambda = \mathrm{Tr}(V)$ *and* $\|\xi\|_{L^1(\mathbb{R})} \leq \|V\|_1$,

(iii) *for every function $\phi \in \mathcal{K}$ (defined in* (v) *of Theorem* 2.2*), $\phi(H) - \phi(H_0) \in \mathcal{B}_1$ and* $\mathrm{Tr}\{\phi(H) - \phi(H_0)\} = \int_{-\infty}^{\infty} \phi'(\lambda)\xi(\lambda)d\lambda$,

(iv) *the function $\lambda \to (\lambda - z)^{-1}$ (with* $\mathrm{Im}z \neq 0$*) belongs to the class $\mathcal{K}$ and hence* $\mathrm{Tr}(R_z - R_z^0) = -\int (\lambda - z)^{-2}\xi(\lambda)d\lambda$.

Proof. At first we let $V \equiv \tau|g\rangle\langle g|;\ \tau > 0$ and $\|g\| = 1$. Hence by Lemma 2.4, we conclude that there exists a sequence $\{P_n\}$ of finite rank projections such that $P_n g \longrightarrow g$ as $n \longrightarrow \infty$ and

$$\mathrm{Tr}\{\mathrm{e}^{\mathrm{i}tH} - \mathrm{e}^{\mathrm{i}tH_0}\} = \lim_{n\longrightarrow\infty} \mathrm{Tr}\{P_n\left[\mathrm{e}^{\mathrm{i}tH_n} - \mathrm{e}^{\mathrm{i}tH_{0,n}}\right]P_n\},$$

where $H_n = P_n H P_n$ and $H_{0.n} = P_n H_0 P_n$, and the convergence is uniform in t for $|t| \leq T$. Note that by construction $P_n\mathcal{H} \subseteq \mathrm{Dom}(H_0) = \mathrm{Dom}(H)$ (see the proof of Theorem 2.3) and hence both H_n and $H_{0,n}$ are self-adjoint operators in the finite-dimensional Hilbert space $P_n\mathcal{H}$. By Theorem 2.1 (iv), we get a $\{0,1\}$-valued $L^1(\mathbb{R})$-function ξ_n such that

$$\mathrm{Tr}\{P_n\left[\mathrm{e}^{\mathrm{i}tH_n} - \mathrm{e}^{\mathrm{i}tH_{0,n}}\right]P_n\} = \mathrm{i}t\int_{\mathbb{R}} \mathrm{e}^{\mathrm{i}t\lambda}\ \xi_n(\lambda)\ d\lambda, \tag{2.7}$$

and hence

$$\mathrm{Tr}\{\mathrm{e}^{\mathrm{i}tH} - \mathrm{e}^{\mathrm{i}tH_0}\} = \mathrm{i}t\lim_{n\longrightarrow\infty}\int_{\mathbb{R}} \mathrm{e}^{\mathrm{i}t\lambda}\ \xi_n(\lambda)\ d\lambda, \tag{2.8}$$

the convergence being uniform in t for $|t| \leq T$. Since $t \longrightarrow \mathrm{e}^{\mathrm{i}tH_n},\ \mathrm{e}^{\mathrm{i}tH_{0,n}}$ are norm continuous in $P_n\mathcal{H}$ and P_nVP_n is rank one, we have from (2.7), by using the bounded convergence theorem that

$$\begin{aligned}
&\int_{\mathbb{R}} \xi_n(\lambda)\ d\lambda = \lim_{t\longrightarrow 0}\int_{\mathbb{R}} \mathrm{e}^{\mathrm{i}t\lambda}\ \xi_n(\lambda)\ d\lambda = \lim_{t\longrightarrow 0}\frac{1}{\mathrm{i}t}\ \mathrm{Tr}\left\{P_n\left[\mathrm{e}^{\mathrm{i}tH_n} - \mathrm{e}^{\mathrm{i}tH_{0,n}}\right]P_n\right\}\\
&= \lim_{t\longrightarrow 0}\frac{1}{\mathrm{i}t}\ \mathrm{Tr}\left\{P_n\int_0^t ds\ \frac{d}{ds}\left(\mathrm{e}^{\mathrm{i}sH_n}.\mathrm{e}^{\mathrm{i}(t-s)H_{0,n}}\right)P_n\right\}\\
&= \lim_{t\longrightarrow 0}\frac{1}{t}\int_0^t ds\ \mathrm{Tr}\{P_n\mathrm{e}^{\mathrm{i}sH_n}\ P_nVP_n\ \mathrm{e}^{\mathrm{i}(t-s)H_{0,n}}P_n\} = \mathrm{Tr}\{P_nVP_n\}\\
&= \tau\|P_ng\|^2 = \tau(1 - \|P_n^{\perp}g\|^2) > \tau(1-\epsilon^2) > 0,
\end{aligned} \tag{2.9}$$

where for a given $\epsilon > 0$, we have chosen a natural number $N \in \mathbb{N}$ such that $\|P_n^{\perp}g\| < \epsilon\ \forall\ n \geq N$. Setting

$$\mu_n(\Delta) = \frac{1}{\tau\|P_ng\|^2}\int_{\Delta} \xi_n(\lambda)\ d\lambda$$

for every Borel set $\Delta \subseteq \mathbb{R}$, we have a family $\{\mu_n\}$ of probability measure by (2.9), the Fourier transform of which by (2.8)

$$\hat{\mu}_n(t) = \int_{\mathbb{R}} \mathrm{e}^{\mathrm{i}t\lambda}\ \mu_n(d\lambda) = \frac{1}{\tau\|P_n g\|^2} \int_{\mathbb{R}} \mathrm{e}^{\mathrm{i}t\lambda}\ \xi_n(\lambda)\ d\lambda,$$

converges to $\frac{1}{it\tau}\mathrm{Tr}\{\mathrm{e}^{\mathrm{i}tH} - \mathrm{e}^{\mathrm{i}tH_0}\} \equiv \hat{\mu}(t)$ uniformly in t in compact sets in $\mathbb{R}\backslash\{0\}$, as $n \longrightarrow \infty$.

On the other hand $\hat{\mu}(0) = \frac{1}{\tau\|P_n g\|^2}\int_{\mathbb{R}} \xi_n(\lambda)\ d\lambda = 1$ for all $n \in \mathbb{N}$ and thus

$$\begin{aligned}
\lim_{t\longrightarrow 0} \hat{\mu}(t) &= \lim_{t\longrightarrow 0} \frac{1}{\mathrm{i}t\tau}\ \mathrm{Tr}\ \left\{\left[\mathrm{e}^{\mathrm{i}tH} - \mathrm{e}^{\mathrm{i}tH_0}\right]\right\} \\
&= \lim_{t\longrightarrow 0} \frac{1}{t\tau} \int_0^t ds\ \mathrm{Tr}\{\mathrm{e}^{\mathrm{i}sH}\ V\ \mathrm{e}^{\mathrm{i}(t-s)H_0}\} = \frac{1}{\tau}\ \mathrm{Tr}\{V\} = 1 \equiv \hat{\mu}(0),
\end{aligned}$$

by definition.

Thus by the Lévy–Cramér continuity theorem [30], there exists a probability measure μ on $\mathbb{R}$ such that $\mu_n \longrightarrow \mu$ weakly, i.e.,

$$\int_{\mathbb{R}} \phi(\lambda)\ \mu_n(d\lambda) \longrightarrow \int_{\mathbb{R}} \phi(\lambda)\ \mu(d\lambda)$$

for every bounded continuous function ϕ.

Let $\Delta = (a,b] \subseteq \mathbb{R}$ and let $\{\phi_n\}$ be a sequence of smooth functions of support in $(a - \frac{1}{n}, b + \frac{1}{n}]$ such that $0 \leq \phi_n \leq 1$ and $\|\chi_\Delta - \phi_n\|_1 \longrightarrow 0$ as $n \longrightarrow \infty$, where χ_Δ is the characteristic function of Δ. Choosing a subsequence if necessary and using the bounded convergence theorem, we have that

$$\lim_{n\longrightarrow\infty} \lim_{m\longrightarrow\infty} \int_{\mathbb{R}} \phi_n(\lambda)\ \mu_m(d\lambda) = \lim_{n\longrightarrow\infty} \int_{\mathbb{R}} \phi_n(\lambda)\ \mu(d\lambda) = \mu(\Delta).$$

Thus

$$\begin{aligned}
\mu(\Delta) &= \lim_{n\longrightarrow\infty} \lim_{m\longrightarrow\infty} \frac{1}{\tau\|P_m g\|^2} \int_{\mathbb{R}} \phi_n(\lambda)\ \xi_m(\lambda)\ (d\lambda) \\
&= \frac{1}{\tau} \lim_{n\longrightarrow\infty} \lim_{m\longrightarrow\infty} \int_{\mathbb{R}} \phi_n(\lambda)\ \xi_m(\lambda)\ (d\lambda) \\
&\leq \tau^{-1} \lim_{n\longrightarrow\infty} \int_{a-\frac{1}{n}}^{b+\frac{1}{n}} \phi_n(\lambda)\ d\lambda = \tau^{-1}(b-a),
\end{aligned}$$

since $0 \leq \xi_m(\lambda) \leq 1$ for all m and all λ.

This shows that μ is absolutely continuous and we set $\xi(\lambda) = \tau\frac{\mu(d\lambda)}{d\lambda}$ so that ξ is a non-negative L^1-function and

$$\hat{\mu}(t) = \int_{\mathbb{R}} \mathrm{e}^{\mathrm{i}t\lambda}\ \mu(d\lambda) = \tau^{-1} \int_{\mathbb{R}} \mathrm{e}^{\mathrm{i}t\lambda}\ \xi(\lambda)\ d\lambda;\ \mathrm{Tr}\{\mathrm{e}^{\mathrm{i}tH} - \mathrm{e}^{\mathrm{i}tH_0}\} = (\mathrm{i}t) \int_{\mathbb{R}} \mathrm{e}^{\mathrm{i}t\lambda}\ \xi(\lambda)\ d\lambda. \tag{2.10}$$

Also dividing both sides of (2.10) by it and taking the limit as $t \longrightarrow 0$, we get that

$$\begin{aligned}\int_{\mathbb{R}} \xi(\lambda)\, d\lambda &= \lim_{t\longrightarrow 0} \int_{\mathbb{R}} \mathrm{e}^{\mathrm{i}t\lambda}\, \xi(\lambda)\, d\lambda = \lim_{t\longrightarrow 0} \frac{1}{\mathrm{i}t} \operatorname{Tr}\left\{\left[\mathrm{e}^{\mathrm{i}tH} - \mathrm{e}^{\mathrm{i}tH_0}\right]\right\} \\ &= \lim_{t\longrightarrow 0} \frac{1}{\mathrm{i}t} \operatorname{Tr}\left\{\int_0^t ds\, \frac{d}{ds}\left(\mathrm{e}^{\mathrm{i}sH}.\mathrm{e}^{\mathrm{i}(t-s)H_0}\right)\right\} \\ &= \lim_{t\longrightarrow 0} \frac{1}{t} \int_0^t ds\, \operatorname{Tr}\{\mathrm{e}^{\mathrm{i}sH}\, V\, \mathrm{e}^{\mathrm{i}(t-s)H_0}\} = \operatorname{Tr}\{V\} = \tau \geq 0.\end{aligned}$$

That is $\|\xi\|_{L^1} = |\tau|$, since ξ is non-negative and $\tau \geq 0$.

If V is rank one and negative, i.e., if $V = \tau|g\rangle\langle g|$ ($\tau < 0$ and $\|g\| = 1$), then $H_0 - H = -\tau|g\rangle\langle g|$ with $-\tau > 0$ and we obtain, similarly as above, a non-negative L^1-function η such that

$$\operatorname{Tr}\{\mathrm{e}^{\mathrm{i}tH_0} - \mathrm{e}^{\mathrm{i}tH}\} = (\mathrm{i}t) \int_{\mathbb{R}} \mathrm{e}^{\mathrm{i}t\lambda}\, \eta(\lambda)\, d\lambda \quad \text{and} \quad \int_{\mathbb{R}} \eta(\lambda)\, d\lambda = \operatorname{Tr}\left(-\tau|g\rangle\langle g|\right) = -\tau.$$

Defining $\xi(\lambda) = -\eta(\lambda)$, we get

$$\operatorname{Tr}\{\mathrm{e}^{\mathrm{i}tH} - \mathrm{e}^{\mathrm{i}tH_0}\} = (\mathrm{i}t) \int_{\mathbb{R}} \mathrm{e}^{\mathrm{i}t\lambda}\, \xi(\lambda)\, d\lambda \quad \text{and} \quad \int_{\mathbb{R}} \xi(\lambda)\, d\lambda = \tau; \ \|\xi\|_{L^1} = \|\eta\|_{L^1} = |\tau|,$$

hence the relation (2.10) is valid for all V rank one with some real-valued L^1-function ξ such that $\int_{\mathbb{R}} \xi(\lambda)\, d\lambda = \operatorname{Tr}(V)$ and $\|\xi\|_{L^1} \leq \|V\|_1$.

Now let $V \in \mathcal{B}_1(\mathcal{H})$, and write $V = \sum_{k=1}^{\infty} \tau_k |g_k\rangle\langle g_k|$ with $\sum_{k=1}^{\infty} |\tau_k| < \infty$; $\|g_k\| = 1$ for each $k \in \mathbb{N}$. Set $V_k \equiv \sum_{j=1}^{k} \tau_j |g_j\rangle\langle g_j|$ and $H_k \equiv H_0 + V_k$ for $k = 1, 2, 3 \ldots$. Then $\|V - V_k\|_1 \longrightarrow 0$ as $k \longrightarrow \infty$ and hence $\|H - H_k\|_1 = \|V - V_k\|_1 \longrightarrow 0$ as $k \longrightarrow \infty$, since $\sum_{k=1}^{\infty} |\tau_k| < \infty$. Therefore

$$\left\|\mathrm{e}^{\mathrm{i}tH} - \mathrm{e}^{\mathrm{i}tH_k}\right\|_1 = \left\|(\mathrm{i}t) \int_0^1 d\alpha\, \mathrm{e}^{\mathrm{i}t\alpha H}[H - H_k]\mathrm{e}^{\mathrm{i}t(1-\alpha)H_k}\right\|_1 \leq |t| \|H - H_k\|_1,$$

which converges to 0 as $k \longrightarrow \infty$, uniformly in t for $|t| \leq T$. But on the other hand

$$\mathrm{e}^{\mathrm{i}tH_k} - \mathrm{e}^{\mathrm{i}tH_0} = \sum_{m=1}^{k} \left(\mathrm{e}^{\mathrm{i}tH_m} - \mathrm{e}^{\mathrm{i}tH_{m-1}}\right)$$

and hence

$$\operatorname{Tr}\{\mathrm{e}^{\mathrm{i}tH_k} - \mathrm{e}^{\mathrm{i}tH_0}\} = \sum_{m=1}^{k} \operatorname{Tr}\{\mathrm{e}^{\mathrm{i}tH_m} - \mathrm{e}^{\mathrm{i}tH_{m-1}}\} = \sum_{m=1}^{k} (\mathrm{i}t) \int_{\mathbb{R}} \mathrm{e}^{\mathrm{i}t\lambda}\, \xi_m(\lambda)\, d\lambda, \quad (2.11)$$

where $\xi_m(\lambda)$ is a real-valued L^1-function as obtained in (2.10) corresponding to the pair (H_m, H_{m-1}) such that $\int_{\mathbb{R}} \xi_m(\lambda)\, d\lambda = \tau_m$ and $\|\xi_m\|_{L^1} = |\tau_m|$. If we set $\xi(\lambda) = \sum_{k=1}^{\infty} \xi_m(\lambda)$, then since $\sum_{m=1}^{\infty} |\tau_m| = \|V\|_1 < \infty$, it is easy to see that ξ is real valued since each ξ_m is and $\xi \in L^1(\mathbb{R})$. Moreover by Fubini's theorem we have that $\|\xi\|_{L^1} \leq \|V\|_1$ and

$$\int_{\mathbb{R}} \xi(\lambda)\, d\lambda = \sum_{m=1}^{\infty} \int_{\mathbb{R}} \xi_m(\lambda)\, d\lambda = \sum_{m=1}^{\infty} \tau_m = \mathrm{Tr}(V).$$

Finally, by taking limit as $k \longrightarrow \infty$ on both sides of (2.11),

$$\begin{aligned}\mathrm{Tr}\,\{\mathrm{e}^{\mathrm{i}tH} - \mathrm{e}^{\mathrm{i}tH_0}\} &= \lim_{k\longrightarrow\infty} \mathrm{Tr}\,\{\mathrm{e}^{\mathrm{i}tH_k} - \mathrm{e}^{\mathrm{i}tH_0}\} = \sum_{m=1}^{\infty} \mathrm{i}t \int_{\mathbb{R}} \mathrm{e}^{\mathrm{i}t\lambda}\xi_m(\lambda)\, d\lambda \\ &= \mathrm{i}t \int_{\mathbb{R}} \mathrm{e}^{\mathrm{i}t\lambda} \sum_{m=1}^{\infty} \xi_m(\lambda)\, d\lambda = \mathrm{i}t \int_{\mathbb{R}} \mathrm{e}^{\mathrm{i}t\lambda}\xi(\lambda)\, d\lambda.\end{aligned}$$

The uniqueness of ξ follows easily from the Fourier Inversion theorem of L^1-functions.

(iii) By functional calculus

$$\phi(H) - \phi(H_0) = \int_{-\infty}^{\infty} \frac{\mathrm{e}^{\mathrm{i}tH} - \mathrm{e}^{\mathrm{i}tH_0}}{it} \nu(dt), \tag{2.12}$$

where the integral in (2.12) exists as a $\mathcal{B}_1$-valued Bochner integral (page 30 of [2]). It also follows that

$$||\varphi(H) - \varphi(H_0)||_1 \leq ||V||_1 \int_{-\infty}^{\infty} |\nu|(dt) < \infty\,,$$

and we have by (i)

$$\begin{aligned}\mathrm{Tr}\{\phi(H) - \phi(H_0)\} &= \int_{-\infty}^{\infty} \frac{\nu(dt)}{\mathrm{i}t}\, \mathrm{Tr}\{\mathrm{e}^{\mathrm{i}tH} - \mathrm{e}^{\mathrm{i}tH_0}\} = \int_{-\infty}^{\infty} \frac{\nu(dt)}{\mathrm{i}t} it \int_{-\infty}^{\infty} \mathrm{e}^{\mathrm{i}t\lambda}\xi(\lambda)d\lambda \\ &= \int_{-\infty}^{\infty} d\lambda\, \xi(\lambda) \int_{-\infty}^{\infty} \mathrm{e}^{\mathrm{i}t\lambda}\nu(dt) = \int_{-\infty}^{\infty} \phi'(\lambda)\xi(\lambda)d\lambda.\end{aligned}$$

(iv) For this part we just note that $(\lambda - z)^{-1} + z^{-1} = \int \frac{\mathrm{e}^{\mathrm{i}t\lambda} - 1}{\mathrm{i}t} \nu(dt)$ with $\nu(dt) = -t\chi_{\mp}(t)\mathrm{e}^{-\mathrm{i}zt}dt$, according as $\mathrm{Im}\ z \gtrless 0$, where $\chi_{\pm}$ are the indicator functions of the intervals $[0, \infty)$ and $(-\infty, 0]$ respectively. □

Here we mention some other authors who have also dealt with this subject, in particular Clancy [14], Kuroda [28]. There is also the interesting approach of Birman and Solomyak ([4], [5], [6], [7], [8], [9]) using the theory of double spectral integrals, they obtain a trace formula of the type (2.1) for a function class somewhat larger than what we have described in this section.

Next we discuss the trace formula for the case when the perturbation V is not necessarily of trace class but is such that the difference of resolvents $R_z - R_z^0$ is trace-class for some $z \in \rho(H) \cap \rho(H_0)$. It is not difficult to see that if $R_z - R_z^0 \in \mathcal{B}_1(\mathcal{H})$ for some such z, then it is so for all such z and hence we shall, in this section, take $z = \mathrm{i}$ as the reference point and assume that $R_\mathrm{i} - R_\mathrm{i}^0 \in \mathcal{B}_1(\mathcal{H})$. Also it is worth noting that in $L^2(\mathbb{R}^3)$, if $H_0 = -\Delta$ and V is the multiplication operator by a function $V \in L^2(\mathbb{R}^3) \cap L^1(\mathbb{R}^3)$, then $R_\mathrm{i} - R_\mathrm{i}^0 \in \mathcal{B}_1(\mathcal{H})$ (see p. 546 of [24]).

We set $U_0 = \frac{H_0+\mathrm{i}}{H_0-\mathrm{i}} = I + 2\mathrm{i}R_i^0$ and $U = \frac{H+\mathrm{i}}{H-\mathrm{i}} = I + 2\mathrm{i}R_\mathrm{i}$ so that $U - U_0 = 2\mathrm{i}(R_\mathrm{i} - R_\mathrm{i}^0) \in \mathcal{B}_1(\mathcal{H})$. If we write $U - U_0 = U_0 T$ then it is clear that T is a normal trace-class operator such that $I + T$ is unitary. Let $T = \sum_{j=1}^{\infty} \tau_j \; |g_j\rangle\langle g_j|$ be the canonical decomposition for T with $||g_j|| = 1$ and $1 + \tau_j = \exp(\mathrm{i}\theta_j)$, $-\pi < \theta_j \leq \pi$. Then it follows that

$$\sum_{j=1}^{ity} |\theta_j| \sum_{j=1}^{ity} \left| e^{-\mathrm{i}\theta_j/2} \left[\frac{\theta_j/2}{\sin(\theta_j/2)} \right] \tau_j \right| \leq \frac{\pi}{2} \sum_{j=1}^{ity} |\tau_j| = \frac{\pi}{2} ||T||_1 < \infty \tag{2.13}$$

since $\left|\frac{\sin\theta}{\theta}\right| \geq 2/\pi$ for $0 \leq \theta \leq \pi/2$. Thus the determinant

$$\Delta(\omega) \equiv \det[(U-\omega)(U_0-\omega)^{-1}] = \det[I + U_0 T (U_0 - \omega)^{-1}]$$

is analytic for $|\omega| < 1$ and has no zeroes there. Now we are in a position to state Krein's theorem in the case when the difference of resolvents is trace class.

Theorem 2.6. [41] *Let $R_i - R_i^0 \in \mathcal{B}_1(\mathcal{H})$ and U and U_0 be defined as above. Then there exists a real-valued function ξ on $\mathbb{R}$ such that*

(i) $\xi(\lambda)(1+\lambda^2)^{-1} \in L^1(\mathbb{R})$,

(ii) $\displaystyle\int_{-\infty}^{\infty} |\xi(\lambda)|(1+\lambda^2)^{-1} d\lambda \leq (\pi/4)||T||_1$ *and*

$$\int_{-\infty}^{\infty} \xi(\lambda)(1+\lambda^2)^{-1} d\lambda = \frac{-\mathrm{i}}{2} \ln \det(U_0^* U),$$

(iii) $\displaystyle\xi(\lambda) = \frac{1}{\pi} \lim_{\rho \uparrow 1} \operatorname{Im} \ln \left[\exp\left(-\frac{\mathrm{i}}{2} \sum_{j=1}^{\infty} \theta_j \right) \Delta(\rho e^{\mathrm{i}\alpha}) \right]$, *with* $e^{\mathrm{i}\alpha} = (\lambda+\mathrm{i})(\lambda-\mathrm{i})^{-1}$,

(iv) $\displaystyle\mathrm{Tr}(R_z - R_z^0) = -\int_{-\infty}^{\infty} (\lambda - z)^{-2} \xi(\lambda)\, d\lambda$ *for* $\operatorname{Im} z \neq 0$.

(v) *Let $\psi \in \tilde{\mathcal{K}}$ be the modified Krein Class, consisting of functions $\psi : \mathbb{R} \longrightarrow \mathbb{C}$ such that $(1+\lambda^2)\psi(\lambda) \in \mathcal{K}$. Then for $\psi \in \tilde{\mathcal{K}}$,*

$$\psi(H) - \psi(H_0) \in \mathcal{B}_1(\mathcal{H}) \text{ and } \mathrm{Tr}[\psi(H) - \psi(H_0)] = \int \psi'(\lambda)\, \xi(\lambda)\, d\lambda.$$

Furthermore, ξ is unique up to an additive constant function.

3. Higher-order results

Let H and H_0 be two self-adjoint operators in $\mathcal{H}$ such that $H - H_0 \equiv V \in \mathcal{B}_2(\mathcal{H})$.

In this section first we discuss Koplienko's formula in finite dimension and then we prove the same for the bounded self-adjoint case via finite-dimensional approximation [11], and then we state Koplienko's formula for the unbounded self-adjoint case.

Theorem 3.1. *Let H and H_0 be two bounded self-adjoint operators in a Hilbert space $\mathcal{H}$ such that $H - H_0 \equiv V$ and let $p(\lambda) = \lambda^r (r \geq 2)$.*

(i) *Then $D^{(1)}p(H_0) \bullet X = \sum_{j=0}^{r-1} H_0^{r-j-1} X H_0^j$ and $\frac{d}{ds}(p(H_s)) = \sum_{j=0}^{r-1} H_s^{r-j-1} V H_s^j$, where $H_s = H_0 + sV$ $(0 \leq s \leq 1)$ and $X \in \mathcal{B}(\mathcal{H})$.*

(ii) *If furthermore $\dim \mathcal{H} < \infty$, then there exists a unique non-negative $L^1(\mathbb{R})$-function η such that*

$$\mathrm{Tr}\{p(H) - p(H_0) - D^{(1)}p(H_0) \bullet V\} = \int_a^b p''(\lambda)\eta(\lambda)d\lambda, \tag{3.1}$$

for some $-\infty < a < b < \infty$.

Moreover,

$$\eta(\lambda) = \int_0^1 \mathrm{Tr}\{V\,[E_{H_0}(\lambda) - E_{H_s}(\lambda)]\}ds; \quad \text{and} \quad \|\eta\|_1 = \frac{1}{2}\|V\|_2^2. \tag{3.2}$$

(iii) *For $\dim \mathcal{H} < \infty$,*

$$\mathrm{Tr}\{\mathrm{e}^{\mathrm{i}tH} - \mathrm{e}^{\mathrm{i}tH_0} - D^{(1)}(\mathrm{e}^{\mathrm{i}tH_0}) \bullet V\} = (\mathrm{i}t)^2 \int_a^b \mathrm{e}^{\mathrm{i}t\lambda}\eta(\lambda)d\lambda, \tag{3.3}$$

for some $-\infty < a < b < \infty$, $t \in \mathbb{R}$ and η is given by (3.2).

The next proposition is a generalization of Lemma 2.3.

Proposition 3.2. *Let A be a self-adjoint operator (possibly unbounded) in a separable infinite-dimensional Hilbert space $\mathcal{H}$ and let $\{f_l\}_{1 \leq l \leq L}$ be a set of normalized vectors in $\mathcal{H}$ and $\epsilon > 0$.*

(i) *Then there exists a finite rank projection P such that $\|(I-P)f_l\| < \epsilon$ for $1 \leq l \leq L$.*
(ii) *Furthermore, $(I-P)AP \in \mathcal{B}_2(\mathcal{H})$, $\|(I-P)AP\|_2 < \epsilon$ and $\|(I-P)\mathrm{e}^{\mathrm{i}tA}P\|_2 < \epsilon$ uniformly for t with $|t| \leq T$.*

As a consequence of the above proposition we have the following result.

Lemma 3.3. *Let H and H_0 be two self-adjoint operators in a separable infinite-dimensional Hilbert space $\mathcal{H}$ such that $H - H_0 \equiv V \in \mathcal{B}_2(\mathcal{H})$. Then given $\epsilon > 0$, there exists a projection P of finite rank such that for all t with $|t| \leq T$,*

(i) $\|(I-P)H_0P\|_2 < \epsilon$, $\quad \|(I-P)\mathrm{e}^{\mathrm{i}tH_0}P\|_2 < \epsilon$,
(ii) $\|(I-P)V\|_2 < 2\epsilon$, $\quad \|(I-P)HP\|_2 < 3\epsilon$.

Remark 3.1. We can reformulate the statement of Lemma 3.3 by saying that there exists a sequence $\{P_n\}$ of finite rank projections in $\mathcal{H}$ such that

$$\|(I-P_n)H_0P_n\|_2,\ \|(I-P_n)\mathrm{e}^{\mathrm{i}tH_0}P_n\|_2,\ \|(I-P_n)V\|_2,\ \|(I-P_n)HP_n\|_2 \longrightarrow 0$$

as $n \longrightarrow \infty$. It may also be noted that $\{P_n\}$ does not necessarily converge strongly to I.

The next theorem shows how Lemma 3.3 can be used to reduce the relevant problem into a finite-dimensional one, in the case when the self-adjoint pair (H_0, H) are bounded.

Theorem 3.4. *Let H and H_0 be two bounded self-adjoint operators in a separable infinite-dimensional Hilbert space $\mathcal{H}$ such that $H - H_0 \equiv V \in \mathcal{B}_2(\mathcal{H})$. Then there exists a sequence $\{P_n\}$ of finite rank projections in $\mathcal{H}$ such that*

$$\begin{aligned}&\operatorname{Tr}\{p(H) - p(H_0) - D^{(1)}p(H_0) \bullet V\} \qquad (3.4)\\ &= \lim_{n\to\infty} \operatorname{Tr}\{P_n \left[p(P_nHP_n) - p(P_nH_0P_n) - D^{(1)}p(P_nH_0P_n) \bullet P_nVP_n\right] P_n\},\end{aligned}$$

where $p(.)$ is a polynomial.

Proof. It will be sufficient to prove the theorem for $p(\lambda) = \lambda^r$. Note that for $r = 0$ or 1, both sides of (3.4) are identically zero. Using the sequence $\{P_n\}$ of finite rank projections as obtained in Lemma 3.3, we have that

$$\begin{aligned}
&\operatorname{Tr}\Big\{\left[p(H) - p(H_0) - D^{(1)}p(H_0) \bullet V\right]\\
&\quad - P_n \left[p(P_nHP_n) - p(P_nH_0P_n) - D^{(1)}p(P_nH_0P_n) \bullet P_nVP_n\right] P_n\Big\}\\
&= \operatorname{Tr}\Big\{\left[H^r - H_0^r - D^{(1)}(H_0^r) \bullet V\right]\\
&\quad - P_n \left[(P_nHP_n)^r - (P_nH_0P_n)^r - D^{(1)}((P_nH_0P_n)^r) \bullet P_nVP_n\right] P_n\Big\}\\
&= \operatorname{Tr}\left\{\left[\sum_{j=0}^{r-1} \left(H^{r-j-1} - H_0^{r-j-1}\right) V H_0^j\right]\right.\\
&\quad \left.-P_n \left[\sum_{j=0}^{r-1} [(P_nHP_n)^{r-j-1} - (P_nH_0P_n)^{r-j-1}](P_nVP_n)(P_nH_0P_n)^j\right] P_n\right\}\\
&= \operatorname{Tr}\left\{\sum_{j=0}^{r-2}\sum_{k=0}^{r-j-2} H^{r-j-k-2} V H_0^k V H_0^j\right.\\
&\quad \left.-\sum_{j=0}^{r-2}\sum_{k=0}^{r-j-2} P_n(P_nHP_n)^{r-j-k-2}(P_nVP_n)(P_nH_0P_n)^k(P_nVP_n)(P_nH_0P_n)^j P_n\right\}\\
&= \sum_{j=0}^{r-2}\sum_{k=0}^{r-j-2} \operatorname{Tr}\Big\{\left[H^{r-j-k-2}P_n - (P_nHP_n)^{r-j-k-2}\right] P_nVH_0^kVH_0^j\\
&\quad + H^{r-j-k-2}P_n^{\perp}VH_0^kVH_0^j + (P_nHP_n)^{r-j-k-2}P_nVP_n^{\perp}H_0^kVH_0^j\\
&\quad + (P_nHP_n)^{r-j-k-2}(P_nVP_n)\left[P_nH_0^k - (P_nH_0P_n)^k\right] VH_0^j\\
&\quad + (P_nHP_n)^{r-j-k-2}(P_nVP_n)(P_nH_0P_n)^kP_nVP_n^{\perp}H_0^j
\end{aligned}$$

$$+ (P_nHP_n)^{r-j-k-2}(P_nVP_n)(P_nH_0P_n)^k(P_nVP_n)\left[P_nH_0^j - (P_nH_0P_n)^j\right]\Big\}. \tag{3.5}$$

Using the results of Lemma 3.3, the first term of the expression (3.5) leads to

$$\begin{aligned}&\left\|\left[H^{r-j-k-2} - (P_nHP_n)^{r-j-k-2}\right]P_n\right\|_2\\ &\quad\leq (r-j-k-2)\|H\|^{r-j-k-3}\left\|P_n^{\perp}HP_n\right\|_2 \leq r(1+\|H\|)^r\left\|P_n^{\perp}HP_n\right\|_2,\end{aligned}$$

and hence

$$\begin{aligned}&\left\|\left[H^{r-j-k-2}P_n - (P_nHP_n)^{r-j-k-2}\right]P_nVH_0^kVH_0^j\right\|_1\\ &\leq \left\|\left[H^{r-j-k-2} - (P_nHP_n)^{r-j-k-2}\right]P_n\right\|_2\|VH_0^kVH_0^j\|_2\\ &\leq r(1+\|H\|)^r\left\|P_n^{\perp}HP_n\right\|_2\|V\|_2^2\|H_0\|^{k+j},\end{aligned}$$

which converges to 0 as $n \longrightarrow \infty$. Similarly, the fourth and the sixth terms in (3.5) can be seen to converge to 0 as $n \longrightarrow \infty$. For the second term in (3.5) we have

$$\begin{aligned}\left\| H^{r-j-k-2}P_n^{\perp}VH_0^kVH_0^j\right\|_1 &\leq \| H^{r-j-k-2}P_n^{\perp}V\|_2\|H_0^kVH_0^j\|_2\\ &\leq \|P_n^{\perp}V\|_2\|H\|^{r-j-k-2}\|H_0\|^{j+k}\|V\|_2,\end{aligned}$$

which converges to 0 as $n \longrightarrow \infty$ since by Lemma 3.3, $\left\|P_n^{\perp}V\right\|_2 \longrightarrow 0$ as $n \longrightarrow \infty$. An identical set of computations show that the third and fifth terms in (3.5) also converges to 0 as $n \longrightarrow \infty$ and hence the result follows. □

Theorem 3.5 (Koplienko's Trace Formula [25]). *Let H and H_0 be two bounded self-adjoint operators in an infinite-dimensional separable Hilbert space $\mathcal{H}$ such that $H - H_0 \equiv V \in \mathcal{B}_2(\mathcal{H})$. Then for any polynomial $p(.)$, $p(H) - p(H_0) - D^{(1)}p(H_0) \bullet V \in \mathcal{B}_1(\mathcal{H})$ and there exists a unique non-negative $L^1(\mathbb{R})$-function η supported on $[a, b]$ such that*

$$\operatorname{Tr}\{p(H) - p(H_0) - D^{(1)}p(H_0) \bullet V\} = \int_a^b p''(\lambda)\eta(\lambda)d\lambda,$$

where, $a = \inf\sigma(H_0) - \|V\|$, $b = \sup\sigma(H_0) + \|V\|$. Furthermore $\int_a^b |\eta(\lambda)|d\lambda = \frac{1}{2}\|V\|_2^2$.

Proof. By Theorem 3.1, $p(H) - p(H_0) - D^{(1)}p(H_0) \bullet V \in \mathcal{B}_1(\mathcal{H})$ and by Theorem 3.4 we have that

$$\begin{aligned}&\operatorname{Tr}\{p(H) - p(H_0) - D^{(1)}p(H_0) \bullet V\}\\ &= \lim_{n\to\infty}\operatorname{Tr}\{P_n\left[p(P_nHP_n) - p(P_nH_0P_n) - D^{(1)}p(P_nH_0P_n) \bullet P_nVP_n\right]P_n\}\\ &= \lim_{n\to\infty}\int_a^b p''(\lambda)\eta_n(\lambda)d\lambda,\end{aligned}$$

with $\eta_n(\lambda)$ given by (3.2), and $\|\eta_n\|_1 = \frac{1}{2}\|P_n(H-H_0)P_n\|_2^2$, which clearly converges to $\frac{1}{2}\|V\|_2^2$ as $n \longrightarrow \infty$, since $|\|P_nVP_n\|_2 - \|V\|_2| \leq \|P_nVP_n^{\perp}\|_2 + \|P_n^{\perp}V\|_2$, which converges to 0 as $n \longrightarrow \infty$. Set $V_n \equiv P_nVP_n$; $H_n \equiv P_nHP_n$; $H_{0,n} \equiv P_nH_0P_n$ and $E_{H_{0,n}}(.)$, $E_{H_{s,n}}(.)$ are the spectral families of $H_{0,n}$ and $H_{s,n} \equiv P_nH_sP_n$ respectively. Using the expression (3.2), we have for $f \in L^\infty([a,b])$ and $g(\lambda) = \int_a^\lambda f(\mu)d\mu$ that (see also [19])

$$\begin{aligned}
&\int_a^b f(\lambda)\left[\eta_n(\lambda) - \eta_m(\lambda)\right] d\lambda \\
&= \int_a^b g'(\lambda)d\lambda \int_0^1 \operatorname{Tr}\{V_n\left[E_{H_{0,n}}(\lambda) - E_{H_{s,n}}(\lambda)\right]\} ds \\
&\quad - \int_a^b g'(\lambda)d\lambda \int_0^1 \operatorname{Tr}\{V_m\left[E_{H_{0,m}}(\lambda) - E_{H_{s,m}}(\lambda)\right]\} ds \\
&= \int_0^1 ds \int_a^b g'(\lambda) \operatorname{Tr}\{V_n\left[E_{H_{0,n}}(\lambda) - E_{H_{s,n}}(\lambda)\right] \\
&\quad - V_m\left[E_{H_{0,m}}(\lambda) - E_{H_{s,m}}(\lambda)\right]\} d\lambda, \qquad (3.6)
\end{aligned}$$

which after an integration (in λ) by parts and after noting that boundary terms vanish, becomes

$$\begin{aligned}
&-\int_0^1 ds \int_a^b g(\lambda) \operatorname{Tr}\left\{V_n\left[E_{H_{0,n}}(d\lambda) - E_{H_{s,n}}(d\lambda)\right] - V_m\left[E_{H_{0,m}}(d\lambda) - E_{H_{s,m}}(d\lambda)\right]\right\} \\
&= \int_0^1 ds \operatorname{Tr}\{V_n\left[g(H_{s,n}) - g(H_{0,n})\right] - V_m\left[g(H_{s,m}) - g(H_{0,m})\right]\}. \qquad (3.7)
\end{aligned}$$

Next we note that

$$g(H_0) - g(H_s) = -s\int_a^b\int_a^b \frac{g(\alpha)-g(\beta)}{\alpha-\beta}\mathcal{G}(d\alpha\times d\beta)V,$$

where $\mathcal{G}(\Delta\times\delta)X = E_{H_0}(\Delta)XE_{H_s}(\delta)$ ($X \in \mathcal{B}_2(\mathcal{H})$ and $\Delta\times\delta \subseteq \mathbb{R}\times\mathbb{R}$) extends to a spectral measure on $\mathbb{R}^2$ in the Hilbert space $\mathcal{B}_2(\mathcal{H})$. Therefore $\|g(H_s)-g(H_0)\|_2 \leq s\ \|f\|_\infty\ \|V\|_2$, since

$$\sup_{\alpha,\beta\in[a,b];\alpha\neq\beta}\left|\frac{g(\alpha)-g(\beta)}{\alpha-\beta}\right| \leq \|f\|_\infty.$$

We have for $0 \leq s \leq 1$,

$$\begin{aligned}
&P_n\left[g(H_{s,n}) - g(H_s)\right]P_n \\
&= -P_n\{\int_a^b\int_a^b \frac{g(\alpha)-g(\beta)}{\alpha-\beta}\ \mathcal{G}_{(s,n)}(d\alpha\times d\beta)\left[P_nH_0P_n^{\perp} + sP_nVP_n^{\perp}\right]\}P_n,
\end{aligned}$$

where $\mathcal{G}_{(s,n)}(\Delta\times\delta)X = E_{H_{s,n}}(\Delta)XE_{H_s}(\delta)$ ($X \in \mathcal{B}_2(\mathcal{H})$ and $\Delta\times\delta \subseteq \mathbb{R}\times\mathbb{R}$) extends to a spectral measure on $\mathbb{R}^2$ in the Hilbert space $\mathcal{B}_2(\mathcal{H})$ and hence

$$\|P_n\left[g(H_{s,n}) - g(H_s)\right]P_n\|_2 \leq \|f\|_\infty\left(\|P_nH_0P_n^{\perp}\|_2 + s\|P_nVP_n^{\perp}\|_2\right),$$

and in particular for $s = 0$, we have

$$\|P_n [g(H_{0,n}) - g(H_0)] P_n\|_2 \leq \|f\|_\infty \left\|P_n H_0 P_n^\perp\right\|_2 .$$

Therefore from equation (3.7), it follows that

$$\begin{aligned}
&\left|\int_a^b f(\lambda) [\eta_n(\lambda) - \eta_m(\lambda)] \, d\lambda\right| \\
&\quad = \Bigg|\int_0^1 ds \; (\mathrm{Tr}(V_n\{[g(H_{s,n}) - g(H_{0,n})] - [g(H_s) - g(H_0)]\}) \\
&\qquad - \mathrm{Tr}(V_m\{[g(H_{s,m}) - g(H_{0,m})] - [g(H_s) - g(H_0)]\}) \\
&\qquad + \mathrm{Tr}\{(V_n - V_m) [g(H_s) - g(H_0)]\})\Bigg| \\
&\quad \leq \int_0^1 ds \; \{\|V_n\|_2(\|P_n [g(H_{s,n}) - g(H_s)] \|_2 + \|P_n [g(H_{0,n}) - g(H_0)] \|_2) \\
&\qquad - \|V_m\|_2(\|P_m [g(H_{s,m}) - g(H_s)] \|_2 + \|P_m [g(H_{0,m}) - g(H_0)] \|_2) \\
&\qquad + \|(V_n - V_m)\|_2\| [g(H_s) - g(H_0)] \|_2\}.
\end{aligned}$$

Hence

$$\begin{aligned}
\|\eta_n - \eta_m\|_{L^1([a,b])} &\equiv \sup_{f \in L^\infty([a,b])} \frac{\left|\int_a^b f(\lambda) [\eta_n(\lambda) - \eta_m(\lambda)] \, d\lambda\right|}{\|f\|_\infty} \\
&\leq \|f\|_\infty \|V\|_2 \left(\int_0^1 ds\{2 \left(\left\|P_n H_0 P_n^\perp\right\|_2 + \left\|P_m H_0 P_m^\perp\right\|_2\right)\right. \\
&\qquad \left. + s \left(\left\|P_n V P_n^\perp\right\|_2 + \left\|P_m V P_m^\perp\right\|_2\right) + s\|V_n - V_m\|_2\}\right),
\end{aligned}$$

which converges to zero as $m, n \longrightarrow \infty$ and therefore there exists a non-negative $L^1([a,b])$-function η such that $\{\eta_n\}$ converges to η in L^1-norm. Thus

$$\mathrm{Tr}\{p(H) - p(H_0) - D^{(1)} p(H_0) \bullet V\} = \lim_{n \to \infty} \int_a^b p''(\lambda)\eta_n(\lambda) d\lambda = \int_a^b p''(\lambda)\eta(\lambda) d\lambda.$$

The uniqueness of $\eta \in L^1([a,b])$ follows as in the proof of Theorem 2.5. □

Next we state, without proof, Koplienko's trace formula for the unbounded self-adjoint case ([11], [25], [42], [45]). Recently Potapov and Sukochev obtained Koplienko's trace formula in a unitary case [37]. In this context Peller also obtained an extension of the Koplienko–Neidhardt trace formula using multiple operator integrals ([31], [32]).

Theorem 3.6 ([11]). *Let H and H_0 be two self-adjoint operators in an infinite-dimensional separable Hilbert space $\mathcal{H}$ such that $H - H_0 \equiv V \in \mathcal{B}_2(\mathcal{H})$ and $f \in \mathcal{S}(\mathbb{R})$ (the Schwartz class of smooth functions of rapid decrease). Then $f(H) -$*

$f(H_0) - D^{(1)}f(H_0) \bullet V \in \mathcal{B}_1(\mathcal{H})$ and

$$\mathrm{Tr}\{f(H) - f(H_0) - D^{(1)}f(H_0) \bullet V\} = \int_{\mathbb{R}} f''(\lambda)\eta(\lambda)d\lambda, \tag{3.8}$$

where η is a unique non-negative $L^1(\mathbb{R})$-function with $\|\eta\|_1 = \frac{1}{2}\|V\|_2^2$.

Dykema and Skripka ([16], [17], [18], [43]) and earlier Boyadzhiev [10] obtained the formula (3.8) in the semi-finite von Neumann algebra setting and also studied the existence of a higher-order spectral shift function. In ([16],Theorem 5.1), Dykema and Skripka showed that for a self-adjoint operator A (possibly unbounded) and a self-adjoint operator $V \in \mathcal{B}_2(\mathcal{H})$, the following assertions hold:

(i) *There is a unique finite real-valued measure ν_3 on $\mathbb{R}$ such that the trace formula*

$$\begin{aligned}\mathrm{Tr}\left\{\phi(A+V) - \phi(A) - D^{(1)}\phi(A) \bullet V - \frac{1}{2}D^{(2)}\phi(A) \bullet (V,V)\right\}& \\ = \int_{-\infty}^{\infty} \phi'''(\lambda)d\nu_3(\lambda),&\end{aligned} \tag{3.9}$$

holds for suitable functions ϕ, where $D^{(2)}\phi(A)$ is the second-order Fréchet derivative of ϕ at A [3]. The total variation of ν_3 is bounded by $\frac{1}{3!}\|V\|_2^3$.

(ii) *If, in addition, A is bounded, then ν_3 is absolutely continuous.*

In [12] the present authors used the finite-dimensional approximation method to obtain the formula (3.9). They also prove the absolute continuity of the measure ν_3 when the unperturbed operator is self-adjoint (bounded or unbounded, but bounded below). More recently, Potapov, Skripka and Sukochev ([36], [35], [37]) have proven the trace-formula for all orders, obtaining a kind of Taylor's theorem under trace. In fact they have proved, in [36], the existence of $\eta_n \in L^1(\mathbb{R})$ for $n \in \mathbb{N}$ such that

$$\mathrm{Tr}\left(\phi(H_0+V) - \sum_{k=0}^{n-1}\frac{1}{k!}D^{(k)}\phi(H_0) \bullet (\underbrace{V,V,\ldots,V}_{k\text{-times}})\right) = \int_{\mathbb{R}} \phi^{(n)}(\lambda)\eta_n(\lambda)d\lambda,$$

for every sufficiently smooth function ϕ, where H_0 is a self-adjoint operator defined on a Hilbert space $\mathcal{H}$, V is a self-adjoint operator such that $V \in \mathcal{B}_n(\mathcal{H})$, $D^{(k)}\phi(H_0) \bullet (\underbrace{V,V,\ldots,V}_{k\text{-times}})$ denotes the kth-order Fréchet derivative of ϕ at H_0 acting on $(\underbrace{V,V,\ldots,V}_{k\text{-times}})$ (see [3]) and where $\phi^{(n)}$ denotes the nth-order derivative of the function ϕ.

4. Two variables trace formula

In this section we consider the generalization of Krein's theorem to a pair of commuting tuples (H_1^0, H_2^0) and (H_1, H_2) of bounded self-adjoint operators in a separable Hilbert space $\mathcal{H}$ with $H_j - H_j^0 = V_j \in \mathcal{B}_2(\mathcal{H})$ (set of all Hilbert–Schmidt operators on $\mathcal{H}$) for $j = 1, 2$, and prove a Stokes-like formula under trace, using finite-dimensional approximation. In this context, it should be mentioned that recently Skripka [44] has studied a related problem for commuting contractions. Let us start with a lemma (without proof) which will be useful to generalize the Weyl–von Neumann–Berg theorem [15].

Lemma 4.1. *Let $A \in \mathcal{B}(\mathcal{H})$ be such that $0 \leq A \leq I$. Now consider the spectral projections $E_k = E_A\left(\bigcup_{j=1}^{2^{k-1}} (2^{-k}(2j-1), 2^{-k}(2j)]\right)$ for $k \geq 1$. Then*

$$A = \sum_{k=1}^{\infty} 2^{-k} E_k, \tag{4.1}$$

where the right-hand side of (4.1) converges in operator norm.

A result due to Weyl and von Neumann [24] proves that for a self-adjoint operator A, given $\epsilon > 0$, $\exists K \in \mathcal{B}_2(\mathcal{H})$ such that $\|K\|_2 < \epsilon$ and $A + K$ has pure point spectrum. Later Berg extended this to n-tuples of bounded commuting self-adjoint operators $(A_1, A_2, \ldots, A_n)$, which says that given $\epsilon > 0$, $\exists$ $\{K_j\}_{j=1}^n$ of compact operators such that $\|K_j\| < \epsilon$ $\forall j$ and $\{A_j - K_j\}_{j=1}^n$ is a commuting family of bounded self-adjoint operators with pure point spectra. We extend, in the next theorem, the ideas of the proof of Berg's result as given in [15]. It is worth mentioning that Voiculescu [46] had earlier obtained related (though not the same) results.

Theorem 4.2. *Let $\{A_i\}_{1\leq i\leq n}$ be a commuting family of bounded self-adjoint operators in an infinite-dimensional separable Hilbert space $\mathcal{H}$. Then there exists a sequence $\{P_N\}$ of finite-rank projections such that $\{P_N\} \uparrow I$ as $N \longrightarrow \infty$ and such that there exists a commuting family of bounded self-adjoint operators $\{B_i^{(N)}\}_{1\leq i\leq n}$ with the properties that for $p \geq n$ and for each i $(1 \leq i \leq n)$, as $N \longrightarrow \infty$,*

$$\text{(i) } P_N B_i^{(N)} P_N = B_i^{(N)} P_N, \quad \text{(ii) } \left\|A_i - B_i^{(N)}\right\|_p \longrightarrow 0, \quad \text{(iii) } \|[A_i, P_N]\|_p \longrightarrow 0,$$

$$\text{(iv) } \left\|P_N A_i P_N - B_i^{(N)} P_N\right\|_p \longrightarrow 0 \quad \textit{and} \quad \text{(v) } \{B_i^{(N)} P_N\} \uparrow A_i.$$

Proof. One can assume without loss of generality that $0 \leq A_i \leq I$ for all $1 \leq i \leq n$, and therefore for each i, by Lemma 4.1,

$$A_i = \sum_{k=1}^{\infty} 2^{-k} E_k^{(i)},$$

where $E_k^{(i)} = E_{A_i}\left(\bigcup_{j=1}^{2^{k-1}} (2^{-k}(2j-1), 2^{-k}(2j)]\right)$ with E_{A_i} the spectral measure associated to the bounded self-adjoint operator A_i. Next set for $N \in \mathbb{N}$ (the set of natural numbers),

$$\mathcal{L}_N \equiv \operatorname{span}\left\{\left[\prod_{k=1}^{N}\prod_{i=1}^{n}\left(E_k^{(i)}\right)^{\epsilon}\right] f_j \mid 1 \le j \le N;\ \epsilon = \pm 1\right\},$$

where $\{f_1, f_2, \ldots, f_N, \ldots\}$ is a countable orthonormal basis of $\mathcal{H}$ and $\left(E_k^{(i)}\right)^1 = E_k^{(i)}$ and $\left(E_k^{(i)}\right)^{-1} = I - E_k^{(i)}$. Thus $\mathcal{L}_N$ is a finite-dimensional subspace of $\mathcal{H}$ and it has the following properties:

(a) $\mathcal{L}_N \subseteq \mathcal{L}_{N+1}$, (b) $\overline{\left(\bigcup_{N=1}^{\infty} \mathcal{L}_N\right)} = \mathcal{H}$, (c) $\dim(\mathcal{L}_N) \le N\left(2^n - 1\right)^N + N$.

Of these, we only give the proof of (c).

(c): According to the definition of $E_K^{(i)}$, it follows that for each fixed $k \in \mathbb{N}$,

$$\sum_{\epsilon=\pm 1}\prod_{i=1}^{n}\left(E_k^{(i)}\right)^{\epsilon} = I. \tag{4.2}$$

We claim that for any fixed vector $f \in \mathcal{H}$, the span $\left\{\left[\prod_{k=1}^{N}\prod_{i=1}^{n}\left(E_k^{(i)}\right)^{\epsilon}\right] f : \epsilon = \pm 1\right\}$ contains at most $\left(2^n - 1\right)^N$ linearly independent vectors, without counting f. We prove our claim by induction on N. For $N = 1$, because of the identity (4.2) we conclude that the span $\left\{\left[\prod_{i=1}^{n}\left(E_1^{(i)}\right)^{\epsilon}\right] f : \epsilon = \pm 1\right\}$ contains at most $(2^n - 1)$ linearly independent vectors besides f. Since $\{A_i\}_{1 \le i \le n}$ is a commuting family, we have the following:

$$\begin{aligned}
&\operatorname{span}\left\{\left[\prod_{k=1}^{N+1}\prod_{i=1}^{n}\left(E_k^{(i)}\right)^{\epsilon}\right] f : \epsilon = \pm 1\right\} \\
&:= \operatorname{span}\left\{\prod_{i=1}^{n}\left(E_{N+1}^{(i)}\right)^{\epsilon}\left[\prod_{k=1}^{N}\prod_{i=1}^{n}\left(E_k^{(i)}\right)^{\epsilon}\right] f : \epsilon = \pm 1\right\}
\end{aligned}$$

and thus by the induction hypothesis, span $\left\{\left[\prod_{k=1}^{N}\prod_{i=1}^{n}\left(E_k^{(i)}\right)^{\epsilon}\right] f : \epsilon = \pm 1\right\}$ contains at most $\left(2^n - 1\right)^N$ linearly independent vectors, other than f. Therefore using equation (4.2) we conclude that the span $\left\{\left[\prod_{k=1}^{N+1}\prod_{i=1}^{n}\left(E_k^{(i)}\right)^{\epsilon}\right] f : \epsilon = \pm 1\right\}$ contains at most

$$2^n\left(2^n - 1\right)^N - \left(2^n - 1\right)^N = \left(2^n - 1\right)^{N+1}$$

number of linearly independent vectors, other than f itself, completing the induction. Hence for any fixed vector $f \in \mathcal{H}$, the span $\left\{\left[\prod_{k=1}^{N}\prod_{i=1}^{n}\left(E_k^{(i)}\right)^{\epsilon}\right] f : \epsilon = \pm 1\right\}$ contains the maximum of possible $\left\{\left(2^n - 1\right)^N + 1\right\}$ linearly independent vectors, including the vector f. This implies that $\mathcal{L}_N$ contains at most $N\{(2^n-1)^N+1\}$ number of linearly independent vectors and therefore $\dim(\mathcal{L}_N) \leq N(2^n-1)^N + N$. Now we set P_N to be the finite rank projection associated with the finite-dimensional subspace $\mathcal{L}_N$. Then by (a) and (b) the sequence $\{P_N\}$ increases to I. Next we define

$$B_i^{(N)} = \sum_{k=1}^{N} 2^{-k} E_k^{(i)} + \sum_{k=N+1}^{\infty} 2^{-k} E_k^{(i)} (I - P_k),$$

and observe that since $\{E_k^{(i)}\}_{1\leq k\leq N; 1\leq i\leq n}$ is a commuting family and since each member of that family for fixed k commutes with P_l for $1 \leq k \leq l$, it is easy to verify that

$$\begin{aligned} E_k^{(i)}(I-P_k)E_{k'}^{(i)}(I-P_{k'}) &= (I-P_k)(I-P_{k'})E_{k'}^{(i)}E_k^{(i)} \\ &= (I-P_{k'})E_{k'}^{(i)}E_k^{(i)} = E_{k'}^{(i)}(I-P_{k'})E_k^{(i)}(I-P_k), \end{aligned}$$

where we have assumed without loss of generality that $k \leq k'$. Thus $\{B_i^{(N)}\}_{1\leq i\leq n}$ is a commuting family of positive self-adjoint contractions and since $(I-P_k)P_N = 0$ for $k \geq N+1$, it follows that

$$P_N B_i^{(N)} P_N = B_i^{(N)} P_N = \sum_{k=1}^{N} 2^{-k} E_k^{(i)} P_N, \tag{4.3}$$

and hence $B_i^{(N)} P_N$ is a finite-dimensional self-adjoint operator in the Hilbert space $P_N(\mathcal{H})$. Furthermore, $A_i - B_i^{(N)} = \sum\limits_{k=N+1}^{\infty} 2^{-k} E_k^{(i)} P_k$ and

$$\begin{aligned} \left\| A_i - B_i^{(N)} \right\|_n &\leq \sum_{k=N+1}^{\infty} 2^{-k} \|P_k\|_n \leq \sum_{k=N+1}^{\infty} 2^{-k} \left[k\{1 + (2^n-1)^k\} \right]^{\frac{1}{n}} \\ &\leq \sum_{k=N+1}^{\infty} k^{\frac{1}{n}} 2^{-k} + \sum_{k=N+1}^{\infty} k^{\frac{1}{n}} \left[(1-2^{-n})^{\frac{1}{n}} \right]^k. \end{aligned}$$

Since for fixed n, $(1-2^{-n})^{\frac{1}{n}} < 1$, and since $\sum\limits_{k=1}^{\infty} k^{\frac{1}{n}} \alpha^k < \infty$ for $\alpha < 1$, it follows that for each $i (1 \leq i \leq n)$, $\left\| A_i - B_i^{(N)} \right\|_n \longrightarrow 0$ as $N \longrightarrow \infty$. Therefore for any $p \geq n$ we get

$$\left\| A_i - B_i^{(N)} \right\|_p \leq 2^{(1-\frac{n}{p})} \left\| A_i - B_i^{(N)} \right\|_n^{\frac{n}{p}} \longrightarrow 0$$

as $N \longrightarrow \infty$ and hence

$$\|[A_i, P_N]\|_p = \left\| [A_i - B_i^{(N)}, P_N] \right\|_p \tag{4.4}$$

and

$$\left\| P_N A_i P_N - P_N B_i^{(N)} P_N \right\|_p \leq \left\| A_i - B_i^{(N)} \right\|_p \longrightarrow 0$$

as $N \longrightarrow \infty$ for any $p \geq n$.

From (4.3) we have that

$$B_i^{(N+1)}P_{N+1} - B_i^{(N)}P_N = \sum_{k=1}^{N} 2^{-k}\left(P_{N+1} - P_N\right)E_k^{(i)}\left(P_{N+1} - P_N\right) + 2^{-(N+1)}P_{N+1}E_{N+1}^{(i)}P_{N+1} \geq 0,$$

since $\{P_N\} \uparrow I$. Finally by using (4.4) and the fact that $P_N A_i P_N \longrightarrow A_i$ strongly as $N \longrightarrow \infty$, we conclude that $B_i^{(N)}P_N \longrightarrow A_i$ strongly as $N \longrightarrow \infty$. This completes the proof. □

Remark 4.1. The choice that $0 \leq A_i \leq I$ does not materially affect the calculations of Theorem 4.2. For if $C_i \in \mathcal{B}(\mathcal{H})$ $(1 \leq i \leq n)$, then we can set

$$A_i = (2\|C_i\|)^{-1}C_i + \frac{1}{2}I$$

so that $0 \leq A_i \leq I$ and thus $C_i = 2\|C_i\|(\sum 2^{-k}E_k^{(i)} - \frac{1}{2}I)$. Thus choosing

$$B_i^{(N)} = 2\|C_i\|\left\{\sum_{k=1}^{N} 2^{-k}E_k^{(i)} + \sum_{k=N+1}^{\infty}(I - P_k)E_k^{(i)} - \frac{1}{2}I\right\}$$

one has $\|[C_i, B_i^{(N)}]\|_p = 2\|C_i\|\|[A_i, B_i^{(N)}]\|_p \to 0$ as $N \to \infty$ for $p \geq n$.

Now we are going to define spectral integrals of operator functions [1] in the next few lemmas and for details of the proof see [13].

Lemma 4.3. *Let H be a bounded self-adjoint operator in $\mathcal{H}$ with spectrum in $[a, b]$ and let $A : [a, b] \longrightarrow \mathcal{B}(\mathcal{H})$ be operator norm Hölder continuous with Hölder index $k > \frac{1}{2}$, that is,*

$$\|A(\alpha_1) - A(\alpha_2)\| \leq C|\alpha_1 - \alpha_2|^k,$$

where C is some positive constant and $k > \frac{1}{2}$. Then

$$\int_a^b A(\alpha)E_H(d\alpha)$$

is well defined as an operator norm Riemann–Stieltjes integral, where $E_H(.)$ is the spectral measure corresponding to the bounded self-adjoint operator H.

Lemma 4.4. *Let A, B, C be three bounded self-adjoint operators in an infinite-dimensional Hilbert space $\mathcal{H}$ such that $\sigma(A), \sigma(B), \sigma(C) \subseteq [a, b]$. Let $\phi : [a, b]^2 \longrightarrow \mathbb{C}$ be a bounded measurable function. Then the symbol $\int_A^B \phi(x, C)dx$, defined as:*

$$\int_A^B \phi(x, C)dx \equiv \int_a^b \left(\int_a^\alpha \phi(x, C)dx\right)[E_B(d\alpha) - E_A(d\alpha)],$$

(where $E_A(.), E_B(.)$ are the spectral measures of the operators A, B respectively), exists as a bounded operator.

Next we derive two formulae for the trace of a Stokes-like expression, one in terms of a spectral function and the other in terms of divided differences. First we need a simple lemma.

Lemma 4.5. *Let $\psi \in L^\infty([a,b]^2)$. Then there exist two measurable functions ϕ_1, ϕ_2 on $[a,b] \times [a,b]$ such that ϕ_1 and ϕ_2 are differentiable (almost everywhere) with respect to the second and first variable respectively with bounded derivatives such that,*

$$\frac{\partial \phi_2}{\partial x}(x,y) - \frac{\partial \phi_1}{\partial y}(x,y) = \psi(x,y).$$

Moreover ϕ_1 and ϕ_2 are Lipschitz in the second and first variable respectively, uniformly with respect to the other variable. Conversely, if ϕ_1 and ϕ_2 are two measurable functions differentiable with respect to the second and first variable respectively with bounded measurable derivatives, then

$$\psi(x,y) = \frac{\partial \phi_2}{\partial x}(x,y) - \frac{\partial \phi_1}{\partial y}(x,y) \in L^\infty([a,b]^2).$$

Proof. Let $\psi \in L^\infty([a,b]^2)$ and ϕ_1, ϕ_2 be defined as:

$$\begin{aligned} \phi_1(x,y) &= -\frac{1}{2}\int_a^y \psi(x,t)dt + \psi_1(x) = \widetilde{\phi}_1(x,y) + \psi_1(x) \\ \text{and} \qquad & \\ \phi_2(x,y) &= \frac{1}{2}\int_a^x \psi(t,y)dt + \psi_2(y) = \widetilde{\phi}_2(x,y) + \psi_2(y), \end{aligned} \tag{4.5}$$

where ψ_1, ψ_2 are two measurable functions on $[a,b]$. The rest of the proof follows easily. □

The following is a theorem about the trace formula for two variables in finite dimension (see [13]).

Theorem 4.6. *Let P and Q be two finite-dimensional projections in $\mathcal{H}$ and let (H_1^0, H_2^0) and (H_1, H_2) be two commuting pairs of self-adjoint operators acting in the reducing subspaces $P(\mathcal{H})$ and $Q(\mathcal{H})$ respectively. Let*

$$\psi \in L^\infty([a,b]^2),$$

where

$$\sigma(H_1), \sigma(H_2), \sigma(H_1^0), \sigma(H_2^0) \subseteq [a,b].$$

Then

$$\begin{aligned} \mathcal{I} \equiv \operatorname{Tr}\Big\{ &\int_{H_1^0}^{H_1} P\phi_1(x, H_2^0)Q dx + \int_{H_2^0}^{H_2} Q\phi_2(H_1, y)P dy \\ &+ \int_{H_1}^{H_1^0} P\phi_1(x, H_2)Q dx + \int_{H_2}^{H_2^0} Q\phi_2(H_1^0, y)P dy \Big\} \end{aligned}$$

$$
\begin{aligned}
&= \operatorname{Tr}\left\{\int_{H_1^0}^{H_1} P\left[\phi_1\left(x, H_2^0\right) - \phi_1\left(x, H_2\right)\right] Q dx \right.\\
&\quad \left. + \int_{H_2^0}^{H_2} Q\left[\phi_2\left(H_1, y\right) - \phi_2\left(H_1^0, y\right)\right] P dy\right\} \\
&= \int_a^b \int_a^b \left[\frac{\partial \phi_2}{\partial x}(x,y) - \frac{\partial \phi_1}{\partial y}(x,y)\right] \xi(x,y) dx dy \\
&= \int_a^b \int_a^b \psi(x,y)\xi(x,y) dx dy, \qquad (4.6)
\end{aligned}
$$

where

$$\xi(x,y) = \operatorname{Tr}\{Q\left[E_{H_1}(x) - E_{H_1^0}(x)\right] P \left[E_{H_2}(y) - E_{H_2^0}(y)\right] Q\}$$

and $E_{H_1}(.)$, $E_{H_2}(.)$, $E_{H_1^0}(.)$, $E_{H_2^0}(.)$ *are the spectral measures of the operators* H_1, H_2, H_1^0, H_2^0 *respectively and* ϕ_1, ϕ_2 *are the same as in* (4.5).

The next theorem gives another formula for the above Stokes-like expression $\mathcal{I}$ of operator functions under its trace in terms of divided differences, which is useful to control the measure generated by ξ [13].

Theorem 4.7. *Under the hypotheses of Theorem* 4.6,

$$
\begin{aligned}
\mathcal{I} = \int_{[a,b]^2} \int_{[a,b]^2} & \frac{\int_{x_2}^{x_1}\int_{y_2}^{y_1} \psi(x,y) dx dy}{(x_1 - x_2)(y_1 - y_2)} \\
& \times \left\langle \left(H_1 - H_1^0\right), P E_{\underline{H}^0}(dx_2 \times dy_1) \left(H_2 - H_2^0\right) E_{\underline{H}}(dx_1 \times dy_2) Q\right\rangle_2,
\end{aligned}
$$

where $\underline{H}^0 = (H_1^0, H_2^0)$, $\underline{H} = (H_1, H_2)$ *and* $E_{\underline{H}^0}(.)$ *and* $E_{\underline{H}}(.)$ *are the spectral measures of the operator tuples* $\underline{H}^0$ *and* $\underline{H}$ *respectively on the Borel sets of* $[a,b]^2$ *and where* $\langle .,.\rangle_2$ *denotes the inner product of the Hilbert space* $\mathcal{B}_2(\mathcal{H})$.

The next theorem shows how Theorem 4.2 can be used for reduction to a finite dimension (for $n = 2$). In the statement of the theorem below we apply Theorem 4.2 to the pairs (H_1^0, H_2^0) and (H_1, H_2) to get two commuting pairs of finite-dimensional self-adjoint operators $\left(H_1^{0(N)}, H_2^{0(N)}\right)$ and $\left(H_1^{(N)}, H_2^{(N)}\right)$ in $P_N^0(\mathcal{H})$ and $P_N(\mathcal{H})$ respectively, such that

$$\left\|[H_j^0, P_N^0]\right\|_p, \left\|P_N^0 H_j^0 P_N^0 - H_j^{0(N)} P_N^0\right\|_p \longrightarrow 0 \text{ as } N \longrightarrow \infty \text{ for } p \geq 2,\ j = 1,2 \qquad (4.7)$$

and

$$\left\|[H_j, P_N]\right\|_p, \left\|P_N H_j P_N - H_j^{(N)} P_N\right\|_p \longrightarrow 0 \text{ as } N \longrightarrow \infty \text{ for } p \geq 2,\ j = 1,2, \qquad (4.8)$$

where P_N^0, P_N are projections increasing to I (i.e., $P_N^0, P_N \uparrow I$).

Theorem 4.8. *Let (H_1^0, H_2^0) and (H_1, H_2) be two commuting pairs of bounded self-adjoint operators in a separable Hilbert space $\mathcal{H}$ such that $H_j - H_j^0 \equiv V_j \in \mathcal{B}_2(\mathcal{H})$ and such that $\sigma(H_j)$, $\sigma(H_j^0) \subseteq [a,b]$ for $j = 1,2$. Furthermore let*

$$p_1(x,y) = \sum_{0\leq i+j\leq n} c(i,j)x^i y^j, \qquad p_2(x,y) = \sum_{0\leq r+s\leq m} d(r,s)x^r y^s$$

be two polynomials in $[a,b]^2$ with complex coefficients. Then

$$\begin{aligned}
\mathcal{J} \equiv \operatorname{Tr}&\Big\{\int_{H_1^0}^{H_1} p_1(x, H_2^0)dx + \int_{H_2^0}^{H_2} p_2(H_1, y)dy \\
&+ \int_{H_1}^{H_1^0} p_1(x, H_2)dx + \int_{H_2}^{H_2^0} p_2(H_1^0, y)dy\Big\} \\
= \lim_{N\longrightarrow\infty} \operatorname{Tr}&\Big\{\int_{H_1^{0(N)}}^{H_1^{(N)}} P_N^0 \left[p_1\left(x, H_2^{0(N)}\right) - p_1\left(x, H_2^{(N)}\right)\right] P_N dx \\
&+ \int_{H_2^{0(N)}}^{H_2^{(N)}} P_N \left[p_2\left(H_1^{(N)}, y\right) - p_2\left(H_1^{0(N)}, y\right)\right] P_N^0 dy\Big\} \\
= \lim_{N\longrightarrow\infty} &\int_a^b\int_a^b \left[\frac{\partial p_2}{\partial x}(x,y) - \frac{\partial p_1}{\partial y}(x,y)\right] \xi_N(x,y)dxdy, \qquad (4.9)
\end{aligned}$$

where

$$\xi_N(x,y) = \operatorname{Tr}\left\{P_N\left[E_{H_1^{(N)}}(x) - E_{H_1^{0(N)}}(x)\right] P_N^0 \left[E_{H_2^{(N)}}(y) - E_{H_2^{0(N)}}(y)\right] P_N\right\}$$

and

$$E_{H_1^{0(N)}}(.),\ E_{H_2^{0(N)}}(.),\ E_{H_1^{(N)}}(.),\ E_{H_2^{(N)}}(.)$$

are the spectral measures of the operators $H_1^{0(N)}$, $H_2^{0(N)}$, $H_1^{(N)}$, $H_2^{(N)}$ respectively and P_N^0, P_N are the projections obtained by applying Theorem 4.2 to the pairs (H_1^0, H_2^0) and (H_1, H_2) respectively, as mentioned above.

Lemma 4.9. *Let H_j^0, H_j, $H_j^{0(N)}$, $H_j^{(N)}$, P_N^0, P_N be as above for $j = 1,2$. Then*

(i) $\left\|P_N^0\left(H_j^0\right)^k - \left(P_N^0 H_j^0 P_N^0\right)^k\right\|_2$ *and* $\left\|P_N\left(H_j\right)^k - \left(P_N H_j P_N\right)^k\right\|_2 \longrightarrow 0$ *as $N \longrightarrow \infty$ for $j = 1,2$ and $k \geq 1$.*

(ii) $\left\|\left(H_j^0\right)^k P_N^0 - \left(P_N^0 H_j^0 P_N^0\right)^k\right\|_2$ *and* $\left\|\left(H_j\right)^k P_N - \left(P_N H_j P_N\right)^k\right\|_2 \longrightarrow 0$ *as $N \longrightarrow \infty$ for $j = 1,2$ and $k \geq 1$.*

(iii) $\left\|P_N^0\left[\left(P_N^0 H_j^0 P_N^0\right)^k - \left(P_N H_j P_N\right)^k\right] P_N\right\|_2$ *is uniformly bounded in N for $j = 1,2$ and $k \geq 1$.*

(iv) $\lim_{N\longrightarrow\infty} \operatorname{Tr}\left\{P_N^0\left[(H_2^0)^k - (H_2)^k\right]P_N\left[(H_1)^l - (H_1^0)^l\right]P_N^0\right\}$

$= \lim_{N\longrightarrow\infty} \operatorname{Tr}\left\{P_N^0\left[\left(H_2^{0(N)}\right)^k - \left(H_2^{(N)}\right)^k\right]P_N\left[\left(H_1^{(N)}\right)^l - \left(H_1^{0(N)}\right)^l\right]P_N^0\right\}$

for $k, l \geq 1$.

(v) $\left\|P_N\left(H_j^{(N)} - H_j^{0(N)}\right)P_N^0 - P_N V_j P_N^0\right\|_2 \longrightarrow 0$ *as* $N \longrightarrow \infty$ *for* $j = 1, 2$.

Proof of Theorem 4.8. Using the above lemma and applying Theorem 4.6 appropriately we can achieve the conclusion of the theorem. For details of the proof see [13]. □

Now we are in a position to state and sketch the proof of our main result.

Theorem 4.10. *Let* $\underline{H}^0 = (H_1^0, H_2^0)$ *and* $\underline{H} = (H_1, H_2)$ *be two commuting tuples of bounded self-adjoint operators in a separable Hilbert space* $\mathcal{H}$ *such that* $H_j - H_j^0 \equiv V_j \in \mathcal{B}_2(\mathcal{H})$ *for* $j = 1, 2$. *Then there exists a unique complex Borel measure* μ *on* $[a, b]^2$ *such that*

$$\operatorname{Tr}\left\{\int_{H_1^0}^{H_1} p_1(x, H_2^0)dx + \int_{H_2^0}^{H_2} p_2(H_1, y)dy + \int_{H_1}^{H_1^0} p_1(x, H_2)dx + \int_{H_2}^{H_2^0} p_2(H_1^0, y)dy\right\}$$

$$= \operatorname{Tr}\left\{\int_{H_1^0}^{H_1}\left[p_1(x, H_2^0) - p_1(x, H_2)\right]dx + \int_{H_2^0}^{H_2}\left[p_2(H_1, y) - p_2(H_1^0, y)\right]dy\right\}$$

$$= \int_{[a,b]^2}\left[\frac{\partial p_2}{\partial x}(x, y) - \frac{\partial p_1}{\partial y}(x, y)\right]\mu(dx \times dy),$$

where p_1 *and* p_2 *are two polynomials in* $[a, b]^2$ *and* $\bigcup_{j=1}^{2}\{\sigma(H_j)\bigcup\sigma(H_j^0)\} \subseteq [a, b]$.

Proof. By Theorems 4.8 and 4.6 corresponding to the tuples $\underline{H}^0$, $\underline{H}$, we conclude that

$$\mathcal{J} = \lim_{N\longrightarrow\infty}\int_a^b\int_a^b\left[\frac{\partial p_2}{\partial x}(x, y) - \frac{\partial p_1}{\partial y}(x, y)\right]\xi_N(x, y)dxdy. \tag{4.10}$$

For a Borel subset Δ of $[a, b]^2$, set

$$\mu_N(\Delta) = \int_\Delta \xi_N(x, y)dxdy$$

and observe that $\|\xi_N\|_\infty \leq 4\|P_N\|_2\|P_N^0\|_2$ and therefore μ_N is a complex Borel measure on $[a, b]^2$. Next we want to show that there exists a complex Borel measure μ on $[a, b]^2$ such that for a suitable subsequence $\{N_k\}$, μ_{N_k} converges weakly to μ, i.e.,

$$\lim_{k\longrightarrow\infty}\int_{[a,b]\times[a,b]}\psi(x, y)\mu_{N_k}(dx \times dy) = \int_{[a,b]\times[a,b]}\psi(x, y)\mu(dx \times dy)$$

for all $\psi(x,y) \in C([a,b]^2)$. Let $\psi(x,y) \in C([a,b]^2)$ and let $\phi_j (j=1,2)$ be given as in (4.5). Then by applying Theorem 4.7, for the pairs $\left(H_1^{(N)}, H_2^{(N)}\right)$ and $\left(H_1^{0(N)}, H_2^{0(N)}\right)$, we have that

$$\begin{aligned}
\mathcal{J}_N \equiv \operatorname{Tr}&\left\{\int_{H_1^{0(N)}}^{H_1^{(N)}} P_N^0 \phi_1\left(x, H_2^{0(N)}\right) P_N dx + \int_{H_2^{0(N)}}^{H_2^{(N)}} P_N^0 \phi_2\left(H_1^{(N)}, y\right) P_N dy\right.\\
&\left.+ \int_{H_1^{(N)}}^{H_1^{0(N)}} P_N^0 \phi_1\left(x, H_2^{(N)}\right) P_N dx + \int_{H_2^{(N)}}^{H_2^{0(N)}} P_N^0 \phi_2\left(H_1^{0(N)}, y\right) P_N dy\right\}\\
&= \int_{[a,b]^2}\int_{[a,b]^2} \frac{\int_{x_2}^{x_1}\int_{y_2}^{y_1} \psi(x,y)dxdy}{(x_1-x_2)(y_1-y_2)}\\
&\times \left\langle P_N^0\left(H_1^{(N)} - H_1^{0(N)}\right) P_N, P_N^0 E_{H_1^{0(N)}}(dx_2) E_{H_2^{0(N)}}(dy_1)\right.\\
&\quad\left.\times \left(H_2^{(N)} - H_2^{0(N)}\right) E_{H_1^{(N)}}(dx_1) E_{H_2^{(N)}}(dy_2) P_N\right\rangle_2,
\end{aligned} \tag{4.11}$$

Next we recall from Lemma 4.9 (v) that

$$\sup_N \left\| P_N\left(H_j^{(N)} - H_j^{0(N)}\right) P_N^0 \right\|_2 < C_j < \infty \quad \text{for } j = 1,2.$$

Thus by the property of a spectral measure in a Hilbert space, one has that

$$\begin{aligned}
&\Big\| \Big\langle P_N^0\left(H_1^{(N)} - H_1^{0(N)}\right) P_N, P_N^0 E_{H_1^{0(N)}}(\bullet) E_{H_2^{0(N)}}(\bullet)\\
&\qquad \times \left(H_2^{(N)} - H_2^{0(N)}\right) E_{H_1^{(N)}}(\bullet) E_{H_2^{(N)}}(\bullet) P_N \Big\rangle_2 \Big\|_{\text{var}}\\
&\leq \left\| P_N^0\left(H_1^{(N)} - H_1^{0(N)}\right) P_N \right\|_2 \left\| P_N^0\left(H_2^{(N)} - H_2^{0(N)}\right) P_N \right\|_2 < C_1 C_2.
\end{aligned}$$

On the other hand, by Theorems 4.6 and 4.7, we have that

$$\begin{aligned}
\mathcal{J}_N &\equiv \int_a^b\int_a^b \left[\frac{\partial \phi_2}{\partial x}(x,y) - \frac{\partial \phi_1}{\partial y}(x,y)\right] \xi_N(x,y) dxdy \equiv \int_{[a,b]^2} \psi(x,y) \mu_N(dx \times dy)\\
&= \int_{[a,b]^2}\int_{[a,b]^2} \frac{\int_{x_2}^{x_1}\int_{y_2}^{y_1} \psi(x,y)dxdy}{(x_1-x_2)(y_1-y_2)}\\
&\quad \cdot \left\langle P_N^0\left(H_1^{(N)} - H_1^{0(N)}\right) P_N, P_N^0 E_{H_1^{0(N)}}(dx_2) E_{H_2^{0(N)}}(dy_1)\right.\\
&\qquad \left. \cdot \left(H_2^{(N)} - H_2^{0(N)}\right) E_{H_1^{(N)}}(dx_1) E_{H_2^{(N)}}(dy_2) P_N\right\rangle_2,
\end{aligned} \tag{4.12}$$

for all $\psi(x,y) \in C([a,b]^2)$. Thus

$$\left| \int_{[a,b]^2} \psi(x,y)\mu_N(dx \times dy) \right| < C_1 C_2 \|\psi\|_\infty. \tag{4.13}$$

Thus, one can apply Helley's theorem (page 171, [38]) to conclude that there exists a subsequence μ_{N_k} of μ_N such that μ_{N_k} converges weakly to a unique complex Borel measure μ on $[a,b]^2$, i.e.,

$$\lim_{k \longrightarrow \infty} \int_{[a,b]^2} \psi(x,y)\mu_{N_k}(dx \times dy) = \int_{[a,b]^2} \psi(x,y)\mu(dx \times dy) \ \forall\ \psi \in C([a,b]^2). \tag{4.14}$$

This completes the proof, by applying this conclusion to the right-hand side of the equation (4.9). □

5. Trace formula for Toeplitz operators and Helton–Howe theorem

This section deals with traces of commutators of Toeplitz operators [29]. Consider the circle $\mathbb{T}$ with its normalized arc length measure (= Haar measure), denoted by dz, and write $L^p(\mathbb{T})$ for $L^p(\mathbb{T}, dz)$. Thus, if $f \in L^1(\mathbb{T})$, then $\int f(z)dz = \frac{1}{2\pi}\int_0^{2\pi} f(e^{it})dt$.

For $n \in \mathbb{Z}_+$, the set of non-negative integers, we define $\varepsilon_n : \mathbb{T} \longrightarrow \mathbb{T}$ by $\varepsilon_n(z) = (2\pi)^{-\frac{1}{2}} z^n$ so that $\{\varepsilon_n\}_{n \in \mathbb{Z}_+}$ is an orthonormal set in $L^2(\mathbb{T})$ and constitutes a complete orthonormal basis of the Hilbert subspace $H^2(\mathbb{T})$.

It may be noted that $H^2(\mathbb{T})$ is known as the Hardy space of the circle – the space of all square integrable boundary values of functions holomorphic on the unit disc $\mathbb{D}$. It is easy to verify that for $\phi \in L^\infty(\mathbb{T})$ the inclusion $\phi H^2(\mathbb{T}) \subseteq H^2(\mathbb{T})$ holds if and only if $\phi \in H^\infty(\mathbb{T})$.

A prime example of an operator with trace class "self-commutator" is the unilateral shift T_z: for the Hilbert space $H^2(\mathbb{T})$, defined by $T_z(\varepsilon_n) = \varepsilon_{n+1}$ $(n \in \mathbb{Z}_+)$. Then T_z is an isometry and $[T_z^*, T_z] = |\varepsilon_0\rangle\langle\varepsilon_0|$, a rank one projection.

Recall that $\phi \in L^\infty(\mathbb{T})$, then the Toeplitz operator T_ϕ with symbol ϕ is defined by

$$T_\phi(f) = P(\phi f) \text{ so that } T_\phi = PM_\phi P \text{ on } L^2(\mathbb{T}),$$

for $f \in H^2(\mathbb{T}) \subseteq L^2(\mathbb{T})$, where P is the projection in $L^2(\mathbb{T})$ onto $H^2(\mathbb{T})$ and M_ϕ is the operator of multiplication by ϕ in $L^2(\mathbb{T})$. Clearly $\|T_\phi\| \leq \|\phi\|_\infty$. Moreover, the map

$$L^\infty(\mathbb{T}) \ni \phi \mapsto T_\phi \in \mathcal{B}(H^2(\mathbb{T}))$$

is linear and preserves adjoints; that is $T_\phi^* = T_{\bar{\phi}}$. Therefore, if $\phi = \bar{\phi}$, then T_ϕ is self-adjoint. Let us start with a lemma which will be useful in the sequel.

Lemma 5.1. *For $z \in \mathbb{T}$ and $m, n \in \mathbb{Z}$, a Toeplitz operator as discussed above has the following properties:*

(i) $[T_{z^n}, T_{z^m}] = 0$ *if* $mn \geq 0$,

(ii) $[T_{z^n}, T_{\bar{z}}] = -|\varepsilon_{n-1}\rangle\langle\varepsilon_0|$ *for* $n \geq 1$,

(iii) $[T_{z^n}, T_{\bar{z}^m}] = -\sum_{k=0}^{n-1} |\varepsilon_k\rangle\langle\varepsilon_{m-n+k}|$ *for* $m \geq n \geq 1$,

(iv) $[T_{z^n}, T_{\bar{z}^m}] = -\sum_{k=0}^{m-1} |\varepsilon_{n-m+k}\rangle\langle\varepsilon_k|$ *for* $n \geq m \geq 1$,

(v) $\mathrm{Tr}\{[T_{z^n}, T_{\bar{z}^m}]\} = -m\delta_{mn}$.

Proof. Note that $T_{\bar{z}} = T_z^*$ and for $z \in \mathbb{T}$ and $n \in \mathbb{Z}_+$, we have $\bar{z}^n = z^{-n}$.

(i) Statement (i) is a simple consequence of the definitions.

(ii) $[T_{z^n}, T_{\bar{z}}] = [T_z^n, T_{\bar{z}}] = \sum_{r=0}^{n-1} T_z^{n-1-r}[T_z, T_{\bar{z}}]T_z^r = -\sum_{r=0}^{n-1} T_z^{n-1-r}|\varepsilon_0\rangle\langle\varepsilon_0|T_z^r = -T_z^{n-1}|\varepsilon_0\rangle\langle\varepsilon_0| = -|\varepsilon_{n-1}\rangle\langle\varepsilon_0|$,
where we have used the fact that $T_{\bar{z}}^r|\varepsilon_0\rangle = 0$ for all $r > 0$.

(iii) For $m \geq n$, we have $[T_{z^n}, T_{\bar{z}^m}] = [T_{z^n}, T_{\bar{z}}^m] = \sum_{r=0}^{m-1} T_{\bar{z}}^r[T_{z^n}, T_{\bar{z}}]T_{\bar{z}}^{m-r-1} = -\sum_{r=0}^{m-1} T_{\bar{z}}^r|\varepsilon_{n-1}\rangle\langle\varepsilon_0|T_{\bar{z}}^{m-r-1} = -\sum_{r=0}^{n-1} T_{\bar{z}}^r|\varepsilon_{n-1}\rangle\langle\varepsilon_0|T_{\bar{z}}^{m-r-1} = -\sum_{r=0}^{n-1} |\varepsilon_{n-r-1}\rangle\langle\varepsilon_{m-r-1}| = -\sum_{k=0}^{n-1} |\varepsilon_k\rangle\langle\varepsilon_{m-n+k}|$,
where we have used the fact that $T_{\bar{z}}^r|\varepsilon_{n-1}\rangle = 0$ for $n < r \leq m-1$ and in the last equality we have changed the summation index by setting $n - r - 1 = k$.

(iv) Property (iv) follows from (iii) by taking adjoint and interchanging m and n.

(v) It is easy to see that for $m > n$,

$$\mathrm{Tr}\{[T_{z^n}, T_{\bar{z}^m}]\} = -\sum_{k=0}^{n-1} \mathrm{Tr}\{|\varepsilon_k\rangle\langle\varepsilon_{m-n+k}|\} = -\sum_{k=0}^{n-1} \langle\varepsilon_k, \varepsilon_{m-n+k}\rangle = 0$$

and similarly

$$\mathrm{Tr}\{[T_{z^n}, T_{\bar{z}^m}]\} = 0 \text{ for } n > m.$$

On the other hand for $n = m$, we have

$$\mathrm{Tr}\{[T_{z^n}, T_{\bar{z}^n}]\} = -\sum_{k=0}^{n-1} \mathrm{Tr}\{|\varepsilon_k\rangle\langle\varepsilon_k|\} = -n$$

and hence the conclusion (v) follows. □

Now let us assume that $f, g \in C^2(\mathbb{T})$, of which $\tilde{f}$ and $\tilde{g}$ are a C^2-extensions respectively, to the unit disc $\mathbb{D}$. The following theorem is a Helton–Howe type theorem in this simple case ([21], [22], [23], [33]).

Theorem 5.2. *Let $f, g \in C^2(\mathbb{T})$ with $\tilde{f}, \tilde{g} \in C^2(\mathbb{D})$ as above. Then $[T_f, T_g]$ is a trace class operator and*

$$\mathrm{Tr}\{[T_f, T_g]\} = \frac{1}{2\pi \mathrm{i}} \int_{\mathbb{D}} J(\tilde{f}, \tilde{g}) dz d\bar{z},$$

where $J(\tilde{f}, \tilde{g}) = \frac{\partial \tilde{f}}{\partial z}\frac{\partial \tilde{g}}{\partial \bar{z}} - \frac{\partial \tilde{g}}{\partial z}\frac{\partial \tilde{f}}{\partial \bar{z}}$ is the Jacobian of $\tilde{f}$ and $\tilde{g}$ in $\mathbb{D}$ and $dzd\bar{z}$ is the Lebesgue measure on $\mathbb{D}$.

Proof. Since $f, g \in C^2(\mathbb{T}) \subseteq L^2(\mathbb{T})$, then the Fourier series expansions of f and g are as follows:

$$\psi(z, \bar{z}) = \sum_{n=1}^{\infty} \langle \psi, \varepsilon_n \rangle z^n + \sum_{n=1}^{\infty} \langle \psi, \varepsilon_{-n} \rangle \bar{z}^n + \langle \psi, \varepsilon_0 \rangle$$

where $\langle \psi, \varepsilon_n \rangle = \frac{1}{2\pi} \int_0^{2\pi} \psi(t) \mathrm{e}^{-\mathrm{i}nt} dt$ $(n \in \mathbb{Z})$ and $\sum\limits_{n \in \mathbb{Z}} |\langle \psi, \varepsilon_n \rangle|^2 < \infty$ for $\psi = f$ and g. From the assumption $f, g \in C^2(\mathbb{T})$ it follows that

$$\sum_{n \in \mathbb{Z}} |\langle \psi, \varepsilon_n \rangle| < \infty \text{ for } \psi = f \text{ and } g. \tag{5.1}$$

Indeed, by doing integration by parts twice we get the following for $n \in \mathbb{Z} \setminus \{0\}$,

$$\begin{aligned}
\langle f, \varepsilon_n \rangle &= \frac{1}{2\pi} \int_0^{2\pi} f(t) \mathrm{e}^{-\mathrm{i}nt} dt = \frac{1}{2\pi} \left\{ f(t) \frac{\mathrm{e}^{-\mathrm{i}nt}}{-\mathrm{i}n} \Big|_{t=0}^{2\pi} - \int_0^{2\pi} f'(t) \frac{\mathrm{e}^{-\mathrm{i}nt}}{-\mathrm{i}n} dt \right\} \\
&= \frac{1}{2\mathrm{i}\pi n} \int_0^{2\pi} f'(t) \mathrm{e}^{-\mathrm{i}nt} dt = \frac{1}{2\mathrm{i}\pi n} \left\{ f'(t) \frac{\mathrm{e}^{-\mathrm{i}nt}}{-\mathrm{i}n} \Big|_{t=0}^{2\pi} - \int_0^{2\pi} f''(t) \frac{\mathrm{e}^{-\mathrm{i}nt}}{-\mathrm{i}n} dt \right\} \\
&= \frac{1}{2\pi (\mathrm{i}n)^2} \int_0^{2\pi} f''(t) \mathrm{e}^{-\mathrm{i}nt} dt
\end{aligned}$$

and therefore

$$|\langle \psi, \varepsilon_n \rangle| \leq \frac{C_\psi}{n^2}, \text{ which implies that } \sum_{n \in \mathbb{Z}} |\langle \psi, \varepsilon_n \rangle| < \infty \text{ with } C_\psi = \frac{1}{2\pi} \int_0^{2\pi} |\psi''(t)| dt \tag{5.2}$$

for $\psi = f$ and g. Next we define

$$f_{j,k}(z, \bar{z}) = \sum_{n=1}^{j} \langle f, \varepsilon_n \rangle z^n + \sum_{n=1}^{k} \langle f, \varepsilon_{-n} \rangle \bar{z}^n + \langle f, \varepsilon_0 \rangle$$

and

$$g_{l,m}(z, \bar{z}) = \sum_{n=1}^{l} \langle g, \varepsilon_n \rangle z^n + \sum_{n=1}^{m} \langle g, \varepsilon_{-n} \rangle \bar{z}^n + \langle g, \varepsilon_0 \rangle.$$

Using the linearity of the map $L^\infty(\mathbb{T}) \ni \phi \mapsto T_\phi \in \mathcal{B}(H^2(\mathbb{T}))$, we get that

$$T_{f_{j,k}} = \sum_{n=1}^{j} \langle f, \varepsilon_n \rangle T_{z^n} + \sum_{n=1}^{k} \langle f, \varepsilon_{-n} \rangle T_{\bar{z}^n} + \langle f, \varepsilon_0 \rangle$$

and

$$T_{g_{l,m}} = \sum_{n=1}^{l} \langle g, \varepsilon_n \rangle T_{z^n} + \sum_{n=1}^{m} \langle g, \varepsilon_{-n} \rangle T_{\bar{z}^n} + \langle g, \varepsilon_0 \rangle.$$

Next by using the estimate (5.2) and the fact that $\|T_f\| \leq \|f\|_\infty$, we conclude that $T_{f_{j,k}} \longrightarrow T_f$ in operator norm as $j, k \longrightarrow \infty$ and

$$T_\psi = \sum_{n=1}^{\infty} \langle \psi, \varepsilon_n \rangle T_{z^n} + \sum_{n=1}^{\infty} \langle \psi, \varepsilon_{-n} \rangle T_{\bar{z}^n} + \langle \psi, \varepsilon_0 \rangle \in \mathcal{B}(H^2(\mathbb{T}))$$

for $\psi = f$ and g. By (ii) of Lemma 5.1, we have that

$$\left[T_{f_{j,k}}, T_{g_{l,m}}\right] = \sum_{n,n'=1}^{j,m} \langle f, \varepsilon_n \rangle \langle g, \varepsilon_{-n'} \rangle \left[T_{z^n}, T_{\bar{z}^{n'}}\right] + \sum_{n,n'=1}^{k,l} \langle f, \varepsilon_{-n} \rangle \langle g, \varepsilon_{n'} \rangle \left[T_{\bar{z}^n}, T_{z^{n'}}\right]$$

and

$$\left[T_f, T_g\right] = \sum_{n,n'=1}^{\infty} \langle f, \varepsilon_n \rangle \langle g, \varepsilon_{-n'} \rangle \left[T_{z^n}, T_{\bar{z}^{n'}}\right] + \sum_{n,n'=1}^{\infty} \langle f, \varepsilon_{-n} \rangle \langle g, \varepsilon_{n'} \rangle \left[T_{\bar{z}^n}, T_{z^{n'}}\right].$$

Note that $\left[T_{f_{j,k}}, T_{g_{l,m}}\right]$ is a finite rank operator and therefore

$$\left[T_{f_{j,k}}, T_{g_{l,m}}\right] \in \mathcal{B}_1(\mathcal{H}) \quad \text{for all } j, k, l, m \in \mathbb{N}.$$

Next we want to show that

$$\left[T_f, T_g\right] = \lim_{j,k,l,m \longrightarrow \infty} \left[T_{f_{j,k}}, T_{g_{l,m}}\right] \text{ in trace norm } \|.\|_1.$$

To prove the above claim first note that

$$\begin{aligned} &\left[T_f, T_g\right] - \left[T_{f_{j,k}}, T_{g_{l,m}}\right] \\ &= \sum_{\substack{n=j+1 \\ n'=m+1}}^{\infty} \langle f, \varepsilon_n \rangle \langle g, \varepsilon_{-n'} \rangle \left[T_{z^n}, T_{\bar{z}^{n'}}\right] + \sum_{\substack{n=k+1 \\ n'=l+1}}^{\infty} \langle f, \varepsilon_{-n} \rangle \langle g, \varepsilon_{n'} \rangle \left[T_{\bar{z}^n}, T_{z^{n'}}\right]. \end{aligned} \tag{5.3}$$

The two terms in the right-hand side of equation (5.3) are similar and therefore it is enough to deal with only the first term.

Thus by Lemma 5.1,

$$
\begin{aligned}
&\sum_{\substack{n=j+1\\ n'=m+1}}^{\infty} \langle f,\varepsilon_n\rangle\langle g,\varepsilon_{-n'}\rangle\,[T_{z^n},T_{\bar{z}^{n'}}]\\
&= \sum_{\substack{n=n'\\ =\min\{j,m\}+1}}^{\infty} \langle f,\varepsilon_n\rangle\langle g,\varepsilon_{-n}\rangle\,[T_{z^n},T_{\bar{z}^n}] + \sum_{\substack{n>n'\\ n\geq j+1\\ n'\geq m+1}}^{\infty} \langle f,\varepsilon_n\rangle\langle g,\varepsilon_{-n'}\rangle\,[T_{z^n},T_{\bar{z}^{n'}}]\\
&\quad + \sum_{\substack{n<n'\\ n\geq j+1\\ n'\geq m+1}}^{\infty} \langle f,\varepsilon_n\rangle\langle g,\varepsilon_{-n'}\rangle\,[T_{z^n},T_{\bar{z}^{n'}}].
\end{aligned}
\tag{5.4}
$$

Next using the conclusion (iv) of Lemma 5.1, we compute the trace norm of the first term in the right-hand side of equation (5.4), we get

$$
\begin{aligned}
&\left\|\sum_{n=n'=\min\{j,m\}+1}^{\infty} \langle f,\varepsilon_n\rangle\langle g,\varepsilon_{-n}\rangle\,[T_{z^n},T_{\bar{z}^n}]\right\|_1 \leq \sum_{n=\min\{j,m\}+1}^{\infty} n|\langle f,\varepsilon_n\rangle||\langle g,\varepsilon_{-n}\rangle|\\
&\leq \left(\sum_{n=\min\{j,m\}+1}^{\infty} n^2|\langle f,\varepsilon_n\rangle|^2\right)^{\frac{1}{2}} \left(\sum_{n=\min\{j,m\}+1}^{\infty} |\langle g,\varepsilon_{-n}\rangle|^2\right)^{\frac{1}{2}} \longrightarrow 0
\end{aligned}
\tag{5.5}
$$

as $j,m\longrightarrow\infty$, by equation (5.2).

Again by using the conclusion (iv) of Lemma 5.1, we compute the trace norm of the second term in the right-hand side of equation (5.4), we have that

$$
\begin{aligned}
&\left\|\sum_{n>n';\ n\geq j+1,n'\geq m+1}^{\infty} \langle f,\varepsilon_n\rangle\langle g,\varepsilon_{-n'}\rangle\,[T_{z^n},T_{\bar{z}^{n'}}]\right\|_1\\
&\leq \sum_{n>n';\ n\geq j+1,n'\geq m+1}^{\infty} |\langle f,\varepsilon_n\rangle||\langle g,\varepsilon_{-n'}\rangle| \sum_{k=0}^{n'-1} \||\varepsilon_{n-n'+k}\rangle\langle\varepsilon_k|\|_1\\
&= \sum_{n>n';\ n\geq j+1,n'\geq m+1}^{\infty} |\langle f,\varepsilon_n\rangle||\langle g,\varepsilon_{-n'}\rangle| n' \leq C_g \sum_{n=j+1}^{\infty} |\langle f,\varepsilon_n\rangle| \sum_{n'=1}^{n-1}\frac{1}{n'}\\
&= C_g \sum_{n=j+1}^{\infty} |\langle f,\varepsilon_n\rangle| \left\{\left(\sum_{n'=1}^{n-1}\frac{1}{n'} - \log(n)\right) + \log(n)\right\}\\
&\leq C_f C_g \left\{\sum_{n=j+1}^{\infty}\frac{\gamma}{n^2} + \sum_{n=j+1}^{\infty}\frac{\log(n)}{n^2}\right\} \longrightarrow 0 \text{ as } j\longrightarrow\infty,
\end{aligned}
\tag{5.6}
$$

where we have used the fact that the sequence $\{v_n = 1 + \frac{1}{2} + \cdots + \frac{1}{n} - \log(n+1)\}$ is an increasing sequence, converging to the Euler constant γ. The last term in the right-hand side of (5.4) can be obtained from the second by taking the adjoint and by interchanging n and n', as well as $\bar{f}$ and g. This proves that

$$\left\| [T_f, T_g] - \left[T_{f_{j,k}}, T_{g_{l,m}}\right] \right\|_1 \longrightarrow 0 \text{ as } j,k,l,m \longrightarrow \infty.$$

Therefore

$$\begin{aligned}
\mathrm{Tr}\{[T_f, T_g]\} &= \lim_{j,k,l,m\longrightarrow\infty} \mathrm{Tr}\left\{\left[T_{f_{j,k}}, T_{g_{l,m}}\right]\right\} \\
&= \lim_{j,k,l,m\longrightarrow\infty} \left(\sum_{n,n'=1}^{j,m} \langle f, \varepsilon_n\rangle\langle g, \varepsilon_{-n'}\rangle \mathrm{Tr}\{[T_{z^n}, T_{\bar{z}^{n'}}]\} \right. \\
&\qquad \left. + \sum_{n,n'=1}^{k,l} \langle f, \varepsilon_{-n}\rangle\langle g, \varepsilon_{n'}\rangle \mathrm{Tr}\{[T_{\bar{z}^n}, T_{z^{n'}}]\} \right) \\
&= - \lim_{j,k,l,m\longrightarrow\infty} \left\{ \sum_{n,n'=1}^{j,m} \langle f, \varepsilon_n\rangle\langle g, \varepsilon_{-n'}\rangle n\delta_{nn'} + \sum_{n,n'=1}^{k,l} \langle f, \varepsilon_{-n}\rangle\langle g, \varepsilon_{n'}\rangle - n\delta_{nn'} \right\} \\
&= - \lim_{j,k,l,m\longrightarrow\infty} \left\{ \sum_{n=1}^{\min\{j,m\}} \langle f, \varepsilon_n\rangle\langle g, \varepsilon_{-n}\rangle n + \sum_{n=1}^{\min\{k,l\}} \langle f, \varepsilon_{-n}\rangle\langle g, \varepsilon_{n}\rangle - n \right\} \\
&= - \sum_{n=1}^{\infty} n\{\langle f, \varepsilon_n\rangle\langle g, \varepsilon_{-n}\rangle - \langle f, \varepsilon_{-n}\rangle\langle g, \varepsilon_n\rangle\}. \qquad (5.7)
\end{aligned}$$

Next consider $\tilde{f}$ and $\tilde{g}$ as C^2-extension of f and g respectively, to the open unit disc $\mathbb{D}$, given by:

$$\tilde{f}(z,\bar{z}) = \sum_{n=1}^{\infty} \langle f, \varepsilon_n\rangle z^n + \sum_{n=1}^{\infty} \langle f, \varepsilon_{-n}\rangle \bar{z}^n + \langle f, \varepsilon_0\rangle \qquad \text{for } z \in \mathbb{D}$$

and

$$\tilde{g}(z,\bar{z}) = \sum_{n'=1}^{\infty} \langle g, \varepsilon_{n'}\rangle z^{n'} + \sum_{n'=1}^{\infty} \langle g, \varepsilon_{-n'}\rangle \bar{z}^{n'} + \langle g, \varepsilon_0\rangle \quad \text{for } z \in \mathbb{D}.$$

Then by changing the indices of the summation appropriately we find that for $|z| < 1$, the Jacobian

$$\begin{aligned}
J(\tilde{f}, \tilde{g}) &= \frac{\partial \tilde{f}}{\partial z}\frac{\partial \tilde{g}}{\partial \bar{z}} - \frac{\partial \tilde{g}}{\partial z}\frac{\partial \tilde{f}}{\partial \bar{z}} \\
&= \left(\sum_{n=1}^{\infty} n\langle f, \varepsilon_n\rangle z^{n-1} \right) \left(\sum_{n'=1}^{\infty} n'\langle g, \varepsilon_{-n'}\rangle \bar{z}^{n'-1} \right) \\
&\quad - \left(\sum_{n'=1}^{\infty} n'\langle f, \varepsilon_{-n'}\rangle \bar{z}^{n'-1} \right) \left(\sum_{n=1}^{\infty} n\langle g, \varepsilon_n\rangle z^{n-1} \right)
\end{aligned}$$

$$= \sum_{n,n'=1}^{\infty} nn'\{\langle f, \varepsilon_n\rangle\langle g, \varepsilon_{-n'}\rangle - \langle f, \varepsilon_{-n'}\rangle\langle g, \varepsilon_n\rangle\} z^{n-1}\bar{z}^{n'-1}.$$

Thus

$$\begin{aligned}
&\int_{\mathbb{D}} J(\tilde{f}, \tilde{g}) dz d\bar{z} \\
&= \int_{\mathbb{D}} \left(\sum_{n,n'=1}^{\infty} nn'\{\langle f, \varepsilon_n\rangle\langle g, \varepsilon_{-n'}\rangle - \langle f, \varepsilon_{-n'}\rangle\langle g, \varepsilon_n\rangle\} \right) z^{n-1}\bar{z}^{n'-1} dz d\bar{z} \qquad (5.8)\\
&= \sum_{n,n'=1}^{\infty} nn'\{\langle f, \varepsilon_n\rangle\langle g, \varepsilon_{-n'}\rangle - \langle f, \varepsilon_{-n'}\rangle\langle g, \varepsilon_n\rangle\} \int_{\mathbb{D}} z^{n-1}\bar{z}^{n'-1} dz d\bar{z},
\end{aligned}$$

where we have used the estimate (5.2) and Fubini's theorem to interchange the summation and integration because

$$\begin{aligned}
&\sum_{n,n'=1}^{\infty} |nn'\{\langle f, \varepsilon_n\rangle\langle g, \varepsilon_{-n'}\rangle - \langle f, \varepsilon_{-n'}\rangle\langle g, \varepsilon_n\rangle\}| \left| \int_{\mathbb{D}} z^{n-1}\bar{z}^{n'-1} dz d\bar{z} \right| \\
&\quad \leq 4\pi \sum_{n,n'=1}^{\infty} nn'\{|\langle f, \varepsilon_n\rangle||\langle g, \varepsilon_{-n'}\rangle| + |\langle f, \varepsilon_{-n'}\rangle||\langle g, \varepsilon_n\rangle|\} \int_0^1 r^{n+n'-2} r dr \\
&\quad \leq 4C_f C_g \pi \sum_{n,n'=1}^{\infty} \frac{nn'}{n+n'} \left(\frac{1}{n^2}\right)\left(\frac{1}{n'^2}\right) < \infty.
\end{aligned}$$

Finally we get from (5.8) that

$$\begin{aligned}
\int_{\mathbb{D}} J(\tilde{f}, \tilde{g}) dz d\bar{z} = &-2\mathrm{i} \sum_{n,n'=1}^{\infty} nn'\{\langle f, \varepsilon_n\rangle\langle g, \varepsilon_{-n'}\rangle - \langle f, \varepsilon_{-n'}\rangle\langle g, \varepsilon_n\rangle\} \\
&\times \int_0^1 \int_0^{2\pi} r^{n+n'-2} e^{\mathrm{i}(n-n')\theta} r dr d\theta \\
= &-4\pi\mathrm{i} \sum_{n,n'=1}^{\infty} nn'\{\langle f, \varepsilon_n\rangle\langle g, \varepsilon_{-n'}\rangle - \langle f, \varepsilon_{-n'}\rangle\langle g, \varepsilon_n\rangle\} \frac{1}{n+n'} \delta_{nn'} \\
= &-2\pi\mathrm{i} \sum_{n=1}^{\infty} n\{\langle f, \varepsilon_n\rangle\langle g, \varepsilon_{-n}\rangle - \langle f, \varepsilon_{-n}\rangle\langle g, \varepsilon_n\rangle\}\}. \qquad (5.9)
\end{aligned}$$

The conclusion of the theorem follows by combining equations (5.7) and (5.9). □

Remark 5.1. The right-hand side of the equation in Theorem 5.2 is actually independent of the choice of the extension of f and g to the unit disc, since the left-hand side is. Furthermore, a simple calculation would show that for $\tilde{f}$, $\tilde{g}$ any C^2-extension of given $f, g \in C^2(\mathbb{T})$ to $\mathbb{D}$,

$$\int_{\mathbb{D}} d\tilde{f} \wedge d\tilde{g} = \int_{\mathbb{D}} d(\tilde{f} \wedge d\tilde{g}) = \int_{\mathbb{T}} f \wedge dg.$$

Acknowledgement

The first author acknowledges with thanks financial support from the Department of Atomic Energy, India through N.B.H.M. Post Doctoral Fellowship and the second author is grateful to Jawaharlal Nehru Centre for Advanced Scientific Research and SERB-Distinguished Fellowship for support.

References

[1] Amrein, W.O., Jauch, J.M. and Sinha, K.B. *Scattering theory in quantum mechanics*, Massachusetts, W.A. Benjamin, 1977.

[2] Baumgärtel, M.S. and Wollenberg, M. *Mathematical scattering theory*, Berlin, Akademie Verlag, 1983.

[3] Bhatia, R. *Matrix analysis*, Springer, New York, 1997.

[4] Birman, M.S. and Solomyak, M.Z. *On double stiljes operator integrals*, (Russian), Dokl. Akad. Nauk SSSR, 165, 1965, 1223–1226 (English translation in: Soviet Math. Dokl. 6, no. 2(1965), 1567–1571.).

[5] Birman, M.S. and Solomyak, M.Z. *Double Stieltjes operator integrals*, (Russian), 1966, Problems of Mathematical Physics, no. I, Spectral Theory and Wave Processes, pp. 33–67, Izdat, Leningrad Univ., Leningrad (English translation in: Topics in Mathematical Physics, vol. 1, pp. 25–54, Consultants Bureau Plenum Publishing Corporation, New York 1967).

[6] Birman, M.S. and Solomyak, M.Z.: *Double Stieltjes operator integrals. II*, (Russian), 1967 Problems of Mathematical Physics, no. 2, Spectral Theory, Diffraction Problems, pp. 26–60, Izdat, Leningrad Univ., Leningrad (English translation in: Topics in Mathematical Physics, vol. 2, pp. 19–46, Consultants Bureau, New York 1968).

[7] Birman, M.S. and Solomyak, M.Z. Topics in Math. Phys. (English translation) $\underline{1}$, 15 and $\underline{2}$, 19 (1969).

[8] Birman, M.S. and Solomyak, M.Z. *Remarks on the spectral shift function*, Zap. Nauch. Sem. Len. Otdel. Mat. Instt. Steklova, Akad Nauk. SSSR 27 (1972) 33–46 (English translation: J. Sov. Math. 3(4) (1975) 408–419).

[9] Birman, M.S. and Solomyak, M.Z. *Double operator integrals in a Hilbert space*, Integral Equations Operator Theory 47 (2003), 131–168.

[10] Boyadzhiev, K.N. *Mean value theorems for traces*, Math. Japonica, 38, no. 2 (1993), 217–224.

[11] Chattopadhyay, A. and Sinha, K.B. *Koplienko trace formula*, Integral Equations Operator Theory, volume 73, issue 4 (2012), 573–587.

[12] Chattopadhyay, A. and Sinha, K.B. *Third order trace formula*, Proc. Indian Acad. Sci. (Math. Sci.), vol. 123, no. 4, November (2013), 547–575.

[13] Chattopadhyay, A. and Sinha, K.B. *Trace formula for two variables*, arXiv:1402.0792v3 [math.FA] 5th May 2014.

[14] Clancy, K. *Seminormal Operators*, Lecture Notes in Mathematics 742, Springer Verlag, Heidelberg, 1979.

[15] Davidson, K.R. *C^*-Algebras by example*, Field Institute Monographs, American Mathematical Society, 1996.

[16] Dykema, K. and Skripka, A. *Higher order spectral shift*, J. Funct. Anal., 257 (2009), 1092–1132.

[17] Dykema, K. and Skripka, A. *On single commutators in II1 factors*, Proc. Amer. Math. Soc., 140 (2012), no. 3, 931–940.

[18] Dykema, K. and Skripka, A. *Perturbation formulas for traces on normed ideals*, Comm. Math. Phys., 325 (2014), no. 3, 1107–1138.

[19] Gesztesy, F., Pushnitski, A. and Simon, B. *On the Kopleinko spectral shift function*, I. Basics Zh. Mat. Fiz. Anal. Geom. 4, 1 (2008), 63–107.

[20] Gohberg, I.C. and Krein, M.G. *Introduction to the theory of linear non self-adjoint operators*, Translations of Mathematical Monographs, vol. 18, American Mathematical Society, Providence, R.I., 1969.

[21] Helton, J.W. and Howe, R. *Integral operators: commutators, traces, index and homology*, Proceedings of a Conference on Operator Theory (Dalhousie University, Halifax, N.S. 1973), Lecture Notes in Math. 345, Springer, Berlin, 1973, 141–209.

[22] Helton, J.W and Howe, R. *Traces of commutators of integral operators*, Acta Math. 135 (1975), 271–305.

[23] Howe, R. *Traces of commutators of integral operators – the aftermath*, Operator Theory: Advances and Applications, vol. 222, 221–231.

[24] Kato, T. *Perturbation theory of linear operators* (2nd ed.), New York, Springer Verlag, 1976.

[25] Koplienko, L.S. *Trace formula for nontrace-class perturbations*, Sibirsk. Mat. Zh., 25,5 (1984), 6–21 (Russian). English Translation: Siberian Math. J., 25,5 (1984),735–743.

[26] Krein, M.G. *On a trace formula in perturbation theory*, Matem. Sbornik 33 (1953), 597–626 (Russian).

[27] Krein, M.G. *On certain new studies in the perturbation theory for self-adjoint operators*, Topics in Differential and Integral equations and Operator theory, (Ed. I. Gohberg), OT 7 (Basel: Birkhäuser Verlag) 1983, 107–172.

[28] Kuroda, S.T. *On a generalization of Weinstein–Aronszajn formula and infinite determinant*, Sci. papers Coll. Gen. Ed. Univ. Tokyo, 11 (1961), 1–12.

[29] Murphy, G.J. *C^*-algebras and operator theory*, Academic Press, INC, ISBN 0-12-511360-9, 1990.

[30] Parthasarathy, K.R. *Introduction to Probability and Measure* (Delhi: Macmillan) 1977.

[31] Peller, V.V. *An extension of the Koplienko–Neidhardt trace formulae*, J. Funct. Anal. 221 (2005), 456–481.

[32] Peller, V.V. *Multiple operator integrals and higher operator derivatives*, J. Funct. Anal. 223 (2006), 515–544.

[33] Pincus, J.D. and XIA, D. *A trace formula for subnormal operator tuples*, Integral Equations Operator Theory, vol. 14 (1991), 390–398.

[34] Potapov, D., Sukochev, F and Zanin, D. *Krein's trace theorem revisited*, J. Spectr. Theory 4 (2014), no. 2, 415–430.

[35] Potapov, D., Skripka, A. and Sukochev, F. *Spectral shift function of higher order for contractions*, Proc. London Math. Soc., 108 (2014), no. 3, 327–349.

[36] Potapov, D., Skripka, A. and Sukochev, F. *Spectral shift function of higher order*, Invent. Math., 193 (2013), no. 3, 501–538.

[37] Potapov, D. and Sukochev, F. *Koplienko spectral shift function on the unit circle*, Comm. Math. Phys., vol. 309 (2012), no. 3, pp. 693–702.

[38] Royden, H.L. and Fitzpatrick, P.M. *Real analysis*, Fourth Edition, ISBN 978-0-13-143747-0, English reprint edition copyright 2010.

[39] Sinha, K.B. *On the theorem of M.G. Krein*, Preprint, Univ. of Geneva, Geneve, 1975.

[40] Sinha, K.B. and Mohapatra, A.N. *Spectral shift function and trace formula*, Proc. Indian Acad. Sci. (Math. Sci.), vol. 104, no. 4, November 1994, 819–853.

[41] Sinha, K.B. and Mohapatra, A.N. *Spectral shift function and trace formula for unitaries-a new proof*, Integral Equations Operator Theory 24, 1996, 285–297.

[42] Skripka, A. *Trace inequalities and spectral shift*, Oper. Matrices 3 (2009), no. 2, 241–260.

[43] Skripka, A. *Higher order spectral shift, II. Unbounded case*, Indiana Univ. Math. J. 59 (2010), no. 2, 691–706.

[44] Skripka, A. *Trace formulas for tuples of commuting contractions*, arXiv:1309.4797v2 [math.FA], 22 September 2013.

[45] Skripka, A. *Taylor approximations of operator functions* (survey), Operator Theory: Advances and Applications, to appear.

[46] Voiculescu, D. *Some results on norm-ideal perturbations of Hilbert space operators*, J. Operator Theory, 2 (1979), no. 1, 3–37.

[47] Voiculescu, D. *On a trace formula of M.G. Krein*, Operator Theory: Advances and Applications, vol. 24 (1987), 329–332 (Basel, Birkhäuser Verlag).

Arup Chattopadhyay
Indian Statistical Institute
Statistics and Mathematics Unit
8th Mile, Mysore Road
Bangalore, 560059, India
e-mail: `2003arupchattopadhyay@gmail.com`
`arup@isibang.ac.in`

Kalyan B. Sinha
J.N. Centre for Advanced Scientific Research
and Indian Institute of Science
Bangalore, India
e-mail: `kbs@jncasr.ac.in`

Operator Theory:
Advances and Applications, Vol. 247, 37–56
© Springer International Publishing Switzerland 2015

Continuous Minimax Theorems

Madhushree Basu and V.S. Sunder

Abstract. In matrix theory, there exist useful extremal characterizations of eigenvalues and their sums for Hermitian matrices (due to Ky Fan, Courant–Fischer–Weyl and Wielandt) and some consequences such as the majorisation assertion in Lidskii's theorem. In this paper, we extend these results to the context of self-adjoint elements of finite von Neumann algebras, and their distribution and quantile functions. This work was motivated by a lemma in [1] that described such an extremal characterization of the distribution of a self-adjoint operator affiliated to a finite von Neumann algebra – suggesting a possible analogue of the Courant–Fischer–Weyl minimax theorem for Hermitian matrices, for a self-adjoint operator in a finite von Neumann algebra[1].

Mathematics Subject Classification (2010). 46L10, 60B11, 34L15.

Keywords. Minimax, Ky Fan, Wielandt, Lidskii, Courant–Fischer–Weyl, II_1-factor, non-commutative probability.

1. Introduction

This paper is arranged as follows: in Section 2, we prove an extension of the 'classical' minimax theorem of Ky Fan ([5]) in a von Neumann algebraic setting for self-adjoint operators having no atoms in their distributions, and then, give a few applications in Section 3. First we state and prove an exact analogue of the Courant–Fischer–Weyl minimax theorem ([3]) for operators in non-commutative probability spaces satisfying a *continuity condition.* (Specifically we shall say a finite von Neumann algebra (M, τ) is continuous if $\{\tau(q) : q \in \mathcal{P}(M), q \leq p\} = [0, \tau(p)]\ \forall p \in \mathcal{P}(M)$.)

It is interesting to note that for matrices, the Courant–Fischer–Weyl minimax theorem preceded Ky Fan's theorem as is seen from the title of [5], whereas the order of events is reversed in our proofs. Then, as an application of our version of the Courant–Fischer–Weyl minimax theorem, we prove that for self-adjoint operators without eigenvalues in a 'continuous' finite von Neumann algebra (M, τ), the association of quantile functions to self-adjoint operators is an order-preserving one. Finally we discuss a continuous analogue of Lidskii's majorization relation between the eigenvalue-lists of two Hermitian matrices and their sum. Discussions and proofs of the finite-dimensional version can be found in [16], [15], [19]. In Section 4, we state and prove an analogue of Wielandt's minimax theorem ([19]), for $a = a^* \in M$, with both M and $A = W^*(a)$ being in the 'continuous case' in our sense. The matricial (and not 'continuous' in out sense) version of it yields an extremal characterization for arbitrary sums of eigenvalues of Hermitian matrices.

These and other continuous analogues of minimax-type results have been worked out earlier, for example in [4], [7] and [8], at the level of generality of unbounded operators affiliated to semi-finite von Neumann algebras equipped with a semi-finite trace. However in those papers, the emphasis has been on positive operators and the von Neumann algebraic versions of minimax-type results corresponded to singular values of Hermitian matrices. On the other hand, our proofs are simple, independent of the approach of these papers, deal explicitly with self-adjoint (as against positive) operators in certain von Neumann algebras and correspond to eigenvalues (as against singular values) of Hermitian matrices in the finite-dimensional case. Moreover as far as we know, unlike former works on this topic, our formulations, for the particular case of finite-dimensional matrix algebras, give the exact statements of Ky Fan, Courant–Fischer–Weyl's and Wielandt's theorems for matrices. However in the continuous case, our results are restricted to the case when both M and A (as above) are continuous.

In order to describe our results, which are continuous analogues of certain inequalities that appear among the set of inequalities mentioned in Horn's conjecture ([9]), it will be convenient to re-prove the well-known fact that any monotonic function with appropriate one-sided continuity is the distribution function of a random variable X – which can in fact be assumed to be defined on the familiar Lebesgue space $[0, 1)$ equipped with the Borel σ-algebra and Lebesgue measure. (We adopt the convention of [1] that the *distribution function* F_μ of a compactly supported probability measure[1] μ defined on the σ-algebra $\mathcal{B}_{\mathbb{R}}$ of Borel sets in $\mathbb{R}$, is *left-continuous*; thus $F_\mu(x) = \mu((-\infty, x).)$

Proposition 1.1. *If $F : \mathbb{R} \to [0, 1]$ is monotonically non-decreasing and left continuous and if there exists $\alpha, \beta \in \mathbb{R}$ with $\alpha < \beta$ such that*

$$F(t) = 0, \text{ for } t \leq \alpha \text{ and } F(t) = 1 \text{ for } t \geq \beta, \tag{1.1}$$

then there exists a monotonically non-decreasing right-continuous function X : $[0, 1) \to \mathbb{R}$ such that F is the distribution function of X, i.e., $F(t) = m(\{s : X(s) < t\})$, where m denotes the Lebesgue measure on $[0, 1)$. Moreover range$(X) \subset [\alpha, \beta]$.

[1]Actually Bercovici and Voiculescu considered possibly unbounded self-adjoint operators affiliated to M, so as to also be able to handle probability measures which are not necessarily compactly supported, but we shall be content with the case of bounded $a \in M$, having a compactly supported probability measure as its distribution.

Proof. Define $X : [0,1) \to \mathbb{R}$ by

$$\begin{aligned} X(s) &= \inf\{t : F(t) > s\} \\ &= \inf\{t : t \in E_s\}\ , \end{aligned} \tag{1.2}$$

where $E_s = \{t \in \mathbb{R} : F(t) > s\}\ \forall s \in [0,1)$. (The hypothesis (1.1) is needed to ensure that E_s is a non-empty bounded set for every $s \in [0,1)$ so that, indeed $X(s) \in \mathbb{R}$.)

First deduce from the monotonicity of F that

$$\begin{aligned} s_1 \leq s_2 &\Rightarrow E_{s_2} \subset E_{s_1} \\ &\Rightarrow X(s_1) \leq X(s_2) \end{aligned}$$

and hence X is indeed monotonically non-decreasing.

The definition of X and the fact that F is monotonically non-increasing and left continuous are easily seen to imply that $E_s = (X(s), \infty)$, and hence, it is seen that

$$\begin{aligned} X(s) < t &\Leftrightarrow \exists t_0 < t \text{ such that } F(t_0) > s \\ &\Leftrightarrow F(t) > s \text{ (since } F \text{ is left-continuous).} \end{aligned} \tag{1.3}$$

Hence, if $t \in \mathbb{R}$,

$$m(\{s \in [0,1) : X(s) < t\}) = m([0, F(t)) = F(t), \text{ proving the required statement.} \tag{1.4}$$

Moreover, if for any $s \in [0,1)$, $X(s) < \alpha$, then by definition of X, $\exists\, t' < \alpha$ such that $F(t') > s \geq 0$, a contradiction to the first hypothesis in equation (1.1). On the other hand, if for any $s \in [0,1)$, $X(s) > \beta$, then by (1.3), $s \geq F(\beta) = 1$ (by the second hypothesis in (1.1)), a contradiction. Hence indeed range$(X) \subset [\alpha, \beta]$. $\square$

This function X is known as a *quantile function*[2] of the distribution F. If $F = F_\mu$ for a probability measure μ on $\mathbb{R}$, then X is denoted as X_μ. The function X can also be thought of as an element of $L^\infty(\mathbb{R}, \mu)$, where μ is a compactly supported probability measure on $\mathbb{R}$ such that $\mu = m \circ X^{-1}$ and supp $\mu \subset [\alpha, \beta]$. It should be observed that the quantile function $X(s)$ (corresponding to the self-adjoint operator a) here is the non-decreasing version of the generalized s-numbers $\mu_s(a)$ in [4] as well as the spectral scale $\lambda_s(a)$ in [17]. We will elaborate further on this function later in Proposition 2.1.

Given a self-adjoint element a in a von Neumann algebra M and a (usually faithful normal) tracial state τ on M, define

$$\mu_a(E) := \tau(1_E(a)) \tag{1.5}$$

(for the associated scalar spectral measure) to be the *distribution of* a. Since τ is positivity preserving, μ_a indeed turns out to be a probability measure on $\mathbb{R}$.

[2]This function acts as the inverse of the distribution function at every point that is not an atom of the probability measure μ.

For simplicity we write F_a, X_a instead of F_{μ_a}, X_{μ_a} (to be pedantic, one should also indicate the dependence on (M, τ), but the trace τ and the M containing a will usually be clear). Note that only the abelian von Neumann subalgebra A generated by a and $\tau|_A$ are relevant for the definition of F_a and X_a.

For M, a, τ as above, it was shown in [1] that

$$1 - F_{\mu_a}(t) = \max\{\tau(p) : p \in \mathcal{P}(M), pap \geq tp\}. \tag{1.6}$$

Example 1.2. Let $M = M_n(\mathbb{C})$ with τ as the tracial state on this M. If $a = a^* \in M$ has distinct eigenvalues $\lambda_1 < \lambda_2 < \cdots < \lambda_n$, then $F_a(t) = \frac{1}{n}|\{j : \lambda_j < t\}| = \sum_{j=1}^n \frac{j}{n} 1_{(\lambda_j, \lambda_{j+1}]}$. We see that the distinct numbers less than 1 in the range of F_a are attained at the n distinct eigenvalues of a, and further that equation (1.6) for $t = \lambda_j$ says that $n - j + 1$ is the largest possible dimension of a subspace W of $\mathbb{C}^n$ such that $\langle a\xi, \xi\rangle \geq \lambda_j$ for every unit vector $\xi \in W$. In other words equation (1.6) suggests a possible extension of the matricial Courant–Fischer minimax theorem for a self-adjoint operator in a von Neumann algebra, involving its distribution.

It is also true and not hard to see that the right-hand side of equation (1.6) is indeed a maximum (and not just a supremum), and is in fact attained at a spectral projection of a; i.e., the two sides of equation (1.6) are also equal to $\max\{\tau(p) : p \in \mathcal{P}(A), pap \geq ta\}$, where $A = \{a\}''$.

2. Our version of Ky Fan's theorem

In this section we wish to proceed towards obtaining non-commutative counterparts of the matricial Ky Fan minimax theorem formulated for appropriate self-adjoint elements of appropriate finite von Neumann algebras. This result (Theorem 2.3) is not new – Lemma 4.1 of [4] but we give its detailed proof with our language in order to make the exposition of the paper self-contained.

Proposition 2.1. *Let $(\Omega, \mathcal{B}, P)$ be a probability measure space, and suppose $Y : \Omega \to \mathbb{R}$ is an essentially bounded random variable. Let $\sigma(Y) = \{Y^{-1}(E) : E \in \mathcal{B}_{\mathbb{R}}\}$ and let $\mu = P \circ Y^{-1}$ be the distribution of Y. Then, for any $s_0 \in F_\mu(\mathbb{R})$, we have*

$$\begin{aligned}\inf\left\{\int_{\Omega_0} Y dP : \Omega_0 \in \sigma(Y), P(\Omega_0) \geq s_0\right\} &= \inf\left\{\int_E f_0 d\mu : E \in \mathcal{B}_{\mathbb{R}}, \mu(E) \geq s_0\right\}\\ = \inf\left\{\int_G X_\mu dm : G \in \sigma(X_\mu), m(G) \geq s_0\right\} &= \int_0^{s_0} X_\mu dm,\end{aligned} \tag{2.1}$$

where $f_0 = id_{\mathbb{R}}$ and m denotes the Lebesgue measure on $[0, 1)$.

Proof. The version of the change of variable theorem we need says that if $(\Omega_i, \mathcal{B}_i, P_i)$, $i = 1, 2$ are probability spaces and $T : \Omega_1 \to \Omega_2$ is a measurable function such that $P_2 = P_1 \circ T^{-1}$, then

$$\int_{\Omega_2} g dP_2 = \int_{\Omega_1} g \circ T dP_1\ , \tag{2.2}$$

for every bounded measurable function $g : \Omega_2 \to \mathbb{R}$.

For every $\Omega_0 \in \sigma(Y)$ that is of the form $Y^{-1}(E)$ for some $E \in \mathcal{B}_{\mathbb{R}}$, set $G = X_\mu^{-1}(E)$. Notice, from equations (1.3) and (1.4) that

$$\begin{aligned} m \circ X_\mu^{-1}(-\infty, t) &= m(\{s \in [0,1) : X_\mu(s) < t\}) \\ &= m(\{s \in [0,1) : s < F_\mu(t)\}) \\ &= F_\mu(t) = \mu(-\infty, t); \end{aligned}$$

i.e., $m \circ X_\mu^{-1} = \mu = P \circ Y^{-1}$. Now, set $g = 1_E \cdot f_0$. Since $g \circ Y = 1_E \circ Y \cdot Y = 1_{Y^{-1}(E)} Y = 1_{\Omega_0} Y$, and (similarly) $g \circ X_\mu = 1_G X_\mu$, we see that the first two equalities in (2.1) are immediate consequences of two applications of the version stated in equation (2.2) above, of the 'change of variable' theorem.

As for the last equality, if $G \in \mathcal{B}_{[0,1)}$ with $m(G) \geq s_0$, then write $I = G \cap [0, s_0)$, $J = [0, s_0) \setminus I$, $K = G \setminus I$ and note that $G = I \coprod K, [0, s_0) = I \coprod J$ (where $\coprod$ denotes disjoint union, and $K = G \setminus [0, s_0) \subset [s_0, 1)$. So we may deduce that

$$\begin{aligned} \int_G X_\mu dm - \int_0^{s_0} X_\mu dm &= \int_K X_\mu dm - \int_J X_\mu dm \\ &\geq X_\mu(s_0) m(K) - X_\mu(s_0) m(J) \geq 0, \end{aligned}$$

since $s_1 \in J, s_2 \in K \Rightarrow s_1 \leq s_0 \leq s_2 \Rightarrow X_\mu(s_1) \leq X_\mu(s_0) \leq X_\mu(s_2)$ (by the monotonicity of X_μ), and $m(K) \geq m(J)$. Thus, we see that

$$\inf\{\int_G X_\mu dm : G \in \sigma(X_\mu), m(G) \geq s_0\} \geq \int_0^{s_0} X_\mu dm,$$

while conversely,

$$\inf\{\int_G X_\mu dm : G \in \sigma(X_\mu), m(G) \geq s_0\} \leq \int_{[0,s_0)} X_\mu dm = \int_0^{s_0} X_\mu dm,$$

thereby establishing the last equality in (2.1). □

Remark 2.2. With the same notations as in the above proposition, a change of variables gives us the following simple but useful equation that will be applied many times in this paper:

$$\int_0^{F(t)} X_\mu \, dm = \int_{-\infty}^t f_0 \, d\mu_a = \tau(a 1_{(-\infty,t]}(a)),$$

where μ is the distribution of a self-adjoint element a in a von Neumann algebra equipped with a faithful normal tracial state τ.

Theorem 2.3. *Let a be a self-adjoint element of a von Neumann algebra M equipped with a faithful normal tracial state τ. Let A be the von Neumann subalgebra generated by a in M and $\mathcal{P}(M)$ be the set of projections in M. Then, for all $s \in F_a(\mathbb{R})$,*

$$\inf\{\tau(ap) : p \in \mathcal{P}(M), \tau(p) \geq s\} = \inf\{\tau(ap) : p \in \mathcal{P}(A), \tau(p) \geq s\} = \int_0^s X_a dm \tag{2.3}$$

(hence the infima are attained and are actually minima), if either

1. (*'continuous case'*) *μ_a has no atoms, or*
2. (*'finite case'*) *$M = M_n(\mathbb{C})$ for some $n \in \mathbb{N}$ and a has spectrum $\{\lambda_1 < \lambda_2 < \cdots < \lambda_n\}$.*

Proof. We begin by noting that in both cases, the last equality in equation (2.3) is an immediate consequence of Proposition 2.1. Moreover the set $\{\tau(ap) : p \in \mathcal{P}(A), \tau(p) \geq s\}$ being contained in $\{\tau(ap) : p \in \mathcal{P}(M), \tau(p) \geq s\}$, it is clear that

$$\inf\{\tau(ap) : p \in \mathcal{P}(A), \tau(p) \geq s\} \geq \inf\{\tau(ap) : p \in \mathcal{P}(M), \tau(p) \geq s\}.$$

So we just need to prove that

$$\inf\{\tau(ap) : p \in \mathcal{P}(A), \tau(p) \geq s\} \leq \inf\{\tau(ap) : p \in \mathcal{P}(M), \tau(p) \geq s\}. \tag{2.4}$$

1. The continuous case. Due to the assumption of μ_a being compactly supported and having no atoms, it is clear that F_a is continuous and that $F_a(\mathbb{R}) = [0, 1]$.

Under the standing assumption of separability of pre-duals of our von Neumann algebras, the hypothesis of this case implies the existence of a probability space $(\Omega, \mathcal{B}, P)$ and a map $\pi : A \to L^\infty(\Omega, \mathcal{B}, P)$ such that $\int \pi(x)dP = \tau(x)\ \forall x \in A$, $Y := \pi(a)$ is a random variable and π is an isomorphism onto $L^\infty(\Omega, \sigma(Y), P)$.

We shall establish the first equality of (2.3) by showing that if $p' \in \mathcal{P}(M)$ and $\tau(p') = s$, then $\tau(ap') \geq \min\{\tau(ap) : p \in \mathcal{P}(A), \tau(p) \geq s\}$. For this, first note that since τ is a faithful normal tracial state on M, there exists a τ-preserving conditional expectation $\mathcal{E} : M \to A$. Then

$$\tau(ap') = \tau(a\mathcal{E}(p')) = \int YZdP,$$

where $Z = \pi(\mathcal{E}(p'))$. Since $\mathcal{E}$ is linear and positive, it is clear that $0 \leq Z \leq 1$ P-a.e. So it is enough to prove that

$$\begin{aligned} &\inf\left\{\int_\Omega YZdP : 0 \leq Z \leq 1, \int ZdP \geq s\right\} \\ &= \inf\left\{\int_E YdP : E \in \sigma(Y), P(E) \geq s\right\}. \end{aligned}$$

For this, it is enough, thanks to the Krein–Milman theorem (see, e.g., [13]), to note that $K = \{Z \in L^\infty(\Omega, \mathcal{B}, P) : 0 \leq Z \leq 1, \int ZdP \geq s\}$ is a convex set that is compact in the weak* topology inherited from $L^1(\Omega, \mathcal{B}, P)$, and prove that the set $\partial_e(K)$ of its extreme points is $\{1_E : P(E) \geq s\}$.

For this, suppose $Z \in K$ is not a projection, Clearly then $P(\{Z \in (0, 1)\}) > 0$, so there exists $\epsilon > 0$ such that $P(\{\epsilon < Z < 1 - \epsilon\}) > 0$. Since μ_a, and hence P has no atoms, we may find disjoint Borel subsets $E_1, E_2 \subset \{Z \in (\epsilon, 1 - \epsilon)\}$ such that $P(E_1) = P(E_2) > 0$. If we now set $Z_1 = Z + \epsilon(1_{E_1} - 1_{E_2})$ and $Z_2 = Z + \epsilon(1_{E_2} - 1_{E_1})$, it is not hard to see that $Z_1, Z_2 \in K, Z_1 \neq Z_2$ and $Z = \frac{1}{2}(Z_1 + Z_2)$ showing that $Z \notin \partial_e(K)$, thereby proving equation (2.4).

2. The finite case. Since a has distinct eigenvalues $\lambda_1 < \lambda_2 < \cdots < \lambda_n$, A is a maximal abelian self-adjoint subalgebra of $M_n(\mathbb{C})$. Recall that in this case, $F_a(t) = \frac{1}{n}|\{j : \lambda_j < t\}| = \sum_{j=1}^n \frac{j}{n} 1_{(\lambda_j, \lambda_{j+1}]}$. It then follows that $F_a(\mathbb{R}) = \{\frac{j}{n} : 0 \leq j \leq n\}$

and that $X_a = \sum_{j=1}^n \lambda_j 1_{[\frac{j-1}{n},\frac{j}{n})}$ and equation (2.3) is then (after multiplying by n) precisely the statement of Ky Fan's theorem (in the case of self-adjoint matrices with distinct eigenvalues):

For $1 \leq j \leq n$,

$$\begin{aligned}&\inf\{\tau(ap) : p \in \mathcal{P}(M_n(\mathbb{C})), \mathrm{rank}(p) \geq j\}\\ &\quad = \inf\{\tau(ap) : p \in \mathcal{P}(A), \mathrm{rank}(p) \geq j\} = \frac{1}{n}\sum_{i=1}^{j} \lambda_i = \int_0^{\frac{j}{n}} X_a(s)ds.\end{aligned}$$

It suffices to prove the following:

$$\inf\{\tau(ap) : p \in \mathcal{P}(A), \mathrm{rank}(p) \geq j\} \leq \inf\{\tau(ap) : p \in \mathcal{P}(M_n(\mathbb{C})), \mathrm{rank}(p) \geq j\}.$$

For this, begin by deducing from the compactness of $\mathcal{P}(M_n(\mathbb{C}))$ that there exists a $p_0 \in \mathcal{P}(M_n(\mathbb{C}))$ with $\mathrm{rank}(p_0) \geq j$ such that $\tau(ap_0) \leq \tau(ap)$ $\forall p \in \mathcal{P}(M_n(\mathbb{C}))$ with $\mathrm{rank}(p) \geq j$. We assert that any such minimizing p_0 must belong to A. The assumption that A is a masa (maximal abelian self-adjoint algebra) means we only need to prove that $p_0 a = ap_0$. For this pick any self-adjoint $x \in M_n(\mathbb{C})$, and consider the function $f : \mathbb{R} \to \mathbb{R}$ defined by $f(t) = \tau(e^{itx}p_0e^{-itx}a)$. Since clearly $e^{itx}p_0e^{-itx} \in \mathcal{P}(M)$ and $\mathrm{rank}(e^{itx}p_0e^{-itx}) = \mathrm{rank}(p_0) \geq j$, for all $t \in \mathbb{R}$, we find that $f(t) \geq f(0)$ $\forall t$. As f is clearly differentiable, we may conclude that $f'(0) = 0$. Hence,

$$0 = \tau(ixp_0a - ip_0xa) = i(\tau(xp_0a) - \tau(p_0xa)) = i(\tau(xp_0a) - \tau(xap_0)),$$

so that $\tau(x(p_0a - ap_0)) = 0$ for all $x = x^* \in M$, and indeed $ap_0 = p_0a$ as desired. □

Case 1 of Theorem 2.3 is our continuous formulation of Ky Fan's result while Case 2 only captures the classical Ky Fan theorem for the case of distinct eigenvalues. However the general case of non-distinct eigenvalues can also be deduced from our proof, as we show in the following corollary:

Corollary 2.4. *Let a be a Hermitian matrix in $M_n(\mathbb{C})$ with spectrum $\{\lambda_1 \leq \cdots \leq \lambda_n\}$, where not all λ_js are necessarily distinct. Then for all $j \in \{1, \ldots, n\}$,*

$$\min\{\tau(ap) : p \in \mathcal{P}(M_n(\mathbb{C})),\ \mathrm{rank}(p) \geq j\} = \frac{1}{n}\sum_{i=1}^{j} \lambda_i.$$

Proof. We may assume that a is diagonal. Let A_1 be the set of all diagonal matrices, so that $A \subsetneq A_1$. Pick $a^{(m)} = \mathrm{diag}(\lambda_1^{(m)}, \lambda_2^{(m)}, \ldots, \lambda_n^{(m)}) \in A_1$ such that $\lambda_j^{(m)}$s are all distinct and $\lim_{m\to\infty} \lambda_j^{(m)} = \lambda_j$ $\forall 1 \leq j \leq n$. Then the already established case of Theorem 2.3 in the case of distinct eigenvalues shows that for all $p \in \mathcal{P}(M_n(\mathbb{C}))$ with $\mathrm{rank}(p) \geq j$, we have

$$\tau(ap) = \lim_{m\to\infty} \tau(a^{(m)}p) \geq \lim_{m\to\infty} \frac{1}{n}\sum_{i=1}^{j} \lambda_i^{(m)} = \frac{1}{n}\sum_{i=1}^{j} \lambda_i.$$

The above, along with the fact that $\tau(ap_j) = \frac{1}{n}\sum_{i=1}^{j} \lambda_i$, where p_j is the obvious diagonal projection, completes our proof of Ky Fan's theorem for Hermitian matrices in full generality. □

Remark 2.5. It is not difficult to see that equation (2.3) holds even if we replace the inequality $\tau(p) \geq s$ with equality.

Remark 2.6. Notice that the hypothesis and hence the conclusion, of the 'continuous case' of Theorem 2.3 are satisfied by any self-adjoint generator of a masa in a II_1 factor.

3. Applications of our version of Ky Fan's theorem

In this section we discuss three applications[3] of our version of Ky Fan's theorem, the first one being a generalization of the classical Courant–Fischer–Weyl minimax theorem:

Theorem 3.1. *Let a be a self-adjoint element of a von Neumann algebra M equipped with a faithful normal tracial state τ. Let t_0 and $t_1 \in \mathbb{R}$ such that $t_0 < t_1$ and $F_a(t_1) - F_a(t_0) =: \delta > 0$. Then*

$$\int_{F_a(t_0)}^{F_a(t_1)} X_a(s)\, ds = \sup_{\substack{r\in\mathcal{P}(M)\\ \tau(r)\geq 1-F_a(t_0)}} \inf_{\substack{q\in\mathcal{P}(M)\\ q\leq r\\ \tau(q)=\delta}} \tau(aq), \tag{3.1}$$

if either

1. *('continuous case') if $B \in \{M, A\}$ (with A the von Neumann subalgebra generated by a in M as before) and $p \in \mathcal{P}(B)$, then $\{\tau(r) : r \in \mathcal{P}(B), r \leq p\} = [0, \tau(p)]$ (this assumption for $B = A$ amounts to requiring that μ_a has no atoms); or*
2. *('finite case') M is a type I_n factor for some $n \in \mathbb{N}$ and a has spectrum $\{\lambda_1 < \lambda_2 < \cdots < \lambda_n\}$.*

Moreover there exists $r_0 \in \mathcal{P}(A) \subset \mathcal{P}(M)$ with $\tau(r_0) \geq 1 - F(t_0)$ such that

$$\int_{F_a(t_0)}^{F_a(t_1)} X_a(s)\, ds = \min_{\substack{q\in\mathcal{P}(M)\\ q\leq r_0\\ \tau(q)=\delta}} \tau(aq),$$

so that the supremum is actually maximum.

Proof. For simplicity we write F and X for F_a and X_a respectively.

1. The continuous case. For proving "$\leq$", let $r_0 = 1_{[t_0,\infty)}(a)$ and $q_0 = 1_{[t_0,t_1)}(a)$. Then $\tau(r_0) = 1 - F(t_0), \tau(q_0) = \delta$ and $q_0 \leq r_0$.

[3] In the rest of the paper we frequently make use of the equation in Remark 2.2 without mentioning it.

If we consider any other $q \in A, q \leq r_0$ with $\tau(q) = \delta$, then q is of the form $1_E(a)$, such that $E \subset [t_0, \infty)$, $\mu_a(E) = \delta$. Arguing as in the proof of Proposition 2.1,

$$\int_{F(t_0)}^{F(t_1)} X(s)\,ds \leq \int_{F(E)} X(s)\,ds \Rightarrow \int_{t_0}^{t_1} t\,d\mu_a(t) \leq \int_E t\,d\mu_a(t)$$
$$\Rightarrow \tau(aq_0) \leq \tau(aq).$$

To prove the same for any $q \leq r_0$, first we note that since $r_0 \in W^*(\{a\})$, $(M_0, \tau_0) := \left(r_0 M r_0, \frac{\tau(\cdot)}{\tau(r_0)}\right)$ is also a von Neumann algebra (satisfying the same 'continuity hypotheses as M and A) equipped with a faithful normal tracial state and $a_0 := r_0 a r_0$ is a self-adjoint element with a continuous distribution μ_0 (with respect to τ_0) in it.

Let the von Neumann subalgebra generated by a_0 in M_0 be A_0.

Any $q \leq r_0$ with $\tau(q) = \delta$ can be thought of as $q \in \mathcal{P}(M_0)$ with $\tau_0(q) = \delta_0 := \frac{\delta}{\tau(r_0)}$, and conversely.

Now as in the proof of the continuous case of Theorem 2.3 we can assume that there exists a non-atomic probability space $(\Omega_0, \mathcal{B}_0, P_0)$ and a map $\pi_0 : A_0 \to L^\infty(\Omega_0, \mathcal{B}_0, P_0)$ such that $\int \pi_0(x) dP_0 = \tau_0(x)\ \forall x \in A_0$, $Y_0 := \pi_0(a_0)$ and π_0 is an isomorphism onto $L^\infty(\Omega_0, \sigma(Y_0), P_0)$.

It follows from Theorem 2.3 – applied to $a_0, A_0, M_0, \tau_0, Y_0, P_0, \delta_0$ – that there exists $E \in \sigma(Y_0)$ with $P_0(E) = \frac{\delta}{\tau(r_0)}$ such that

$$\min\left\{\int Y_0 Z_0\,dP_0 : Z_0 \in L^\infty(\Omega_0, \mathcal{B}_0, P_0), 0 \leq Z_0 \leq 1, \int Z_0 dP_0 = \frac{\delta}{\tau(r_0)}\right\}$$
$$= \int_E Y_0 dP_0.$$

Thus if $\pi_0(q_0) = 1_E$, we have

$$\tau_0(a_0 q_0) = \min_{\substack{q \in \mathcal{P}(M_0) \\ \tau_0(q) = \delta_0}} \tau_0(a_0 q)$$
$$\Rightarrow \frac{\tau(a_0 q_0)}{\tau(r_0)} = \min_{\substack{q \in \mathcal{P}(M) \\ q \leq r_0 \\ \frac{\tau(q)}{\tau(r_0)} = \frac{\delta}{\tau(r_0)}}} \frac{\tau(a_0 q)}{\tau(r_0)}$$
$$\Rightarrow \tau(aq_0) = \min_{\substack{q \in \mathcal{P}(M) \\ q \leq r_0 \\ \tau(q) = \delta}} \tau(aq), \text{ since } r_0 \text{ commutes with } a \text{ and any } q \leq r_0,$$

$$\Rightarrow \int_{F(t_0)}^{F(t_1)} X(s)\,ds \leq \sup_{\substack{r \in \mathcal{P}(M) \\ \tau(r) \geq 1 - F(t_0)}} \inf_{\substack{q \in \mathcal{P}(M) \\ q \leq r \\ \tau(q) = \delta}} \tau(aq). \tag{3.2}$$

For "$\geq$", let us choose any projection r with $\tau(r) \geq 1 - F(t_0)$.

Let $r_1 = 1_{(-\infty, t_1)}(a)$. Then $\tau(r_1) = F(t_1) \Rightarrow \tau(r_1 \wedge r) \geq F(t_1) - F(t_0) = \delta$.

Hence, by the hypothesis in this continuous case, $\exists\, q_1 \leq r \wedge r_1$ with $\tau(q_1) = \delta$.

Now consider the II_1 factor $(M_1, \tau_1) := \left(r_1 M r_1, \frac{\tau(\cdot)}{\tau(r_1)}\right)$, where τ_1 is a faithful normal tracial state on M_1. Then q_1 can be thought of as a projection in $\mathcal{P}(M_1)$ with $\tau_1(q_1) = \frac{\delta}{\tau(r_1)}$.

Note that $q_0 = 1_{[t_0,t_1)}(a) \leq r_1$.

As above $a_1 := r_1 a r_1$ is a self-adjoint element with continuous distribution in M_1. So we can consider our version of Ky Fan's theorem in M_1 (Theorem 2.3) (also see Remark 2.5):

$$\frac{\int_0^{F(t_0)} X(s)ds}{\tau(r_1)} = \tau_1(a(r_1 - q_0)) = \min_{\substack{q \in \mathcal{P}(M_1) \\ \tau_1(q) = \frac{F(t_0)}{\tau(r_1)}}} \tau_1(aq).$$

(using the fact that a, q_0 and $q \in \mathcal{P}(M_1)$ commute with r_1).

Subtracting both sides from $\tau_1(a_1)$ and writing q' for $r_1 - q$ in the index, we can rewrite it as:

$$\frac{\int_{F(t_0)}^{F(t_1)} X(s)\, ds}{\tau(r_1)} = \max_{\substack{q' \in \mathcal{P}(M_1) \\ \tau_1(q') = \frac{F(t_1) - F(t_0)}{\tau(r_1)} = \frac{\delta}{\tau(r_1)}}} \tau_1(aq'),$$

or equivalently,

$$\int_{F(t_0)}^{F(t_1)} X(s)\, ds = \max_{\substack{q' \in \mathcal{P}(M) \\ q' \leq r_1 \\ \tau(q') = \delta}} \tau(aq').$$

Now using the fact that $q_1 \leq r \wedge r_1$, we have:

$$\int_{F(t_0)}^{F(t_1)} X(s)\, ds = \max_{\substack{q' \in \mathcal{P}(M) \\ q' \leq r_1 \\ \tau(q') = \delta}} \tau(aq') \geq \tau(aq_1) \geq \inf_{\substack{q \in \mathcal{P}(M) \\ q \leq r \\ \tau(q) = \delta}} \tau(aq),$$

thus, and using the fact that our choice of r was arbitrary with $\tau(r) \geq 1 - F(t_0)$, we have:

$$\int_{F(t_0)}^{F(t_1)} X(s)\, ds \geq \sup_{\substack{r \in \mathcal{P}(M) \\ \tau(r) \geq 1 - F(t_0)}} \inf_{\substack{q \in \mathcal{P}(M) \\ q \leq r \\ \tau(q) = \delta}} \tau(aq). \tag{3.3}$$

Equations (3.2) and (3.3) together give us the required equality.

2. The finite case. Notice that if we set $t_0 = \lambda_i, t_1 = \lambda_{i+j}, \delta = \frac{j}{n}$, where $i, j \in \{1, \ldots, n\}$ such that $i + j - 1 \leq n$, equation (3.1) translates to:

$$\lambda_i + \lambda_{i+1} + \cdots + \lambda_{i+j-1} = \sup_{\substack{r \in \mathcal{P}(M_n(\mathbb{C})) \\ Tr(r) \geq n - i + 1}} \inf_{\substack{q \in \mathcal{P}(M_n(\mathbb{C})) \\ q \leq r \\ Tr(q) = j}} Tr(aq),$$

where Tr is the sum of the diagonal entries of matrices.

For the inequality "$\leq$" we prove,

$$\lambda_i + \lambda_{i+1} + \cdots + \lambda_{i+j-1} = Tr(aq_0) = \min_{\substack{q \in \mathcal{P}(M_n(\mathbb{C})) \\ q \leq r_0 \\ Tr(q)=j}} Tr(aq),$$

where $r_0 = 1_{\{\lambda_i, \lambda_{i+1}, \ldots, \lambda_n\}}(a)$ and $q_0 = 1_{\{\lambda_i, \lambda_{i+1}, \ldots, \lambda_{i+j-1}\}}(a)$, by first showing that any minimizing projection below r_0 has to commute with $r_0 a r_0$, and then using the fact that with distinct eigenvalues $r_0 a r_0$ generates a masa in $r_0 M_n(\mathbb{C}) r_0$, concluding that minimizing projections have to be spectral projections (see the exactly similar proof of the finite case of Theorem 2.3).

For proving "$\geq$", we start with an arbitrary projection r with $Tr(r) \geq n-i+1$ and note that if we define $r_1 := 1_{\{\lambda_1, \ldots, \lambda_{i+j-1}\}}(a)$, then $\exists\ q_1 \leq r \wedge r_1$ such that $Tr(q_1) = j$. Now we proceed using Ky Fan's theorem for a finite-dimensional Hermitian matrix $r_1 a r_1$ in $r_1 M_n(\mathbb{C}) r_1$, exactly as in the above proof of the continuous case of this theorem. □

Remark 3.2. Theorem 3.1 can equivalently be stated as:

$$\int_{F(t_0)}^{F(t_1)} X(s)\, ds = \inf_{\substack{p \in \mathcal{P}(M) \\ \tau(p) \geq F(t_1)}} \sup_{\substack{q \leq p \\ \tau(q)=\delta}} \tau(aq).$$

Moreover we can get the classical Courant–Fischer–Weyl minimax theorem for Hermitian matrices in full generality (i.e., involving non-distinct eigenvalues as well) from the above theorem in a similar manner as in Corollary 2.4.

The classical Courant–Fischer–Weyl minimax theorem has a natural corollary that says if a, b are Hermitian matrices in $M_n(\mathbb{C})$ such that $a \leq b$ (i.e., $b - a$ is positive semi-definite), and if $\{\alpha_1 \leq \cdots \leq \alpha_n\}$ and $\{\beta_1 \leq \cdots \leq \beta_n\}$ are their spectra respectively, then $\alpha_j \leq \beta_j$ for all $j \in \{1, \ldots, n\}$. As expected, Theorem 3.1 leads us to the same corollary for the 'continuous case':

Corollary 3.3. *Let M be a II_1 factor equipped with faithful normal tracial state τ. If $a, b \in M$ such that $a = a^*, b = b^*$ and μ_a, μ_b have no atoms. Then*

$$a \leq b \Rightarrow X_a \leq X_b. \tag{3.4}$$

Proof. Notice that since $a \leq b$ and τ is positivity preserving, we have

$$\tau(xax^*) \leq \tau(xbx^*) \tag{3.5}$$

for all $x \in M$.

Fix $0 \leq s_0 < s_1 < 1$.

By our assumptions on a and b, μ_a, μ_b are compactly supported probability measures with no atoms. Hence F_a and F_b are continuous functions with range(F_a) = range$(F_b) = [0,1]$. Thus $\exists\ t_0^a, t_1^a, t_0^b, t_1^b \in \mathbb{R}$ such that $s_0 = F_a(t_0^a) = F_b(t_0^b)$ and $s_1 = F_a(t_1^a) = F_b(t_1^b)$.

Now using Theorem 3.1,

$$\begin{aligned}\int_{s_0}^{s_1} X_a\, dm &= \sup_{\substack{r\in\mathcal{P}(M)\\ \tau(r)\geq 1-F_a(t_0^a)}} \inf_{\substack{q\in\mathcal{P}(M)\\ q\leq r\\ \tau(r)=s_1-s_0}} \tau(aq) = \sup_{\substack{r\in\mathcal{P}(M)\\ \tau(r)\geq 1-F_a(t_0^a)}} \inf_{\substack{q\in\mathcal{P}(M)\\ q\leq r\\ \tau(r)=s_1-s_0}} \tau(qaq)\\ &\leq \sup_{\substack{r\in\mathcal{P}(M)\\ \tau(r)\geq 1-F_b(t_0^b)}} \inf_{\substack{q\in\mathcal{P}(M)\\ q\leq r\\ \tau(r)=s_1-s_0}} \tau(qbq) \text{ by the inequality (3.5)}\\ &= \sup_{\substack{r\in\mathcal{P}(M)\\ \tau(r)\geq 1-F_b(t_0^b)}} \inf_{\substack{q\in\mathcal{P}(M)\\ q\leq r\\ \tau(r)=s_1-s_0}} \tau(bq) = \int_{s_0}^{s_1} X_b\, dm.\end{aligned}$$

This proves that

$$\int_I X_a\, dm \leq \int_I X_b\, dm \tag{3.6}$$

for any interval $I = [s_0, s_1) \subset [0,1)$, and in fact for any $I \in \mathcal{A} := \{\sqcup_{j=1}^k [s_0^j, s_1^j) : 0 \leq s_0^j < s_1^j < 1, k \in \mathbb{N}\}$.

But $\mathcal{A}$ is an algebra of sets which generates the σ-algebra $\mathcal{B}_{[0,1)}$. Thus for any Borel $E \subset [0,1)$, there exists a sequence $\{I_n : n \in \mathbb{N}\} \subset \mathcal{A}$ such that $\mu(I_n \Delta E) \to 0$.

Recall from Proposition 1.1 that our quantile functions of self-adjoint elements of von Neumann algebras are elements of $L^\infty([0,1), \mathcal{B}_{[0,1)}, m)$. We may hence deduce from the sentence following equation (3.6) that if E, I_n are as in the previous paragraph, we have:

$$\int_E X_a dm = \lim_{n\to\infty} \int_{I_n} X_a dm \leq \lim_{n\to\infty} \int_{I_n} X_a dm = \int_E X_b dm.$$

As $E \in \mathcal{B}_{[0,1)}$ was arbitrary, this shows that, $X_a \leq X_B$ m-a.e.; as X_a, X_b are continuous by our hypotheses, this shows that indeed $X_a \leq X_b$. □

The following application of a continuous version of Ky Fan's theorem gives a continuous analogue of a majorization result, which can be seen as a special case of Lidskii–Mirsky–Wielandt's theorem, or more popularly known as Lidskii's theorem. We will discuss this theorem in Section 5 as an application (Theorem 5.1) of Wielandt's theorem.

By Theorem 2.3, we have the following lemma:

Lemma 3.4. *If M is a von Neumann algebra with a faithful normal tracial state τ on it, then for $a = a^*, b = b^* \in M$ with μ_a, μ_b non-atomic and for all $s \in [0,1)$,*

$$\int_0^s X_{a+b}\, dm \geq \int_0^s (X_a + X_b)\, dm.$$

Moreover,

$$\int_0^1 X_{a+b}\, dm = \int_0^1 (X_a + X_b)\, dm.$$

Proof. Recall from our proof of Theorem 2.3 that there exists a projection $q \in \mathcal{P}(M)$ (in fact in the von Neumann algebra generated by $a+b$) such that $\tau(q) \geq s$ and

$$\begin{aligned}\int_0^s X_{a+b}\,dm &= \tau((a+b)q) = \tau(aq) + \tau(bq)\\ &\geq \inf\{\tau(ap) : p \in \mathcal{P}(M), \tau(p) \geq s\} + \inf\{\tau(bp) : p \in \mathcal{P}(M), \tau(p) \geq s\}\\ &= \int_0^s X_a\,dm + \int_0^s X_b\,dm = \int_0^s (X_a + X_b)\,dm.\end{aligned}$$

Finally, it is clear (from our change-of-variable argument in Proposition 2.1 for instance) that for any $c = c^* \in M$, we have $\int_0^1 X_c dm = \tau(c)$ and hence

$$\int_0^1 X_{a+b}\,dm = \tau(a+b) = \tau(a)+\tau(b) = \int_0^1 X_a\,dm + \int_0^1 X_b\,dm = \int_0^1 (X_a+X_b)\,dm.$$

□

The above is an analogue of the fact that for $n \times n$ Hermitian matrices a, b, with their eigenvalues $\lambda_1(a) \leq \cdots \leq \lambda_n(a)$ and $\lambda_1(b) \leq \cdots \leq \lambda_n(b)$, for all $k \in \{1, \ldots, n-1\}$,

$$\sum_{j=1}^{k} \lambda_j(a+b) \geq \sum_{j=1}^{k} \lambda_j(a) + \sum_{j=1}^{k} \lambda_j(b),$$

and

$$\sum_{j=1}^{n} \lambda_j(a+b) = \sum_{j=1}^{n} \lambda_j(a) + \sum_{j=1}^{n} \lambda_j(b),$$

i.e., $\lambda(a) + \lambda(b)$ is majorized by $\lambda(a+b)$ in the sense of [6].

We consider the definition of majorization in the continuous context (see for example, [18]) as follows:

Definition 3.5. For $a = a^*, b = b^*$ in a von Neumann algebra M with a faithful normal tracial state τ on it, a is said to be *majorized* by b if $\int_0^s X_a\,dm \geq \int_0^s X_b\,dm$ for all $s \in [0, 1)$ and $\int_0^1 X_a\,dm = \int_0^1 X_b\,dm$. When this happens, we simply write $X_a \prec X_b$.

Then, Lemma 3.4 can be written as:

$$X_{a+b} \prec X_a + X_b.$$

Majorization is a weaker concept of comparing self-adjoint operators in von Neumann algebras, for example, Corollary 3.3 together with Lemma 3.4 proves that for $a = a^*, b = b^*$ with $\tau(a) = \tau(b)$,

$$a \leq b \Rightarrow \sigma(a) \prec \sigma(b),$$

but the converse is easily seen to be not true. Similarly, it can be seen that an analogue of Lidskii's result does not imply that $X_{a+b} \leq X_a + X_b$. The study of majorization and its von Neumann algebraic analogue is vast (see for example, [10], [11], [12], [7]) and closely related to the minimax-type results but we will not discuss it further within this paper.

4. Continuous version of Wielandt's minimax principle

In this section we state and prove a continuous analogue of Wielandt's minimax theorem. As in the case of Theorem 3.1, our proof for the finite-dimensional version of Ky Fan's theorem would give a new proof for Wielandt's original result for Hermitian matrices too. But in order to avoid repetition we shall be content with the continuous case here. *We make the standing 'continuity assumption' throughout this section that:* (M, τ) is a von Neumann algebra with a faithful normal tracial state on it, $a = a^* \in M$ and $A = W^*(a)$ the generated commutative von Neumann subalgebra, and that: if $B \in \{M, A\}$, $r \in \mathcal{P}(B)$, then $\forall \epsilon \in [0, \tau(r)]$, $\exists r' \leq r$ in $\mathcal{P}(B)$, with $\tau(r') = \epsilon$. Thus our results are valid for any von Neumann algebra that admits of a faithful normal tracial state and has the above-mentioned property.

Our version of Wielandt's theorem is as follows:

Theorem 4.1. *Let F, X be the distribution and quantile function of a. Let $\delta_j \in \mathbb{R}_+$ and t_0^j, t_1^j, $j = 1, \ldots, k$, be points in the spectrum of a such that $t_0^1 < t_1^1 \leq t_0^2 < t_1^2 \leq \cdots \leq t_0^{k-1} < t_1^{k-1} \leq t_0^k < t_1^k$ and $F(t_1^j) - F(t_0^j) = \delta_j$, for all j. Then*

$$\sum_{j=1}^{k} \int_{[F(t_0^j), F(t_1^j))} X(s)\, ds = \inf_{\substack{p_j \in \mathcal{P}(M) \\ p_1 \leq \cdots \leq p_k \\ \tau(p_j) \geq F(t_1^j)}} \sup_{\substack{\hat{q}_j \in \mathcal{P}(M) \\ \hat{q}_j \leq p_j \\ \tau(\hat{q}_j) = \delta_j \\ \hat{q}_j \perp \hat{q}_i \text{ for } j \neq i}} \sum_{j=1}^{k} \tau(a\hat{q}_j).$$

Moreover, $\exists p_1 \leq \cdots \leq p_k$ with $p_j \in \mathcal{P}(A) \subset \mathcal{P}(M)$, for which there exist mutually orthogonal projections $\hat{q}_j \leq p_j, \tau(\hat{q}_j) = \delta_j$, $\forall j$ such that

$$\sum_{j=1}^{k} \int_{[F(t_0^j), F(t_1^j))} X(s)\, ds = \max_{\substack{\hat{q}_j \leq p_j \\ \tau(\hat{q}_j) = \delta_j \\ \hat{q}_j \perp \hat{q}_i}} \sum_{j=1}^{k} \tau(a\hat{q}_j).$$

The following lemmas lead to the proof of the theorem above:

Lemma 4.2. *Let (M, τ) be as above. Consider, for any $k \geq 2$,*

$$\begin{aligned}
&\{r_1, r_2, \ldots, r_k; q_1', \ldots, q_{k-1}'\} \subset \mathcal{P}(M),\\
&r_1 \geq \cdots \geq r_k,\\
&\tau(r_j) \geq \delta_k + \cdots + \delta_j \ \forall\ 1 \leq j \leq k,\\
&q_j' \leq r_j \ \forall 1 \leq j \leq k-1,\\
&q_s' q_t' = 0 \ \forall 1 \leq s < t \leq k-1,\\
&\tau(q_j') = \delta_j \ \forall 1 \leq j < k-1.
\end{aligned}$$

Then there exist mutually orthogonal projections $q_j \leq r_j$ $\forall 1 \leq j \leq k$ in M, such that $\sum_{j=1}^{k} q_j \geq \sum_{j=1}^{k-1} q'_j$, and $\tau(q_j) = \delta_j$ $\forall 1 \leq j \leq k$.

Proof. The proof follows by induction. For $k = 2$, choose $q_2 \leq r_2$ such that $\tau(q_2) = \delta_2$.

Let $e = q_2 \vee q'_1$. Then $\tau(e) \leq \tau(q_2) + \tau(q'_1) = \delta_2 + \delta_1$ and $e \leq r_1$.

But by the hypothesis for $k = 2$, $\tau(r_1) \geq \delta_2 + \delta_1$. Hence by the 'standing continuity assumption', there exists $f \in \mathcal{P}(M)$ such that $e \leq f \leq r_1$ and $\tau(f) = \delta_2 + \delta_1$. In particular $q_2 \leq e \leq f$; thus $f - q_2 \in \mathcal{P}(M)$ with trace δ_1.

Choose $q_1 = f - q_2$. Then $q_j \leq r_j$ with trace δ_j for $j = 1, 2$ and $q_1 + q_2 = f \geq e \geq q'_1$, as required.

Suppose now, for the inductive step, that this result holds with k replaced by $k-1$, and that $r_1, \ldots, r_k, q_1, \ldots, q_{k-1}$ are as in the statement of the Lemma.

By induction hypothesis – applied to $\{r_2, \ldots, r_k; q'_2, \ldots, q'_{k-1}\} \subset \mathcal{P}(M)$ – there exist mutually orthogonal projections $q_2, \ldots, q_k$ in M such that $q_j \leq r_j$ and $\tau(q_j) = \delta_j, \forall 2 \leq j \leq k$ and

$$\sum_{j=2}^{k} q_j \geq \sum_{j=2}^{k-1} q'_j. \tag{4.1}$$

Let $e_2 = q_2 + \cdots + q_k$ and $e = e_2 \vee q'_1$. Then $\tau(e) \leq \tau(e_2) + \tau(q'_1) = (\delta_k + \cdots + \delta_2) + \delta_1$ and $e \leq r_1$.

But $\tau(r_1) \geq \delta_k + \cdots + \delta_1$; thus (by the 'standing continuity assumption') there exists $f \in \mathcal{P}(M)$ such that $e \leq f \leq r_1$ and $\tau(f) = \delta_k + \cdots + \delta_1$. In particular $e_2 \leq e \leq f$; thus $f - e_2 \in \mathcal{P}(M)$ with trace δ_1.

Choose $q_1 = f - e_2$. Then $q_1 \leq r_1$ and $q_1 \perp q_j$ for $2 \leq j \leq k$.

Moreover,

$$\begin{aligned} q_1 + q_2 + \cdots + q_k = f \geq e = e_2 \vee q'_1 &= \left(\sum_{j=2}^{k} q_j\right) \vee q'_1 \\ &\geq \left(\sum_{2}^{k-1} q'_j\right) \vee q'_1 \text{ by equation (4.1)} \\ &= \sum_{1}^{k-1} q'_j, \end{aligned}$$

thus completing the proof of the inductive step. □

Lemma 4.2 can be rewritten as:

Lemma 4.3. *Let (M, τ) be as above. Suppose $\delta_j \in \mathbb{R}_+$, and $\{r_1 \geq \cdots \geq r_k\} \subset \mathcal{P}(M)$ such that $\tau(r_j) \geq \delta_k + \cdots + \delta_j, \forall j = 1, \ldots, k$ and suppose we are given $(k-1)$ mutually orthogonal projections q'_j such that $q'_j \leq r_j$ and $\tau(q'_j) = \delta_j$ $\forall$ $j = 1, \ldots, k-1$. Let*

$$e' = q'_1 + \cdots + q'_{k-1} \leq r_1.$$

Then there exist projections $q \leq r_1 - e', q_j \leq r_j \ \forall\ j = 1, \ldots, k$, *such that* $\tau(q) = \delta_k$ *and* $\tau(q_j) = \delta_j \ \forall\ j$, $\{q_j : 1 \leq j \leq k\}$ *pairwise mutually orthogonal and*

$$q + e' = q_1 + \cdots + q_k,$$

which is also a projection below r_1.

Proof. Use Lemma 4.2 and choose $q = (q_1 + \cdots + q_k) - e'$. □

Before proceeding further, we state a short but useful result:

Lemma 4.4. *For* (M, τ) *as above and* $r, e \in \mathcal{P}(M)$,

$$\tau(r \wedge e^{\perp}) \geq \tau(r) - \tau(e)$$

where, of course, $e^{\perp} = 1 - e$.

Proof.
$$1 + \tau(r \wedge e^{\perp}) \geq \tau(r \vee e^{\perp}) + \tau(r \wedge e^{\perp}) = \tau(r) + 1 - \tau(e)$$
as required. □

The above results lead to the following lemma:

Lemma 4.5. *Let* (M, τ), t_0^j, t_1^j, δ_j *be as in Wielandt's theorem. Let* $\{r_1 \geq \cdots \geq r_k\}$ *and* $\{p_1 \leq \cdots \leq p_k\}$ *be sets of projections in* M *such that* $\tau(p_j) \geq F(t_1^j)$, $\tau(r_j) \geq 1 - F(t_0^j)$ *for all* $1 \leq j \leq k$. *Then there exist mutually orthogonal projections* $q_j \leq r_j$ *and mutually orthogonal projections* $\tilde{q}_j \leq p_j$ *such that* $\tau(q_j) = \tau(\tilde{q}_j) = \delta_j \ \forall\ j$ *and* $q_1 + \cdots + q_k = \tilde{q}_1 + \cdots + \tilde{q}_k$.

Proof. The proof is by induction.

For $k = 1$, deduce from Lemma 4.4 that

$$\begin{aligned}\tau(p_1 \wedge r_1) &\geq \tau(p_1) - \tau(r_1^{\perp}) = \tau(p_1) - 1 + \tau(r_1) \\ &\geq F(t_1^1) - 1 + 1 - F(t_0^1) = F(t_1^1) - F(t_0^1) = \delta_1,\end{aligned}$$

and thus (by our standing 'continuity assumption') there exists a projection $q_1 = \tilde{q}_1 \leq p_1 \wedge r_1$ of trace δ_1.

For the inductive step, assume $p_1 \leq \cdots \leq p_k, r_1 \geq \cdots \geq r_k$ are as in the lemma and that the lemma is valid with k replaced by $k - 1$. By the induction hypothesis applied to $p_1 \leq \cdots \leq p_{k-1}, r_1 \geq \cdots \geq r_{k-1}$, there are mutually orthogonal projections $q_j' \leq r_j$ and mutually orthogonal projections $\tilde{q}_j \leq p_j$ such that $\tau(q_j') = \tau(\tilde{q}_j) = \delta_j$ for all $j = 1, \ldots, k - 1$ and $\sum_{j=1}^{k-1} q_j' = \sum_{j=1}^{k-1} \tilde{q}_j =: e'$, say.

Then $e' \leq p_{k-1} \leq p_k$.

Let $\ell_j = r_j \wedge p_k, \ \forall\ j = 1, \ldots, k$.

Then $\ell_k \leq \cdots \leq \ell_1$. An application of Lemma 4.4, as seen above in the $k = 1$ case, gives:

$$\begin{aligned}\tau(\ell_j) &\geq F(t_1^k) - F(t_0^j) \\ &\geq F(t_1^k) - F(t_0^k) + F(t_1^{k-1}) - F(t_0^{k-1}) + \cdots + F(t_1^j) - F(t_0^j) \\ &= \delta_k + \cdots + \delta_j \ \forall\ j = 1, \ldots, k.\end{aligned}$$

Now by Lemma 4.3 – applied with ℓ_j in place of r_j – we may conclude that $\exists\, q \leq \ell_1 - e'$, $q_j \leq \ell_j$ $(\leq r_j)$ with $\tau(q) = \delta_k$, $\tau(q_j) = \delta_j$ $\forall\, j$ and $q_j \perp q_i$ $\forall\, j \neq i$, such that $q + e' = q_1 + \cdots + q_k$.

But $q + e' = q + \tilde{q}_1 + \cdots + \tilde{q}_{k-1}$, where $\tilde{q}_j \leq p_j$ $\forall\, j = 1, \ldots, k-1$ and $q \leq \ell_1 - e' \leq \ell_1 = r_1 \wedge p_k$.

Choosing $\tilde{q}_k = q$, the proof of the inductive step is complete. □

Now we are ready to prove Theorem 4.1.

Proof. For "$\geq$", we take $p_j := 1_{(-\infty, t_1^j)}(a)$ and $\tilde{q}_j := 1_{[t_0^j, t_1^j)}(a) \leq p_j$.

For proving "$\leq$" here, let us choose any $p_1 \leq \cdots \leq p_k$ such that $p_j \in \mathcal{P}(M)$ and $\tau(p_j) \geq F(t_1^j)$.

Let $r_j = 1_{[t_0^j, \infty)}(a)$ $\forall\, j = 1, \ldots, k$. Then $r_1 \geq \cdots \geq r_k$ with $\tau(r_j) = 1 - F(t_0^j)$.

Now by Lemma 4.5, there exist mutually orthogonal projections $q_j \leq r_j$ and mutually orthogonal projections $\tilde{q}_j \leq p_j$ with $\tau(q_j) = \tau(\tilde{q}_j) = \delta_j$ such that $q_1 + \cdots + q_k = \tilde{q}_1 + \cdots + \tilde{q}_k$.

Notice that by our version of Ky Fan's theorem,

$$\tau(aq_j) \geq \inf_{\substack{q \in \mathcal{P}(M) \\ q \leq r_j \\ \tau(q) = \delta_j}} \tau(aq) = \int_{F(t_0^j)}^{F(t_1^j)} X(s)\, ds.$$

Hence,

$$\sum_{j=1}^{k} \int_{F(t_0^j)}^{F(t_1^j)} X(s)\, ds \leq \sum_{j=1}^{k} \tau(aq_j) = \sum_{j=1}^{k} \tau(a\tilde{q}_j)$$

(since $q_1 + \cdots + q_k = \tilde{q}_1 + \cdots + \tilde{q}_k$), where $\tilde{q}_j \in \mathcal{P}(M), \tilde{q}_j \leq p_j$ with $\tau(\tilde{q}_j) = \delta_j$ and $\tilde{q}_j \perp \tilde{q}_i$.

Hence,

$$\sum_{j=1}^{k} \int_{F(t_0^j)}^{F(t_1^j)} X(s)\, ds \leq \sup_{\substack{\hat{q}_j \in \mathcal{P}(M) \\ \hat{q}_j \leq p_j \\ \tau(\hat{q}_j) = \delta_j \\ \hat{q}_j \perp \hat{q}_i}} \sum_{j=1}^{k} \tau(a\hat{q}_j).$$

Now the theorem follows from the fact that $p_1 \leq \cdots \leq p_k$ were chosen arbitrarily. □

Remark 4.6. For $\delta_1 = \cdots = \delta_k = \delta$, the theorem can be written as:

$$\sum_{j=1}^{k} \int_{[F(t_0^j), F(t_1^j))} X(s)\, ds = \min_{\substack{p_j \in \mathcal{P}(M) \\ p_1 \leq \cdots \leq p_k \\ \tau(p_j) \geq F(t_1^j)}} \sup_{\substack{q \in \mathcal{P}(M) \\ q \leq p_k \\ \tau(q) = k\delta \\ \tau(q \wedge p_j) \geq j\delta}} \tau(aq).$$

5. Continuous version of Lidskii's theorem

The continuous analogue of Lidskii's majorization theorem is a majorization result similar to Lemma 3.4 above, but a strictly stronger one. The matricial version of this result states that given $1 \leq i_1 < \cdots < i_k \leq n$, for $n \times n$ Hermitian matrices a and b with eigenvalues given as $\lambda_1(a) \leq \cdots \leq \lambda_n(a)$ and $\lambda_1(b) \leq \cdots \leq \lambda_n(b)$,

$$\sum_{j=1}^{k} \lambda_{i_j}(a+b) \geq \sum_{j=1}^{k} \lambda_{i_j}(a) + \sum_{j=1}^{k} \lambda_j(b).$$

In this section we state and prove a continuous version of the above. However, we would like to mention here that continuous versions of Lidskii's result have been discussed and proved in several other places, e.g., in [8]. But it is a natural application of Theorem 4.1, so we would like to present it for the sake of completion of our article.

Theorem 5.1. *Let $a = a^*, b = b^* \in M$ be such that μ_a, μ_b, μ_{a+b} are non-atomic. Let F_a, F_b, F_{a+b} and X_a, X_b, X_{a+b} be the distribution and quantile functions of a, b and $(a+b)$ respectively. Let $\delta_j \in \mathbb{R}_+$. Let us choose points $\{t_0^j, t_1^j,\ j = 1, \dots, k\}$ and $\{u_0^j, u_1^j,\ j = 1, \dots, k\}$ in the spectra of $(a+b)$ and a respectively such that $t_0^1 < t_1^1 \leq t_0^2 < t_1^2 \leq \cdots \leq t_0^{k-1} < t_1^{k-1} \leq t_0^k < t_1^k$, $u_0^1 < u_1^1 \leq u_0^2 < u_1^2 \leq \cdots \leq u_0^{k-1} < u_1^{k-1} \leq u_0^k < u_1^k$ and $F_{a+b}(t_1^j) - F_{a+b}(t_0^j) = F_a(u_1^j) - F_a(u_0^j) = \delta_j$ for all j. Then*

$$\sum_{j=1}^{k} \int_{F_{a+b}(t_0^j)}^{F_{a+b}(t_1^j)} X_{a+b}\, dm \geq \sum_{j=1}^{k} \int_{F_a(u_0^j)}^{F_a(u_1^j)} X_a\, dm + \int_0^{\sum_{i=1}^k \delta_i} X_b\, dm.$$

Proof. We know by Theorem 4.1 that M contains projections p_j^{a+b} with $\tau(p_j^{a+b}) = F_{a+b}(t_1^j)$ for all $j = 1, \dots, k$ such that

$$\sum_{j=1}^{k} \int_{F_{a+b}(t_0^j)}^{F_{a+b}(t_1^j)} X_{a+b}\, dm = \sup_{\substack{\hat{q}_j \in \mathcal{P}(M) \\ \hat{q}_j \leq p_j^{a+b} \\ \tau(\hat{q}_j) = \delta_j \\ \hat{q}_j \perp \hat{q}_i \text{ for } j \neq i}} \sum_{j=1}^{k} \tau((a+b)\hat{q}_j).$$

Note that if $\{\alpha_q, \beta_q : q \in \mathcal{Q}\} \subset \mathbb{R}$ for some index set $\mathcal{Q}$, then $\sup_{\{q\in\mathcal{Q}\}}\{\alpha_q + \beta_q\} \geq \sup_{\{q\in\mathcal{Q}\}}\{\alpha_q\} + \inf_{\{q\in\mathcal{Q}\}}\{\beta_q\}$, since we clearly have $\alpha_{q'} + \beta_{q'} \geq \alpha_{q'} + \inf_{\{q\in\mathcal{Q}\}}\{\beta_q\}$ for all $q' \in \mathcal{Q}$, and we may now take $\sup_{q'\in\mathcal{Q}}$ of both sides.

Thus,

$$\sum_{j=1}^{k} \int_{F_{a+b}(t_0^j)}^{F_{a+b}(t_1^j)} X_{a+b}\, dm = \sup_{\substack{\hat{q}_j \in \mathcal{P}(M) \\ \hat{q}_j \leq p_j^{a+b} \\ \tau(\hat{q}_j) = \delta_j \\ \hat{q}_j \perp \hat{q}_i \text{ for } j \neq i}} \sum_{j=1}^{k} \tau((a+b)\hat{q}_j)$$

$$\geq \sup_{\substack{\hat{q}_j \in \mathcal{P}(M) \\ \hat{q}_j \leq p_j^{a+b} \\ \tau(\hat{q}_j)=\delta_j \\ \hat{q}_j \perp \hat{q}_i \text{ for } j \neq i}} \sum_{j=1}^{k} \tau(a\hat{q}_j) + \inf_{\substack{\hat{q}_j \in \mathcal{P}(M) \\ \hat{q}_j \leq p_j^{a+b} \\ \tau(\hat{q}_j)=\delta_j \\ \hat{q}_j \perp \hat{q}_i \text{ for } j \neq i}} \sum_{j=1}^{k} \tau(b\hat{q}_j)$$

$$\geq \sum_{j=1}^{k} \int_{F_a(u_0^j)}^{F_a(u_1^j)} X_a \, dm + \int_0^{\sum_{i=1}^{k} \delta_i} X_b \, dm \text{ by Theorem 4.1 and Theorem 2.3,}$$

proving the continuous analogue of Lidskii's theorem. □

Acknowledgment

It is a pleasure to record our appreciation of the very readable [2] whose proof of the matrix case of Wielandt's theorem we could assimilate and adapt to the continuous case. We also wish to thank Manjunath Krishnapur and Vijay Kodiyalam for helpful discussions. The second author also wishes to gratefully acknowledge the generous support of the J.C. Bose Fellowship.

We would like to thank the referee for pointing out that [14] contains short proofs of Wielandt's and Lidskii's theorems for matrices as well as multiplicative analogues. We intend to see in the future if those can be extended in our context.

References

[1] Hari Bercovici and Dan Voiculescu. Free convolution of measures with unbounded support. *Indiana Univ. Math. J.*, 42(3):733–773, 1993.

[2] Rajendra Bhatia. *Matrix analysis*, volume 169 of *Graduate Texts in Mathematics*. Springer-Verlag, New York, 1997.

[3] R. Courant and D. Hilbert. *Methods of mathematical physics. Vol. II.* Wiley Classics Library. John Wiley & Sons Inc., New York, 1989. Partial differential equations, Reprint of the 1962 original, A Wiley-Interscience Publication.

[4] Thierry Fack and Hideki Kosaki. Generalized s-numbers of τ-measurable operators. *Pacific J. Math.*, 123(2):269–300, 1986.

[5] Ky Fan. On a theorem of Weyl concerning eigenvalues of linear transformations. I. *Proc. Nat. Acad. Sci. USA*, 35:652–655, 1949.

[6] G.H. Hardy, J.E. Littlewood, and G. Pólya. Some simple inequalities satisfied by convex functions. *Messenger Math.*, 58:145–152, 1929.

[7] Fumio Hiai. Majorization and stochastic maps in von Neumann algebras. *J. Math. Anal. Appl.*, 127(1):18–48, 1987.

[8] Fumio Hiai and Yoshihiro Nakamura. Majorizations for generalized s-numbers in semifinite von Neumann algebras. *Math. Z.*, 195(1):17–27, 1987.

[9] Alfred. Horn. Eigenvalues of sums of Hermitian matrices. *Pacific J. Math*, 12:225–241, 1962.

[10] Eizaburo Kamei. Majorization in finite factors. *Math. Japon.*, 28(4):495–499, 1983.

[11] Eizaburo Kamei. On the von Neumann ordering of statistical operators. *Sūrikaisekikenkyūsho Kōkyūroku*, (582):19–30, 1985. Linear operator theory (Japanese) (Kyoto, 1985).

[12] Eizaburo Kamei. Erratum to: "Double stochasticity in finite factors". *Math. Japon.*, 31(5):825, 1986.

[13] M. Krein and D. Milman. On extreme points of regular convex sets. *Studia Math.*, 9:133–138, 1940.

[14] Chi-Kwong Li and Roy Mathias. The Lidskii–Mirsky–Wielandt theorem – additive and multiplicative versions. *Numer. Math.*, 81(3):377–413, 1999.

[15] B.V. Lidskii. A polyhedron of the spectrum of the sum of two Hermitian matrices. *Funktsional. Anal. i Prolozhen.*, 16(2):76–77, 1982.

[16] V.B. Lidskii. On the characterization numbers of the sum and product of symmetric matrices. *Doklady Akad. Nauk SSSR (N.S.)*, 75:769–772, 1950.

[17] Dénes Petz. Spectral scale of selfadjoint operators and trace inequalities. *J. Math. Anal. Appl.*, 109(1):74–82, 1985.

[18] Yūji Sakai. Weak spectral order of Hardy, Littlewood and Pólya. *J. Math. Anal. Appl.*, 108(1):31–46, 1985.

[19] Helmut Wielandt. An extremum property of sums of eigenvalues. *Proc. Amer. Math. Soc.*, 6:106–110, 1955.

Madhushree Basu
Indian Statistical Institute
8th Mile, Mysore Road
Gnana Bharathi, Bangalore 560059, India
e-mail: madhushree@isibang.ac.in

V.S. Sunder
Institute of Mathematical Sciences
Taramani, Chennai 600113, India
e-mail: sunder@imsc.res.in

Operator Theory:
Advances and Applications, Vol. 247, 57–67
© Springer International Publishing Switzerland 2015

Deformation by Dual Unitary Cocycles and Generalized Fixed Point Algebra for Quantum Group Actions

Debashish Goswami and Soumalya Joardar

Abstract. Given a compact quantum group action on a von Neumann algebra and a dual unitary 2-cocycle on the quantum group, we show that the definitions of the deformed algebras given in [3] and [2] are equivalent. Moreover, the deformed von Neumann algebra coincides with the generalized fixed point subalgebra in the sense of ([3]).

Mathematics Subject Classification (2010). 46L65, 46L89, 81R50.

Keywords. Deformation, compact quantum group, cocycle.

1. Introduction

In the theory of operator algebras and related fields, including noncommutative geometry in particular, deformation of a C^*- or von Neumann algebra by suitable 2-cocycles is one of the most important tools for producing a rich source of noncommutative examples. It was Marc Rieffel who did the pioneering construction of cocycle-deformation by actions of $\mathbb{R}^n$ or $\mathbb{T}^n$ ([5]). This was generalized in several directions by many other mathematicians ([1], [3] etc., just to name a few) for actions by Lie groups or even quantum groups. In [1] an interesting description of the deformation of a C^*-algebra by a $\mathbb{T}^n$-action, has been given by identifying with the fixed point algebra of some kind of 'diagonal' action of $\mathbb{T}^n$ on the tensor product of $\mathcal{A}$ with the noncommutative n-torus obtained by deforming $C(\mathbb{T}^n)$ by its own action. This result has been extended to the more general setting of action by a compact group in [3] where the authors have left open the interesting question: is it possible to extend it further for a compact quantum group action, where the fixed point algebra should be replaced by an appropriate 'generalized'

D.G. was partially supported by Swarnajayanti Fellowship from D.S.T. (Govt. of India). S.J. acknowledges support from CSIR..

analogue? The aim of the present note is to give an affirmative answer to a similar question in the (somewhat easier) von Neumann algebraic framework. On our way to prove this, we also compare the definition of deformed algebra given by Neshveyev and Tuset in [3] with our approach in [2] and establish the equivalence of the two approaches for compact quantum group action.

2. Notations and preliminaries

Throughout this note, $\hat{\otimes}, \bar{\otimes}, \otimes$ will denote the injective tensor product between two C^* algebras, tensor product between two von Neumann algebras and tensor product between two Hilbert spaces respectively. $\otimes_{\rm alg}$ will denote the algebraic tensor product. $\otimes$ shall also denote the tensor product between a Hilbert space and a C^*-algebra to get a Hilbert module. For a Hilbert space $\mathcal{H}$, $\mathcal{B}(\mathcal{H})$ and $\mathcal{K}(\mathcal{H})$ will denote the C^*-algebra of bounded operators and compact operators on the Hilbert space $\mathcal{H}$ respectively. We'll also use the standard leg-numbering notation. For a subset A of a vector space V, SpA denotes the linear span of A. We'll denote by $R(T)$ the range of a linear map T.

As in [3] and references therein, let G denote a locally compact quantum group and let $C_0(G) \equiv C_r^*(\hat{G})$, $L^2(G)$ and $L^\infty(G)$ be the underlying C^*-algebra, the GNS Hilbert space of the Haar state and the von Neumann algebraic quantum group generated by the image of $C(G)$ in the GNS representation respectively. We caution the reader that $C_0(G)$ is just a notation for a possibly noncommutative C^*-algebra and should not be confused with the commutative algebra of functions on a classical group G which is a point-set in particular. When G is compact quantum group (CQG from now on), we use the notation $C(G)$ instead of $C_0(G)$ to denote the underlying C^*-algebra. We also refer the reader to [6] for the definition and properties of CQGs.

For a CQG G, Rep(G) and $\hat{G}$ will denote the set of all inequivalent irreducible representations of G and the dual discrete quantum group respectively. We refer to [6] for the definition and properties of unitary representations of CQGs. Given a unitary representation V of G on $\mathcal{H}$ we view it in two ways: as a Hilbert space isometry $V : \mathcal{H} \to \mathcal{H} \otimes L^2(G)$ as well as the corresponding adjointable unitary map $\widetilde{V}$ on the Hilbert module $\mathcal{H} \otimes C(G)$ defined by $\widetilde{V}(\sum_i \xi_i \otimes q_i) := \sum_i V(\xi_i).q_i$ for $\xi_i \in \mathcal{H}$, $q_i \in C(G)$. Observe that this is a unitary element of $\mathcal{M}(\mathcal{K}(\mathcal{H})\hat{\otimes}C(G))$ and extends as a unitary operator on the Hilbert space $\mathcal{H} \otimes L^2(G)$ satisfying $(\mathrm{id} \otimes \Delta)\widetilde{V} = \widetilde{V}_{12}\widetilde{V}_{13}$. For any unitary representation V of G on a Hilbert space $\mathcal{H}$, $R(V)$ will denote the range of V. There is a distinguished unitary representation, called the left regular representation, W of G on $L^2(G)$.

We usually denote by Δ and ϵ the coproduct and counit of a CQG and by $\mathcal{Q}_0$ or $C(G)_0$ the dense Hopf $*$ algebra consisting of matrix coefficients of irreducible representations. For $q \in \mathcal{Q}_0$, we use Sweedler's notation, i.e., we write $\Delta(q) = q_{(1)} \otimes q_{(2)}$.

We say that the compact quantum group $C(G)$ has a right coaction on a unital C^*-algebra $\mathcal{A}$ if there is a unital C^*-homomorphism (called an action) $\alpha : \mathcal{A} \to \mathcal{A}\hat{\otimes}C(G)$ satisfying the following:

(i) $(\alpha \otimes \mathrm{id}) \circ \alpha = (\mathrm{id} \otimes \Delta) \circ \alpha$, and
(ii) the linear span of $\alpha(\mathcal{A})(1 \otimes C(G))$ is norm-dense in $\mathcal{A}\hat{\otimes}C(G)$.

We similarly define a left coaction which is a homomorphism from $\mathcal{A}$ to $C(G)\hat{\otimes}\mathcal{A}$. In this note the word 'action' will usually mean a right coaction unless stated otherwise. We shall have at least one occasion where we have to consider a left coaction. Given a unitary representation V of G on a Hilbert space $\mathcal{H}$, we denote by ad_V the normal $*$-homomorphism from $\mathcal{B}(\mathcal{H})$ to $\mathcal{B}(\mathcal{H} \otimes L^2(G))$ given by $\mathrm{ad}_V(x) = \widetilde{V}(x \otimes 1)\widetilde{V}^*$. We say an action α on a C^* subalgebra $\mathcal{A}$ of $\mathcal{B}(\mathcal{H})$ to be implemented by V if α coincides with ad_V on $\mathcal{A}$.

We say ad_V leaves a C^* or von Neumann subalgebra $\mathcal{C}$ invariant if $(\mathrm{id} \otimes \phi)\mathrm{ad}_V(\mathcal{C}) \subset \mathcal{C}$ for every state ϕ of $C(G)$. For $\pi \in \mathrm{Rep}(G)$, let us recall from [2] the definition of the linear functional ρ^π. For that let d_π and $\{q^\pi_{jk} : j, k = 1, \ldots, d_\pi\}$ be the dimension and matrix coefficients (see [6]) of the corresponding finite-dimensional representation, say U_π, respectively. For each $\pi \in \mathrm{Rep}(G)$, we have a unique $d_\pi \times d_\pi$ complex matrix F_π such that

(i) F_π is positive and invertible with $Tr(F_\pi) = Tr(F_\pi^{-1}) = M_\pi > 0$(say).
(ii) $h(q^\pi_{ij} q^{\pi^*}_{kl}) = \frac{1}{M_\pi}\delta_{ik}F_\pi(j, l)$,

where h is the Haar state of the CQG $C(G)$. Corresponding to $\pi \in \mathrm{Rep}(\mathcal{Q})$, let ρ^π_{sm} be the linear functional on $\mathcal{Q}$ given by $\rho^\pi_{sm}(x) = h(x^\pi_{sm}x), s, m = 1, \ldots, d_\pi$ for $x \in \mathcal{Q}$, where $x^\pi_{sm} = (M_\pi)q^{\pi *}_{km}(F_\pi(k, s))$. Then ρ^π is the linear functional given by $\sum_{s=1}^{d_\pi} \rho^\pi_{ss}$. Now we can define the spectral projection $P^{\mathcal{C}}_\pi := (\mathrm{id} \otimes \rho^\pi)\mathrm{ad}_V$ and denote the image of $\mathcal{C}$ under $P^{\mathcal{C}}_\pi$ by $\mathcal{C}^\pi$ (see [4] for details). We define $\mathcal{C}_0 := Sp\{\mathcal{C}^\pi; \pi \in \mathrm{Rep}(G)\}$, which is called the spectral subalgebra corresponding to $\mathcal{C}$. Then it is known that $\mathrm{ad}_V(\mathcal{C}_0) \subset \mathcal{C}_0 \otimes_{\mathrm{alg}} \mathcal{Q}_0$ (see [4]). For $a \in \mathcal{C}_0$, we shall write $\mathrm{ad}_V(a) = a_{(0)} \otimes a_{(1)}$ (Sweedler's notation). For a C^* action condition (ii) of the definition is equivalent to saying that $\mathcal{C}_0$ is norm dense in $\mathcal{C}$. For a von Neumann algebraic action we have the following (see Proposition 2.3 of [2]):

Proposition 2.1. *If $\mathcal{C}$ is a von Neumann algebra, then $\mathcal{C}_0$ is dense in $\mathcal{C}$ in any of the natural locally convex topologies of $\mathcal{C}$, i.e., $\mathcal{C}_0'' = \mathcal{C}$.*

The left regular representation W implements the coproduct on $C(G)$, i.e., $\Delta(.) = \mathrm{ad}_W(.)$ and denote the left and right spectral projections corresponding to $\pi \in \mathrm{Rep}(G)$ as $P^{G,l}_\pi$ and $P^{G,r}_\pi$ respectively given by $(\rho^\pi \otimes \mathrm{id})\mathrm{ad}_W$ and $(\mathrm{id} \otimes \rho^\pi)\mathrm{ad}_W$ respectively. Clearly $\mathrm{Sp}\{P^{G,l}_\pi(C(G)) : \pi \in \mathrm{Rep}(G)\}$ and $\mathrm{Sp}\{P^{G,r}_\pi(C(G)) : \pi \in \mathrm{Rep}(G)\}$ coincides with $\mathcal{Q}_0$.

3. Main results about deformation by dual unitary cocycles

3.1. Two different approaches of deformation

We refer the reader to [3] where a general construction of deformation of a C^*-algebra $\mathcal{A}$ by a regular dual unitary 2-cocycle σ on a locally compact quantum group acting on $\mathcal{A}$ has been given. Let us recall that a dual unitary 2-cocycle σ of a compact quantum group $C(G)$ is a unitary element of $\mathcal{M}(C_0(\hat{G})\hat{\otimes}C_0(\hat{G}))$ satisfying

$$(1\otimes\sigma)(\mathrm{id}\otimes\hat{\Delta})\sigma=(\sigma\otimes 1)(\hat{\Delta}\otimes\mathrm{id})\sigma,$$

where $\hat{\Delta}$ denotes the coproduct of $C_0(\hat{G})$. As in [3] let us denote the deformed algebra by $\mathcal{A}_\sigma$. However, we work with compact quantum groups only, so every dual unitary 2-cocycle is regular and moreover, the construction can be done in a more algebraic way, as in [2]. Let σ be a dual unitary 2-cocycle on G and let G_σ and $C_r^*(\hat{G},\sigma)$ be the deformed or twisted quantum group and the reduced twisted group C^*-algebra as in [3]. We denote by $L^\infty(G;\sigma)$ the weak closure of $C_r^*(G;\sigma)$ in $\mathcal{B}(L^2(G))$. Suppose furthermore that there is a unitary representation V of G on a Hilbert space $\mathcal{H}$ and $\mathcal{A}\subseteq\mathcal{B}(\mathcal{H})$ is a unital C^*-algebra such that $\alpha:=\mathrm{ad}_V$ gives a C^* action of G on $\mathcal{A}$. Then we have the deformed C^*-algebra $\mathcal{A}_\sigma$ constructed in [3], which is viewed there as a subalgebra of $\mathcal{B}(\mathcal{H}\otimes L^2(G))$. We also denote by $\mathcal{M}$ the weak closure of $\mathcal{A}$ in $\mathcal{B}(\mathcal{H})$ and let $\mathcal{M}_\sigma$ be the weak closure of $\mathcal{A}_\sigma$ in $\mathcal{B}(\mathcal{H}\otimes L^2(G))$. Recall $P_\pi^{G,r}$ for $\pi\in\mathrm{Rep}(G)$. Then it is clear that $(\mathrm{id}\otimes P_\pi^{G,r})X\in\mathcal{B}(\mathcal{H})\otimes_{\mathrm{alg}}\mathcal{Q}_0^\sigma$ for any $X\in(\mathcal{B}(\mathcal{H})\bar{\otimes}L^\infty(G;\sigma))$. Also we denote by $p_\pi^{G,l}$ and $p_\pi^{G,r}$ the Hilbert space projections on $L^2(G)$ corresponding to the spectral projections $P_\pi^{G,l}$ and $P_\pi^{G,r}$ respectively. Then both $p_\pi^{G,r}$s and $p_\pi^{G,l}$s are mutually orthogonal projections whose sum converges to the identity operator on $L^2(G)$ strongly.

Lemma 3.1. *Given any injective C^* (respectively von Neumann algebraic) action β on a C^* (respectively von Neumann) algebra $\mathcal{C}\subset\mathcal{B}(\mathcal{K})$ and $X\in\mathcal{C}$ such that $X_\pi:=P_\pi^{\mathcal{C}}(X)=0$ for all $\pi\in\mathrm{Rep}(G)$, one has $X=0$.*

Proof. We use γ for the ad_W action. We have $X_\pi=0$, i.e., $(\mathrm{id}\otimes\rho^\pi)\beta(X)=0$. So $\beta(\mathrm{id}\otimes\rho^\pi)\beta(X)=0$. Hence by using the fact that $(\beta\otimes\mathrm{id})\beta=(\mathrm{id}\otimes\gamma)\beta$, we obtain,

$$(\mathrm{id}\otimes\mathrm{id}\otimes\rho^\pi)(\mathrm{id}\otimes\gamma)\beta(X)=0.$$

But then $(\mathrm{id}\otimes\rho^\pi)\gamma:=P_\pi^{G,r}$. So $(\mathrm{id}\otimes P_\pi^{G,r})\beta(X)=0$. $\beta(X)\in(\mathcal{B}(\mathcal{K})\bar{\otimes}L^\infty(G))$. Then for any $u\in\mathcal{K}$, $((\mathrm{id}\otimes P_\pi^{G,r})\beta(X))(u\otimes 1_{C(G)})=0$, i.e., $(\mathrm{id}\otimes p_\pi^{G,r})(\beta(X)(u\otimes 1_{C(G)})=0$. But since $\sum_{\pi\in\mathrm{Rep}(G)}p_\pi^{G,r}$ converges strongly to identity operator on $L^2(G)$, we get $\beta(X)(u\otimes 1_{C(G)})=0$ for all $u\in\mathcal{H}$. So $1_{C(G)}$ being a separating vector for $L^\infty(G)\in\mathcal{B}(L^2(G))$, we conclude that $\beta(X)=0$ and hence $X=0$ as β is one one. □

Remark 3.2. *The above lemma clearly holds true if the right action is replaced by the left action.*

In [2] we have given a definition of deformation in $\mathcal{B}(\mathcal{H})$. To distinguish this version of deformation from that of [3] let us denote the deformation a la [2] by $\mathcal{A}^\sigma$. We quickly recall the definition of $\mathcal{A}^\sigma$ without going into the details.

We have a dense subspace $\mathcal{H}_0 \subset \mathcal{H}$ on which V is algebraic, i.e., $V(\mathcal{H}_0) \subset \mathcal{H}_0 \otimes_{\mathrm{alg}} \mathcal{Q}_0$, where $\mathcal{Q}_0$ is the canonical dense Hopf $*$ algebra of $C(G)$ as mentioned before. Then the spectral subalgebras $\mathcal{A}_0$ and $\mathcal{M}_0$ are dense in $\mathcal{A}$ in norm and in $\mathcal{M}$ in SOT respectively. Also $\alpha := \mathrm{ad}_V$ is algebraic over $\mathcal{A}_0$ as well as $\mathcal{M}_0$. Using the dual unitary 2-cocycle $\sigma \in \mathcal{M}(\hat{\mathcal{Q}}\hat{\otimes}\hat{\mathcal{Q}})$, we can define a new representation of $\mathcal{M}_0$ on $\mathcal{H}_0$ by,

$$\rho_\sigma(b)(\xi) := b_{(0)}\xi_{(0)}\sigma^{-1}(b_{(1)}, \xi_{(1)}), \ for\ \xi \in \mathcal{H}_0,$$

where $\alpha(b) = b_{(0)} \otimes b_{(1)}$ and $V(\xi) = \xi_{(0)} \otimes \xi_{(1)}$. We have from Lemma 4.7 of [2] $\rho_\sigma(\mathcal{M}_0) \subset \mathcal{B}(\mathcal{H})$. When $\mathcal{A}$ is a C^* (von Neumann) algebra, we define the C^* algebraic deformation of $\mathcal{A}$ as the norm closure of $\rho_\sigma(\mathcal{A}_0)$ in $\mathcal{B}(\mathcal{H})$ and denote it by $\mathcal{A}^\sigma$. The von Neumann algebraic deformation $\mathcal{M}^\sigma$ of $\mathcal{M}$ is similarly defined as the weak closure of $\rho_\sigma(\mathcal{M}_0)$ in $\mathcal{B}(\mathcal{H})$.

3.2. Equivalence of the frameworks of [2] and [3]

Let us briefly compare the two set-ups ([2] and [3]) and show that they are indeed equivalent. For a dual unitary 2-cocycle σ on $C(G)$, we define $\mathcal{Q}_0^\sigma \in \mathcal{B}(L^2(G))$ by the following:
As vector spaces, $\mathcal{Q}_0$ and $\mathcal{Q}_0^\sigma$ are the same. But the new representation of $\mathcal{Q}_0^\sigma$ in $\mathcal{B}(L^2(G))$ is given by $q * \xi := q_{(1)}\xi_{(1)}\sigma^{-1}(q_{(2)}, \xi_{(2)})$, where $q, \xi \in \mathcal{Q}_0$ and $\Delta(q) = q_{(1)} \otimes q_{(2)}$, $\Delta(\xi) = \xi_{(1)} \otimes \xi_{(2)}$. As $\mathcal{Q}_0$ is dense in the Hilbert space $L^2(G)$, the above formula defines the representation uniquely. Here Δ is the coproduct of G. That $\mathcal{Q}_0^\sigma \subset \mathcal{B}(L^2(G))$ can be shown along the same way as Lemma 4.7 of [2].

Identifying $\mathcal{Q}_0^\sigma$ with $\mathcal{Q}_0$ as a vector space, we have the same counit map $\epsilon : \mathcal{Q}_0^\sigma \to \mathbb{C}$. There is a canonical left action, say Δ_σ, of G on $C_r^*(G;\sigma)$, which coincides with the coproduct Δ as a linear map on the dense $*$-subalgebra $\mathcal{Q}_0^\sigma$ identified with $\mathcal{Q}_0$ as vector spaces. The counit map ϵ satisfies $(\mathrm{id} \otimes \epsilon)\Delta_\sigma = (\epsilon \otimes \mathrm{id})\Delta_\sigma = \mathrm{id}$. However, we should mention that ϵ on $\mathcal{Q}_0^\sigma$ need not be a homomorphism.

Now we introduce the generalized fixed point subalgebras and spaces. We use the notations V and W to denote the unitary representation of G on Hilbert spaces $\mathcal{H}$ and $L^2(G)$ respectively as before.

Definition 3.3. *For a subalgebra (not necessarily closed) $\mathcal{B}$ of $\mathcal{B}(\mathcal{H} \otimes L^2(G))$ the generalized fixed point subalgebra $\mathcal{B}^f$ corresponding to $\mathcal{B}$ is given by*

$$\mathcal{B}^f := \{X \in \mathcal{B} : \ (\mathrm{ad}_V \otimes \mathrm{id})(X) = (\mathrm{id} \otimes \mathrm{ad}_W)(X)\}.$$

Similarly, the generalized fixed point subspace $\mathcal{W}^f$ for a Hilbert subspace $\mathcal{W} \subseteq \mathcal{H} \otimes L^2(G)$ is defined to be

$$\mathcal{W}^f := \{\xi \in \mathcal{W} : \ (V \otimes \mathrm{id})(\xi) = (\mathrm{id} \otimes W)(\xi)\}.$$

There are two possible G-actions of $\mathcal{B}(\mathcal{H})\bar{\otimes}\mathcal{B}(L^2(G))$ given by $\sigma_{23}(\mathrm{ad}_V \otimes \mathrm{id})$ and $\sigma_{23}(\mathrm{id} \otimes \mathrm{ad}_W)$, where σ_{23} flips the second and the third tensor copies and by definition these two actions coincide on $\mathcal{B}^f$. For $\mathcal{B}$ and $\pi \in \mathrm{Rep}(G)$ the corresponding spectral projection $P_\pi^{\mathcal{B}}$ is given by the restriction of $(P_\pi^{\mathcal{B}(\mathcal{H})} \otimes \mathrm{id})$ or $(\mathrm{id} \otimes P_\pi^{G,l})$ on $\mathcal{B}^f$.

The following observation will be crucial.

Lemma 3.4. *Let $\mathcal{L}$ be the range of V viewed as a Hilbert space isometry from $\mathcal{H}$ to $\mathcal{H} \otimes L^2(G)$. Then we have the following:*

(i) *$\mathcal{L} = (\mathcal{H} \otimes L^2(G))^f$, i.e., $V : \mathcal{H} \to (\mathcal{H} \otimes L^2(G))^f$ is a unitary operator.*
(ii) *Any X in $(\mathcal{B}(\mathcal{H}) \otimes \mathcal{B}(L^2(G)))^f$ leaves $\mathcal{L}$ invariant.*

Proof. (i) That V is an isometric operator has been observed previously. So we need to show that $R(V) = (\mathcal{H} \otimes L^2(G))^f$. To that end first note that $\mathcal{H}$ decomposes into spectral subspaces corresponding to the unitary representation V of $C(G)$. More precisely $\mathcal{H}$ is the norm closure of $\mathrm{Sp}\{\mathcal{H}^\pi : \pi \in \mathrm{Rep}(G)\}$, where $\mathcal{H}^\pi = \overline{\mathrm{Sp}}\{e^{\pi,i} : i = 1, \ldots, d_\pi\}$. d_π is the dimension of the πth spectral subspace of $C(G)$. Also $V(e^{\pi,i}) = \sum_j e^{\pi,j} \otimes q_{ji}^\pi$, where q_{ji}^πs are matrix coefficients (see [6]). So we have a basis for the Hilbert space $\mathcal{H} \otimes L^2(G)$ given by the set

$$\{e^{\pi,i} \otimes q_{i'j}^{\pi'} : \pi, \pi' \in \mathrm{Rep}(G); i \in \{1, \ldots, d_\pi\}; i', j \in \{1, \ldots, d_{\pi'}\}\}.$$

So we can write a vector $\xi \in (\mathcal{H} \otimes L^2(G))^f$ as

$$\sum_{\pi,\pi',i,i',j} c_{i,i',j}^{\pi,\pi'} e^{\pi,i} \otimes q_{i'j}^{\pi'}.$$

Then

$$(V \otimes \mathrm{id})(\xi) = \sum_{\pi,\pi',i,i',j,k} c_{i,i',j}^{\pi,\pi'} e^{\pi,k} \otimes q_{ki}^\pi \otimes q_{i'j}^{\pi'},$$

$$(\mathrm{id} \otimes W)(\xi) = \sum_{\pi,\pi',i,i',j,l} c_{i,i',j}^{\pi,\pi'} e^{\pi,i} \otimes q_{i'l}^{\pi'} \otimes q_{lj}^{\pi'}.$$

Now for $\pi \neq \pi'$, and for all l, k, i, i', j, $e^{\pi,k} \otimes q_{ki}^\pi \otimes q_{i'j}^{\pi'}$ and $e^{\pi,i} \otimes q_{i'l}^{\pi'} \otimes q_{lj}^{\pi'}$ are different basis vectors. Hence $c_{i,i',j}^{\pi,\pi'} = 0$ for all i, i', j whenever $\pi \neq \pi'$. So $\xi = \sum_{\pi,i,i',j} c_{i,i',j}^\pi e^{\pi,i} \otimes q_{i',j}^\pi$. Then

$$(V \otimes \mathrm{id})(\xi) = \sum_{\pi,i,i',j,k} c_{i,i',j}^\pi e^{\pi,k} \otimes q_{ki}^\pi \otimes q_{i'j}^\pi. \tag{1}$$

$$(\mathrm{id} \otimes W)(\xi) = \sum_{\pi,i,i',j,l} c_{i,i',j}^\pi e^{\pi,i} \otimes q_{i'l}^\pi \otimes q_{lj}^\pi. \tag{2}$$

Fixing $k = n$ in (1) and $i = n$ in (2)

$$\sum_{\pi,i,i',j} c_{i,i',j}^\pi q_{ni}^\pi \otimes q_{i'j}^\pi = \sum_{\pi,i',j,l} c_{n,i',j}^\pi q_{i'l}^\pi \otimes q_{lj}^\pi.$$

Fixing $i = i_0$ on the left-hand side and $i' = n, l = i_0$ on the right-hand side of the above expression we get,

$$\sum_{\pi,i',j} c^{\pi}_{i_0,i',j} q^{\pi}_{i'j} = \sum_{\pi,j} c^{\pi}_{n,n,j} q^{\pi}_{i_0 j}.$$

So for a fixed i_0 if $i' \neq i_0$, $c^{\pi}_{i_0,i',j} = 0$. Hence the vector ξ can be written as

$$\xi = \sum_{\pi,i,j} c^{\pi}_{i,j} e^{\pi,i} \otimes q^{\pi}_{ij}.$$

Again

$$(V \otimes \mathrm{id})(\xi) = \sum_{\pi,i,j,k} c^{\pi}_{i,j} e^{\pi,k} \otimes q^{\pi}_{ki} \otimes q^{\pi}_{ij}. \tag{3}$$

$$(\mathrm{id} \otimes W)(\xi) = \sum_{\pi,i,j,l} c^{\pi}_{i,j} e^{\pi,i} \otimes q^{\pi}_{il} \otimes q^{\pi}_{lj}. \tag{4}$$

Fix $k = m, i = n$ in (3) and $i = m, l = n$ in (4) to obtain

$$\sum_{\pi,j} c^{\pi}_{n,j} e^{\pi,m} \otimes q^{\pi}_{nj} = \sum_{\pi,j} c^{\pi}_{m,j} e^{\pi,m} \otimes q^{\pi}_{nj}.$$

Hence $c^{\pi}_{m,j} = c^{\pi}_{n,j}$ for all m, n, i.e., ξ takes the form $\sum_{\pi,i,j} c^{\pi}_j e^{\pi,i} \otimes q^{\pi}_{ij}$. So $\xi = V(\sum_{\pi,j} c^{\pi}_j e^{\pi,j})$.

(ii) Let $\xi \in \mathcal{L}$. We have by definition of $\mathcal{L}$, $(V \otimes \mathrm{id})(\xi) = (\mathrm{id} \otimes W)(\xi)$. As $X \in (\mathcal{B}(\mathcal{H}) \otimes \mathcal{B}(L^2(G)))^f$, it follows that

$$(\mathrm{ad}_V \otimes \mathrm{id})(X) = (\mathrm{id} \otimes \mathrm{ad}_W)(X).$$

So

$$\begin{aligned}(V \otimes \mathrm{id})(X\xi) &= [(\mathrm{ad}_V \otimes \mathrm{id})(X)](V \otimes \mathrm{id})(\xi) \\ &= [(\mathrm{id} \otimes \mathrm{ad}_W)(X)](\mathrm{id} \otimes W)(\xi) \\ &= (\mathrm{id} \otimes W)(X\xi).\end{aligned}$$

Hence by definition $X\xi \in \mathcal{L}$. □

Recall the definition of X_π for all $\pi \in \mathrm{Rep}(G)$. For $X \in (\mathcal{B}(\mathcal{H})\bar{\otimes}L^\infty(G;\sigma))^f$, $X_\pi \in (\mathcal{B}(\mathcal{H}) \otimes_{\mathrm{id}} \mathcal{Q}^{\sigma}_0)$. Also X leaves $\mathcal{L}$ invariant. Then we have

Lemma 3.5. *If $X \in (\mathcal{B}(\mathcal{H})\bar{\otimes}L^\infty(G;\sigma))^f$ and $X|_{\mathcal{L}} = 0$, then $X_\pi|_{\mathcal{L}} = 0$, where $X_\pi = (\mathrm{id} \otimes \rho^\pi \otimes \mathrm{id})((\mathrm{ad}_V \otimes \mathrm{id})(X)) = (\mathrm{id} \otimes \rho^\pi \otimes \mathrm{id})((\mathrm{id} \otimes \mathrm{ad}_W)(X))$.*

Proof. Let $\xi \in \mathcal{H}_0$. Using Sweedler's notation we shall write $V(\xi) = \xi_{(0)} \otimes \xi_{(1)}$. Let us denote the operator $\xi(\in \mathcal{H}) \mapsto \xi_{(0)} \otimes \kappa(\xi_{(1)}) \in \mathcal{H} \otimes \mathcal{Q}$ by V'. Then

$$\begin{aligned}
X_\pi(V\xi) &= [(\mathrm{id} \otimes P_\pi^{G,l})X](V\xi) \\
&= [(\mathrm{id} \otimes \rho^\pi \otimes \mathrm{id})(\mathrm{id} \otimes \mathrm{ad}_W)(X)](V\xi) \\
&= [(\mathrm{id} \otimes \rho^\pi \otimes \mathrm{id})\widetilde{W}_{23}(X \otimes 1)\widetilde{W}_{23}^*](V\xi) \\
&= (\mathrm{id} \otimes \rho^\pi \otimes \mathrm{id})[\widetilde{W}_{23}(X \otimes 1)\widetilde{W}_{23}^*(V\xi \otimes 1)] \\
&= (\mathrm{id} \otimes \rho^\pi \otimes \mathrm{id})[\widetilde{W}_{23}(X \otimes 1)\widetilde{W}_{23}^*(\xi_{(0)} \otimes \xi_{(1)} \otimes 1)] \\
&= (\mathrm{id} \otimes \rho^\pi \otimes \mathrm{id})[\widetilde{W}_{23}(X \otimes 1)(\xi_{(0)} \otimes \xi_{(1)(1)} \otimes \kappa(\xi_{(1)(2)}))] \\
&= (\mathrm{id} \otimes \rho^\pi \otimes \mathrm{id})[\widetilde{W}_{23}(X \otimes 1)(\xi_{(0)(0)} \otimes \xi_{(0)(1)} \otimes \kappa(\xi_{(1)}))] \\
&= (\mathrm{id} \otimes \rho^\pi \otimes \mathrm{id})[\widetilde{W}_{23}(X \otimes 1)(V \otimes \mathrm{id})(V'(\xi))],
\end{aligned}$$

where $\widetilde{W}_{23}$ is the corresponding unitary to $(\mathrm{id} \otimes W)$. Since $X|_{\mathcal{L}} = 0$, the above computation implies that $X_\pi(V\xi) = 0$ for all $\xi \in \mathcal{H}_0$. By density of $\mathcal{H}_0$ in $\mathcal{H}$ we can argue that $X_\pi|_{\mathcal{L}} = 0$ for all $\pi \in \mathrm{Rep}(G)$. □

It can easily be seen that $(\mathcal{B}(\mathcal{H})\bar{\otimes}L^\infty(G;\sigma))^f$ is a von Neumann subalgebra of the operator algebra $\mathcal{B}(\mathcal{H} \otimes (L^2(G))$.
Note that by (ii) of Lemma 3.4, any $X \in (\mathcal{B}(\mathcal{H}) \otimes L^\infty(G;\sigma))^f$ leaves $\mathcal{L}$ invariant.

Lemma 3.6. *Let $\Phi : (\mathcal{B}(\mathcal{H}) \otimes L^\infty(G;\sigma))^f \to \mathcal{B}(\mathcal{L})$ be given by*

$$\Phi(X) = X|_{\mathcal{L}}$$

for $X \in (\mathcal{B}(\mathcal{H}) \otimes L^\infty(G;\sigma))^f$. Then Φ is a normal $*$ isomorphism.

Proof. That Φ is a normal $*$-homomorphism is easy to see. So it suffices to show that the map Φ is both one-one and onto. We first prove that Φ is surjective. Fix some operator $T \in \mathcal{B}(\mathcal{L})$. Then $V^*TV \in \mathcal{B}(\mathcal{H})$. Then $\widetilde{V}(V^*TV \otimes 1)\widetilde{V}^* \in \mathcal{B}(\mathcal{H} \otimes L^2(G))$. Let $\eta \in \mathcal{L}$. So we have some $\xi \in \mathcal{H}$ such that $V(\xi) = \eta$. Also we observe that $\widetilde{V}^*V(\xi) = \xi \otimes 1$. Then $TV(\xi) = VV^*TV(\xi) = \widetilde{V}(V^*TV(\xi) \otimes 1)$. That is

$$\widetilde{V}(V^*TV \otimes 1)\widetilde{V}^*V(\xi) = TV(\xi).$$

Hence

$$\widetilde{V}(V^*TV \otimes 1)\widetilde{V}^*|_{\mathcal{L}} = T.$$

It is obvious that $\widetilde{V}(V^*TV \otimes 1)\widetilde{V}^* \in (\mathcal{B}(\mathcal{H})\bar{\otimes}\mathcal{B}(L^2(G))^f$. So the map Φ is onto.

To show Φ is one-one, we have to show that for any $X \in (\mathcal{B}(\mathcal{H})\bar{\otimes}L^\infty(G;\sigma))^f$ such that $X|_{\mathcal{L}} = 0$, $X = 0$ on the Hilbert space $(\mathcal{H} \otimes L^2(G))$. By Remark 3.2 after Lemma 3.1, it suffices to show that $X_\pi = 0$ as an operator on $(\mathcal{H} \otimes L^2(G))$ for all $\pi \in \mathrm{Rep}(G)$, where X_πs are as in Lemma 3.5. As $X|_{\mathcal{L}} = 0$, by Lemma 3.5,

$X_\pi|_{\mathcal{L}} = 0$ for all $\pi \in \mathrm{Rep}(G)$. As for $X \in (\mathcal{B}(\mathcal{H})\bar{\otimes}\mathcal{B}(L^2(G))^f$, $X_\pi \in \mathcal{B}(\mathcal{H}) \otimes_{\mathrm{alg}} \mathcal{Q}_0^\sigma$, $(\mathrm{id} \otimes \epsilon)X_\pi$ makes sense for all $\pi \in \mathrm{Rep}(G)$. Then

$$\begin{aligned} V((\mathrm{id} \otimes \epsilon)X_\pi(\xi)) &= (\mathrm{ad}_V(\mathrm{id} \otimes \epsilon)X_\pi)V(\xi) \\ &= X_\pi(V\xi) \\ &= 0(as\ X_\pi|_{\mathcal{L}} = 0). \end{aligned}$$

Since V is an isometry, $(\mathrm{id} \otimes \epsilon)X_\pi(\xi) = 0$ for all ξ, i.e., $(\mathrm{id} \otimes \epsilon)X_\pi = 0$. Again applying ad_V, we conclude that $X_\pi = 0$ as an element of $\mathcal{B}(\mathcal{H} \otimes L^2(G))$. □

Recall the definition of $\mathcal{Q}_0^\sigma$. If $C(G)$ has a C^* action on $\mathcal{A} \in \mathcal{B}(\mathcal{H})$ implemented by a unitary V, then as before we have a norm dense subalgebra $\mathcal{A}_0$ of $\mathcal{A}$ such that $\alpha(=\mathrm{ad}_V)$ is algebraic over $\mathcal{A}_0$. We denote the vector space isomorphism between $\mathcal{Q}_0$ and $\mathcal{Q}_0^\sigma$ by π^σ. Define $\alpha^\sigma : \mathcal{A}_0 \to \mathcal{A}_0 \otimes_{\mathrm{alg}} \mathcal{Q}_0^\sigma$ by $(\mathrm{id} \otimes \pi^\sigma)\alpha$.

Then we have

Lemma 3.7. *$V^*\alpha^\sigma(a)V = \rho_\sigma(a)$ for all $a \in \mathcal{A}_0$.*

Proof. Let $\xi \in \mathcal{H}_0$ and $V(\xi) = \xi_{(0)} \otimes \xi_{(1)}$ (Sweedler's notation). Also let $\alpha(a) = a_{(0)} \otimes a_{(1)}$. Then we have

$$\begin{aligned} \alpha^\sigma(a)V(\xi) &= a_{(0)}\xi_{(0)} \otimes a_{(1)} * \xi_{(1)} \\ &= a_{(0)}\xi_{(0)} \otimes a_{(1)(1)}\xi_{(1)(1)}\sigma^{-1}(a_{(1)(2)}, \xi_{(1)(2)}). \end{aligned}$$

On the other hand

$$\begin{aligned} V(\rho_\sigma(a)(\xi)) &= V(a_{(0)}\xi_{(0)}\sigma^{-1}(a_{(1)}, \xi_{(1)})) \\ &= a_{(0)(0)}\xi_{(0)(0)} \otimes a_{(0)(1)}\xi_{(0)(1)}\sigma^{-1}(a_{(1)}, \xi_{(1)}) \\ &= a_{(0)}\xi_{(0)} \otimes a_{(1)(1)}\xi_{(1)(1)}\sigma^{-1}(a_{(1)(2)}, \xi_{(1)(2)}). \end{aligned}$$

Hence $V^*\alpha^\sigma(a)V = \rho_\sigma(a)$ for all $a \in \mathcal{A}_0$. □

Now we are ready to prove the main result of this section.

Theorem 3.8. *$\mathcal{A}^\sigma$ is isomorphic with $\mathcal{A}_\sigma$ as a C^*-algebra and $\mathcal{M}_\sigma$ is isomorphic with $\mathcal{M}_\sigma$ as a von Neumann algebra.*

Proof. First observe that according to the definition of $\mathcal{A}_\sigma$ (Definition 3.3 of [3]) for the special case where G $(\mathcal{Q})$ is a CQG, treating $1 \in C(G)$ as a unit vector in $L^2(G)$, we can choose ν according to the notation of Proposition 3.1 of [3] to be the element of $\mathcal{K}(L^2(G))^*$ given by $y \to < 1, y1 >$, i.e., the vector state given by the vector 1 to show that the map T_ν of Definition 3.3 of [3] is nothing but our map π^σ on the dense subspace $\mathcal{Q}_0$. Hence by Lemma 3.6, $\mathcal{A}_\sigma$ is the C^* closure of $\alpha^\sigma(\mathcal{A}_0)$ in $\mathcal{B}(\mathcal{H} \otimes L^2(G))$. By Lemma 3.7 this is isomorphic to the corresponding C^* closure of $\rho_\sigma(\mathcal{A}_0)$ in $\mathcal{B}(\mathcal{H})$, i.e., $\mathcal{A}^\sigma$. Exactly the same reasoning using von Neumann closure in place of C^* closure shows that $\mathcal{M}_\sigma$ is isomorphic to $\mathcal{M}^\sigma$. □

3.3. Identifying $\mathcal{M}^\sigma$ with the generalized fixed point subalgebra

We now give a partial answer to the question asked in [3] by identifying $\mathcal{M}_\sigma$ with the generalized fixed point subalgebra $(\mathcal{M}\bar{\otimes}L^\infty(G;\sigma))^f$ when $\mathcal{M}$ is a von Neumann algebra.

Lemma 3.9. *The map $\rho_\sigma(a) \to \alpha^\sigma(a)$ for $a \in \mathcal{M}_0$ is an isomorphism between $\mathcal{M}_0^\sigma$ and $(\mathcal{M}_0 \otimes_{\text{alg}} \mathcal{Q}_0^\sigma)^f$, where α^σ is as in the previous subsection.*

Proof. It suffices to check that the map is one-one and onto. For injectivity of the map, note that $\rho_\sigma(a) = V^*(\Phi(\alpha^\sigma(a)))V$ in the notation of Lemma 3.7. For any $X \in (\mathcal{M}_0 \otimes_{\text{alg}} \mathcal{Q}_0^\sigma)^f$, $\alpha((\text{id}\otimes\epsilon)X) = (\text{id}\otimes\epsilon\otimes\text{id})(\alpha\otimes\text{id})X = (\text{id}\otimes\epsilon\otimes\text{id})(\text{id}\otimes\Delta_\sigma)X = X$. Hence α is onto. □

Theorem 3.10. $\mathcal{M}^\sigma = (\mathcal{M}\bar{\otimes}L^\infty(G;\sigma))^f$.

Proof. We note that it suffices to show the inclusion

$$(\mathcal{M}\bar{\otimes}L^\infty(G;\sigma))^f \subset \mathcal{M}^\sigma.$$

Let $\mathcal{C} = (\mathcal{M}\bar{\otimes}L^\infty(G;\sigma))^f$. Then by Proposition 2.1, the spectral subalgebra $\mathcal{C}_0$ is weakly dense in $\mathcal{C}$. Moreover $\alpha^\sigma(\mathcal{M}_0) = (\mathcal{M}_0 \otimes_{\text{alg}} \mathcal{Q}_0^\sigma)^f$ is weakly dense in $\mathcal{M}_\sigma$. Hence it is enough to argue that $\mathcal{C}_0 \subset (\mathcal{M}_0 \otimes_{\text{alg}} \mathcal{Q}_0^\sigma)^f$. From the definition of the spectral projections $\{P_\pi^{\mathcal{C}} : \pi \in \text{Rep}(C(G))\}$ corresponding to $\mathcal{C} \in \mathcal{B}(\mathcal{H} \otimes L^2(G))$ and the fact that $(\text{ad}_V \otimes \text{id}) = (\text{id} \otimes \text{ad}_W)$ gives the von Neumann algebraic action on $\mathcal{C}$, we have $P_\pi^{\mathcal{C}} = (P_\pi^{\mathcal{M}} \otimes \text{id}) = (\text{id} \otimes P_\pi^{G,l})$.

Thus on one hand

$$\begin{aligned}(P_\pi^{\mathcal{C}}(\mathcal{C})) &\subset (P_\pi^{\mathcal{M}} \otimes \text{id})(\mathcal{M}\bar{\otimes}L^\infty(G;\sigma))\\ &= \mathcal{M}_0 \otimes_{\text{alg}} L^\infty(G;\sigma).\end{aligned}$$

On the other hand,

$$\begin{aligned}(P_\pi^{\mathcal{C}}(\mathcal{C})) &\subset (\text{id} \otimes P_\pi^{G,l})(\mathcal{M}\bar{\otimes}L^\infty(G;\sigma))\\ &= \mathcal{M} \otimes_{\text{alg}} \mathcal{Q}_0^\sigma,\end{aligned}$$

i.e.,

$$\begin{aligned}(P_\pi^{\mathcal{C}}(\mathcal{C})) &\subset (\mathcal{M}_0 \otimes_{\text{alg}} L^\infty(G;\sigma)) \cap (\mathcal{M} \otimes_{\text{alg}} \mathcal{Q}_0^\sigma)\\ &= (\mathcal{M}_0 \otimes_{\text{alg}} \mathcal{Q}_0^\sigma).\end{aligned}$$

So clearly $P_\pi^{\mathcal{C}}(\mathcal{C}) \subset (\mathcal{M}_0 \otimes_{\text{alg}} \mathcal{Q}_0^\sigma)^f$ for all π. Hence $\mathcal{C}_0 \subset (\mathcal{M}_0 \otimes_{\text{alg}} \mathcal{Q}_0^\sigma)^f$. □

Remark 3.11. *It is an interesting problem to study whether for a CQG action on a C^*-algebra, the deformation coincides with the generalized fixed point subalgebra or not. It is worth mentioning that already for a locally compact quantum group action this is not true. This was already observed in* [3].

References

[1] Alain Connes and Michel Dubois-Violette, Noncommutative finite-dimensional manifolds. I. Spherical manifolds and related examples. Comm. Math. Phys., 230(**3**):539–579, 2002.

[2] Goswami D., Joardar S., Quantum Isometry Groups of Noncommutative Manifolds Obtained by Deformation Using Dual Unitary 2-Cocycles, SIGMA **10** (2014), 076, 18 pages.

[3] Neshveyev S., Tuset L., Deformation of C*-algebras by cocycles on locally compact quantum groups, Adv. Math. **254** (2014), 454–496.

[4] Podles P., Symmetries of Quantum Spaces, subgroups and quotient spaces of $SU(2)$ and $SO(3)$ groups, Comm. Math. Phys., **170(1)** (1995), 1–20.

[5] Rieffel, Mark A.: Deformation Quantization for actions of R^d, Memoirs of the American Mathematical Society, November 1993. Volume **106**. Number 506.

[6] Woronowicz S.L., Compact quantum groups, Symmetries Quantiques (Les Houches, 1995), North-Holland, Amsterdam, 1998, pp. 845–884.

Debashish Goswami and Soumalya Joardar
Stat.-Math. Unit
Indian Statistical Institute
203, B. T. Road
Kolkata 700108, India
e-mail: goswamid@isical.ac.in

Operator Theory:
Advances and Applications, Vol. 247, 69–82
© Springer International Publishing Switzerland 2015

Unbounded Derivations of GB*-algebras

M. Weigt and I. Zarakas

Abstract. Generalized B*-algebras are locally convex $*$-algebras which are generalizations of C*-algebras. We obtain results on unbounded derivations of commutative generalized B*-algebras (GB*-algebras for short) by borrowing some techniques from commutative algebra. We then give an example of a commutative GB*-algebra having a nonzero derivation. Lastly, we also prove that every derivation of a GB*-algebra, with underlying C*-algebra a W*-algebra, is inner.

Mathematics Subject Classification (2010). 46H05, 46H25, 46H35, 46H40, 46K05, 46L05, 46L10.

Keywords. GB*-algebra, topological algebra, derivation.

1. Introduction

GB*-algebras (i.e., generalized B*-algebras) were introduced in [4] by G.R. Allan in 1967, and are locally convex $*$-algebras which are generalizations of C*-algebras. Later, P.G. Dixon extended the notion of a GB*-algebra in [18] to the setting of non-locally convex $*$-algebras. GB*-algebras can also be regarded as abstract algebras of unbounded operators in that the Gelfand–Naimark representation theorem for C*-algebras extends to GB*-algebras [18, Theorem 7.6 and Theorem 7.11]. Also, for every GB*-algebra A, there is a C*-algebra $A[B_0]$ contained in A (see section two), so that we can think of a GB*-algebra as a C*-algebra with "unbounded elements" adjoined to it.

Recall that C*-algebras play a significant role in mathematical physics, and that observables of a quantum mechanical system are regarded as elements of a $*$-algebra consisting of unbounded operators on a Hilbert space. Furthermore, the dynamics of the system can be modeled by one-parameter automorphism groups of the latter algebra, and derivations are the generators of these automorphism groups [30].

If A is an algebra, and B is a subalgebra of A, then a *derivation* is a linear map $\delta : B \to A$ such that $\delta(xy) = x\delta(y) + \delta(x)y$ for all $x, y \in B$. If $B = A$, then we say that δ is a derivation of A. The derivation δ is said to be *inner* if there exists $a \in A$ such that $\delta(x) = ax - xa$ for all $x \in B$. If A is a locally convex $*$-algebra, then we say that the derivation δ is an unbounded derivation if, in addition, B is a $*$-subalgebra dense in A, i.e., the domain of δ is a dense $*$-subalgebra of A.

It is well known that the zero derivation is the only derivation of a commutative C*-algebra [30, Corollary 2.2.8] and that all derivations of a C*-algebra are continuous [30, Theorem 2.3.1]. Also, all derivations of a von Neumann algebra are inner [30, Theorem 2.5.3]. An abundance of automatic continuity results for derivations of Banach algebras can be found in [16].

The article of C. Brödel and G. Lassner [12] is the first article about derivations of unbounded operator algebras to appear in the literature. Much later, in 1992, R. Becker proved that every derivation $\delta : A \to A$ of a pro-C*-algebra $A[\tau_\Gamma]$ is continuous [6, Proposition 2]. By a pro-C*-algebra, we mean a complete topological $*$-algebra $A[\tau_\Gamma]$ for which there exists a directed family of C*-seminorms defining the topology τ_Γ on A [22].

Becker also proved that commutative pro-C*-algebras have no nonzero derivations [6, Corollary 3]. Some other results concerning derivations of non-normed topological $*$-algebras and unbounded operator algebras can be found in [1], [2], [5], [7], [8], [9], [10], [33], [34] and [37]. For a detailed survey of derivations of locally convex $*$-algebras, see [25].

All of the above, together with [25, discussion after Theorem 5.2], provides motivation for a general investigation of derivations of GB*-algebras. In [35], we proved that a complete commutative GB*-algebra having jointly continuous multiplication has no nonzero derivations [35, Theorem 3.3]. In particular, every Fréchet commutative GB*-algebra has no nonzero derivations [35, Corollary 3.4].

Unbounded derivations of C*-algebras are well understood, and have important applications to mathematical physics (see [30]). Up until now, there is nothing in the literature on unbounded derivations of GB*-algebras, and we are therefore motivated to investigate unbounded derivations of GB*-algebras. In Section 3 of this paper, we obtain results about unbounded derivations of commutative GB*-algebras by using techniques in commutative algebra which are generally not used in functional analysis. In particular, we make use of the Kähler module and Kähler derivation of a commutative algebra, along with some of their important properties given in [27] and [28].

Since a commutative C*-algebra has no nonzero derivations, we raised the question in [35] as to whether or not there is a commutative GB*-algebra having a nonzero derivation, and in Section 4 of this paper, we give an example of a commutative GB*-algebra having a nonzero derivation. Motivated by this, we end section four with a characterization of commutative GB*-algebras $A[\tau]$ for which there exists at least one nonzero derivation $\delta : A[B_0] \to A$ (see Proposition 4.1).

In Section 5, we prove that every derivation of a Fréchet GB*-algebra, whose $A[B_0]$ is a W*-algebra, is inner. This generalizes the well-known fact that every derivation of a W*-algebra is inner, as well as results proved in [35] specialized to Fréchet GB*-algebras. Section 2 of our paper contains all preliminary material needed to understand the main results of this paper.

2. Preliminaries

All vector spaces in this paper are over the field $\mathbb{C}$ of complex numbers and all topological spaces are assumed to be Hausdorff. Moreover, all algebras are assumed to have an identity element denoted by 1.

A *topological algebra* is an algebra which is also a topological vector space such that the multiplication is separately continuous in both variables [22]. A topological algebra which is metrizable and complete is called a *Fréchet topological algebra*. A *topological $*$-algebra* is a topological algebra endowed with a continuous involution. A topological $*$-algebra which is also a locally convex space is called a *locally convex $*$-algebra*. The symbol $A[\tau]$ will stand for a topological $*$-algebra A endowed with given topology τ.

Definition 2.1 ([4]). Let $A[\tau]$ be a topological $*$-algebra and $\mathcal{B}^*$ a collection of subsets B of A with the following properties:

(i) B is absolutely convex, closed and bounded,
(ii) $1 \in B$, $B^2 \subseteq B$ and $B^* = B$.

For every $B \in \mathcal{B}^*$, denote by $A[B]$ the linear span of B, which is a normed algebra under the gauge function $\|\cdot\|_B$ of B. If $A[B]$ is complete for every $B \in \mathcal{B}^*$, then $A[\tau]$ is called *pseudo-complete.*

An element $x \in A$ is called (Allan) *bounded* if for some nonzero complex number λ, the set $\{(\lambda x)^n : n = 1, 2, 3, \dots\}$ is bounded in A. We denote by A_0 the set of all bounded elements in A.

A topological $*$-algebra $A[\tau]$ is called *symmetric* if, for every $x \in A$, the element $(1 + x^*x)^{-1}$ exists and belongs to A_0.

In [18], the collection $\mathcal{B}^*$ in the definition above is defined to be the same as above, except that $B \in \mathcal{B}^*$ is no longer assumed to be absolutely convex. The notion of a bounded element is a generalization of the concept of bounded operator on a Banach space, and was introduced by G.R. Allan in [3] in order to develop a spectral theory for general locally convex $*$-algebras.

Definition 2.2 ([4]). A symmetric pseudo-complete locally convex $*$-algebra $A[\tau]$ such that the collection $\mathcal{B}^*$ has a greatest member denoted by B_0, is called a *GB^*-algebra* over B_0.

Every sequentially complete locally convex algebra is pseudo-complete [3, Proposition 2.6]. The Arens algebra $L^\omega([0,1]) = \cap_{p\geq 1} L^p([0,1])$ and every pro-C^*-algebra is a GB*-algebra. In [18], P.G. Dixon extended the notion of GB*-algebras to include topological $*$-algebras which are not locally convex. A typical example is the algebra $M([0,1])$ of all measurable functions (modulo equality almost everywhere) equipped with the topology of convergence in measure [18, p. 696 (3.4)]. For a survey on GB*-algebras, see [23].

Proposition 2.3 ([4, Theorem 2.6]; [11, Theorem 2]). *If $A[\tau]$ is a GB*-algebra, then the Banach $*$-algebra $A[B_0]$ is a C*-algebra sequentially dense in A, and $(1+x^*x)^{-1} \in A[B_0]$ for every $x \in A$. Furthermore, B_0 is the unit ball of $A[B_0]$.*

If A is commutative, then $A_0 = A[B_0]$ [4, p. 94]. In general, A_0 is not a $*$-subalgebra of A, and $A[B_0]$ contains all normal elements of A_0 [4, p. 94]. To be more precise, $A[B_0] \subseteq A_0$ [18, p. 695].

It is well known that every commutative C*-algebra A is topologically and algebraically $*$-isomorphic to $C(X)$ for some compact Hausdorff space X (in fact, X is the maximal ideal space of A and the corresponding algebra isomorphism is the Gelfand isomorphism). This result extends to commutative GB*-algebras, as is seen in the following theorem.

Theorem 2.4 ([4, Theorem 3.9]). *Any commutative GB*-algebra A is algebraically $*$-isomorphic to an algebra $N(X)$ of continuous $\mathbb{C}^*$-valued functions on a compact Hausdorff space X, which are allowed to take the value infinity on at most a nowhere dense subset of X. Here, $\mathbb{C}^*$ denotes the one-point compactification of $\mathbb{C}$. This algebraic $*$-isomorphism extends the Gelfand isomorphism of $A[B_0]$ onto the corresponding C*-algebra $C(X)$.*

Recall that every C*-algebra is topologically-algebraically $*$-isomorphic to a norm closed $*$-subalgebra of $B(H)$ for some Hilbert space H. In general, for every GB*-algebra $A[\tau]$, there exists a faithful $*$-representation $\pi : A \to \pi(A)$ such that $\pi(A)$ is an algebra of closed densely defined operators on a Hilbert space H with B_0 being identified with $\{x \in \pi(A) \cap B(H) : \|x\| \leq 1\}$ [18, Theorem 7.11]. Therefore, for every $a \in A$, it follows that $\|(1+a^*a)^{-1}\| \leq 1$ (see also [4, Theorem 2.6]) and that $a(1+a^*a)^{-1} \in A[B_0]$.

The following concepts of locally measurable operator and EW*-algebra will be needed in Section 5.

Definition 2.5 ([36, Theorem 2.1 and Definition 2.2]). Let M be a von Neumann algebra on a Hilbert space H and x a closed operator affiliated with M.

(i) The operator x is called measurable if the domain of x is dense in H and $1-E_\lambda$ is finite for some $\lambda > 0$, where $|x| = \int_0^\infty \lambda dE_\lambda$ is the spectral decomposition of $|x|$.

(ii) If there exist projections q_n in the centre of M such that $q_n \uparrow 1$ and xq_n is measurable for each n, then x is called locally measurable.

We denote the set of *all locally measurable operators affiliated with a von Neumann algebra M* by $LS(M)$. When equipped with the topology of local convergence in measure τ_{lcm}, one has that $LS(M)$ is a topological $*$-algebra with respect to the usual adjoint, the strong sum and strong product [36, p. 260].

Definition 2.6 ([18, Definitions 7.1], [19, Definition 1.2]). Let A be a set of closed, densely defined operators on a Hilbert space $\mathcal{H}$ which is a $*$-algebra under strong sum, strong product, scalar multiplication (it is understood that $\lambda x = 0$, the zero operator on the whole of $\mathcal{H}$, if $\lambda = 0$) and the usual adjoint of operators. We call A an extended C*-algebra (resp. an EW*-algebra) if the following conditions are satisfied:

(i) $(1 + x^*x)^{-1}$ exists in A for every $x \in A$,
(ii) the $*$-subalgebra A_e of bounded operators in A is a C*-algebra (resp. a W*-algebra).

We sometimes say that *A is an extended C*-algebra (resp. an EW*-algebra) over the C*-algebra (resp. the von Neumann algebra) A_e.*

3. Unbounded derivations of commutative GB*-algebras

Let $A[\tau]$ be a GB*-algebra. We say that a linear map $\delta : D(\delta) \to A$ is an *unbounded derivation* of A if $D(\delta)$ is a dense $*$-subalgebra of A and $\delta(xy) = x\delta(y) + \delta(x)y$ for all $x, y \in D(\delta)$. If, in addition, $\delta(x^*) = \delta(x)^*$ for all $x \in D(\delta)$, then we call δ an *unbounded $*$-derivation* of A. If $D(\delta) = A$, then we say that δ is a *derivation* of A.

The following account, up to and excluding Theorem 3.2, is a summary of [27, pp. 221–224] (alternatively, see [20, Sections 16.1 and 16.8]). Let A be a commutative (abstract) algebra with identity over a field of characteristic zero (for example, K could be the field of complex numbers $\mathbb{C}$). Consider the map $\phi : A \otimes A \to A$, $a \otimes b \mapsto ab$. Let $J = \mathrm{Ker}(\phi)$.

Now consider the map $D : A \to J/J^2$, $a \mapsto (1 \otimes a) - (a \otimes 1) + J^2$. It can be shown that D is a derivation, and is called the *Kähler derivation* of A. It can also be shown that J/J^2 is an A-bimodule, and is called the *Kähler module* of A, also denoted by $E(A)$. It can be proved that $E(A)$ is the A-bimodule generated by the range of D, i.e.,

$$E(A) = \left\{ \sum_{i=1}^{n} b_i D a_i : b_i, a_i \in A, \text{ for all } 1 \le i \le n,\ n \in \mathbb{N} \right\}$$
$$= \left\{ \sum_{i=1}^{n} b_i \big((1 \otimes a_i) - (a_i \otimes 1) + J^2 \big) : b_i, a_i \in A, \text{ for all } 1 \le i \le n,\ n \in \mathbb{N} \right\}.$$

Theorem 3.1 ([27, Proposition 1.6]). *If A is a commutative algebra and F is an A-bimodule, then every derivation $\delta : A \to F$ can be expressed in the form $\delta = h(\delta) \circ D$, where $h(\delta) : E(A) \to F$ is an A-module map.*

If A is a semi-simple commutative algebra over a field K of characteristic zero, in the sense that

$$\cap \{\mathrm{Ker}(\omega) : \omega : A \to K \text{ an algebra homomorphism}\} = \{0\},$$

then A can be realized up to algebra isomorphism as an algebra of K-valued functions on a set $\Omega(A)$, the set of all algebra homomorphisms from A onto K [28,

p. 141]. This is clearly an extension of Gelfand theory for commutative Banach algebras, and we make use of this observation in the following theorem.

Theorem 3.2 ([28, Proposition 3]). *For a commutative and semi-simple algebra A, the K-valued function a on A has finite range if and only if $Da = 0$, where D is the Kähler derivation of A (as above).*

Since a commutative C*-algebra is well known to be semi-simple in the above sense, and is over the field $\mathbb{C}$, which is of characteristic zero, one obtains from Theorem 2.4 the following result concerning unbounded derivations of commutative GB*-algebras, which is an immediate consequence of Theorem 3.1 and Theorem 3.2.

Corollary 3.3. *Let A be a commutative GB*-algebra and $\delta : A[B_0] \to A$ a derivation. If $a \in A[B_0]$ and $\sigma_{A[B_0]}(a)$ is finite, then $\delta(a) = 0$.*

Remarks. (1) The above corollary can, in light of Theorem 2.4, also be deduced as follows: We can regard a as a continuous complex-valued function f on a compact Hausdorff space X with finite range, which is some finite linear combination of characteristic functions χ_{E_i}. It can easily be verified that all E_i are both open and closed, so that all χ_{E_i} are in $C(X)$. Therefore, since A is commutative, it follows that $\delta(\chi_{E_i}) = 0$ for all i, and hence $\delta(a) = 0$.

(2) If, in Theorem 3.1, we take A to be the domain of an arbitrary unbounded derivation of an arbitrary commutative GB*-algebra, then Theorem 3.1 reveals the full structure of unbounded derivations of commutative GB*-algebras. See also Proposition 4.1 in this regard.

If A is a commutative GB*-algebra, then we know from Theorem 2.4 that A is algebraically $*$-isomorphic to a $*$-algebra of functions $N(X)$ on a compact Hausdorff space X, and $A[B_0] \cong C(X)$. Therefore, our next result is about unbounded derivations of commutative GB*-algebras.

Corollary 3.4. *Let X be a compact Hausdorff space and $N(X)$ a $*$-algebra of functions as in the statement of Theorem 2.4. Suppose that for every $f \in C(X)$ and every $x \in X$, there exists a subset E of X having the properties that*

(i) *$x \in E$,*
(ii) *E is an open and closed subset of X, and*
(iii) *$f|_E$ has finite range.*

If $\delta : C(X) \to N(X)$ is a derivation, then $\delta = 0$.

Proof. Let $x \in X$ and $f \in C(X)$. By hypothesis, there exists an open and closed subset E of X such that $x \in E$ and $f|_E$ has finite range. Since E is open and closed, χ_E is continuous, i.e., $\chi_E \in C(X)$. Now $f\chi_E$ has finite range (that is, has finite spectrum relative to $C(X)$), and therefore it follows from the previous corollary that $\delta(f\chi_E) = 0$, and so $0 = \delta(f\chi_E) = \chi_E\delta(f)$. Therefore

$$0 = \big(\chi_E\delta(f)\big)(x) = \chi_E(x)\big(\delta(f)\big)(x) = \big(\delta(f)\big)(x).$$

It follows that $\delta = 0$. □

The example below is an example of a compact Hausdorff space satisfying the conditions of Corollary 3.4. For this, we recall that a totally disconnected space is a compact Hausdorff space X such that for any $x, y \in X$ with $x \neq y$, there exist disjoint open subsets A and B of X such that $x \in A$, $y \in B$ and $X = A \cup B$. Observe that such sets A and B are automatically closed. We recall that a totally disconnected space has a base consisting of open and closed sets [31, Theorem 33.C].

Example 3.5. Let X be a totally disconnected space having the property that for every non-empty open and closed subset F of X, there is a non-empty open and closed subset E of F such that no non-empty open and closed subsets are properly contained in E. We call such an E an *atom* of X, and we call such a space X *atomic* (this is in analogy with an atomic measure space).

Let $x \in X$. Since X is open and closed, there is an atom E in X such that $x \in E$. We now show that f is constant on E, thereby showing that X is an example of a compact Hausdorff space satisfying the three conditions of Corollary 3.4.

Let $f \in C(X)$, and let $a, b \in E$ with $a \neq b$. We show that $f(a) = f(b)$. Suppose that $f(a) \neq f(b)$. Since $\{f(a)\}$ is closed in $\mathbb{C}$ and f is continuous, it follows that $E_1 = f^{-1}(\{f(a)\}) \cap E$ is a closed subset of X. Since $a \in E_1$, we get that E_1 is non-empty. Since $f(a) \neq f(b)$, we get that $b \notin E_1$. Therefore E_1 is properly contained in E. It follows that $E \cap (X \setminus E_1)$ is non-empty, open and properly contained in E (we need the fact that E_1 is non-empty to deduce the proper containment). Since X is totally disconnected, there is a non-empty open and closed subset F of X such that $F \subseteq E \cap (X \setminus E_1)$, and so F is properly contained in E. This contradicts the fact that E is an atom. Therefore $f(a) = f(b)$, implying that f is constant on E.

This implies that X is an example of a compact Hausdorff space satisfying the three conditions of Corollary 3.4.

Corollary 3.6. *Let X be an atomic totally disconnected space, and $N(X)$ a $*$-algebra of functions. If $\delta : C(X) \to N(X)$ is a derivation, then $\delta = 0$.*

We remark that the completeness of the lattice of open and closed subsets of X is not required in Corollary 3.6, so that we have a slight strengthening of one of the implications in the statement of [13, Theorem 4.2].

Recall that a Stonean space is a compact Hausdorff space having the property that the closure of every open set is open. It is well known, and easily verified, that every Stonean space is totally disconnected. However, not every totally disconnected space is Stonean.

Recall that if A is a commutative AW*-algebra, then A is isometrically $*$-isomorphic to $C(X)$, where X is Stonean. This, along with Corollary 3.6, gives us the following result.

Corollary 3.7. *Let A be a commutative GB*-algebra such that $A[B_0]$ is an AW*-algebra having an atomic projection lattice. If $\delta : A[B_0] \to A$ is a derivation, then $\delta = 0$.*

Corollary 3.7 was proved in [35] for the case where $A[B_0]$ is a W*-algebra [35, Proposition 5.12]. Therefore, Corollaries 3.7, 3.6 and 3.4 are extensions of [35, Proposition 5.12] (recall that not every commutative AW*-algebra is a W*-algebra).

4. An example of a commutative GB*-algebra having a nonzero derivation

Let X be a compact Hausdorff space equipped with a positive Radon measure μ assigning nonzero values to non-empty open subsets of X. Then $C(X) \cong \{M_f : f \in C(X)\}$ is a commutative C*-algebra acting on the Hilbert space $L^2(X, \Sigma, \mu)$. Let $N_{\max}(X)$ be the maximal $*$-algebra of functions of X, i.e., the set of *all* continuous functions on X possibly taking the value of infinity on a nowhere dense set. Throughout, we *assume also that each $f \in N_{\max}(X)$ is finite almost everywhere*, so that the nowhere dense set, on which the function can take the value infinity, has measure zero. We also make the convention that $0.\infty = 0$. We show that $N_{\max}(X)$ can be realized as an extended C*-algebra with a common dense domain, thereby proving that it is a locally convex GB*-algebra in some topology.

To begin, first observe that $N_{\max}(X) \cong \{M_f : f \in N_{\max}(X)\}$, where the domain $D(M_f)$ of M_f, for $f \in N_{\max}(X)$, is the set of all $g \in L^2(X, \Sigma, \mu)$ satisfying the property that $fg \in L^2(X, \Sigma, \mu)$. Let $\mathcal{D}_0$ be the set of all (finite-valued) continuous functions on X.

We show that if $f \in N_{\max}(X)$ and $g \in \mathcal{D}_0$, then $fg \in L^2(X, \Sigma, \mu)$. Let

$$D_f = \{x \in X : f(x) < \infty\}.$$

Since $|fg|^2$ is continuous and X is compact, $|fg|^2(X)$ is a compact subset of $\mathbb{C}^*$, the one point compactification of $\mathbb{C}$. Therefore $|fg|^2(X)$ is a bounded subset of $\mathbb{C}^*$, implying that $|fg|^2(D_f)$ is a bounded subset of $\mathbb{C}^*$. Since $|fg|^2(D_f) \subseteq \mathbb{C}$, $|fg|^2(D_f)$ is a bounded subset of $\mathbb{C}$, i.e., there exists $M > 0$ such that $|(fg)(x)|^2 \leq M$ for all $x \in D_f$. Therefore

$$\int_{D_f} |fg|^2 d\mu \leq \int_{D_f} M d\mu = M\mu(D_f) \leq M\mu(X) < \infty.$$

By the assumption above, $\mu(X \setminus D_f) = 0$, and hence $\int_{X \setminus D_f} |fg|^2 = 0$. Therefore $\int |fg|^2 d\mu < \infty$, i.e., $fg \in L^2(X, \Sigma, \mu)$.

It follows immediately that $\mathcal{D}_0 \subseteq D(M_f)$ for all $f \in N_{\max}(X)$. Since $\mathcal{D}_0$ is the set of all finite-valued continuous functions on X having compact support, we get that $\mathcal{D}_0$ is a dense subspace of $L^2(X, \Sigma, \mu)$ [15, Proposition 7.4.2].

Finally, observe that $N_{\max}(X) \cong \{M_f : f \in N_{\max}(X)\}$ is an extended C*-algebra over the C*-algebra $\{M_f : f \in C(X)\} \cong C(X)$, which, from the above, has common dense domain $\mathcal{D}_0$. By [18, Theorem 7.12], $N_{\max}(X)$ is a locally convex GB*-algebra in the weak-operator topology.

If Y is an arbitrary totally disconnected compact Hausdorff space, then $N_{\max}(Y)$ admits a nonzero derivation if and only if Y is not σ-distributive [13, Theorem 3.1 or Theorem 4.2] (and therefore not atomic).

Therefore, if, in addition, X is a totally disconnected space in such a way that the Boolean algebra of closed and open subsets of X is complete and not σ-distributive, then $N_{\max}(X)$ is a commutative locally convex GB*-algebra admitting a nonzero derivation $\delta : N_{\max}(X) \to N_{\max}(X)$, and hence also a nonzero derivation $\delta : C(X) \to N_{\max}(X)$ [35, Proposition 3.1].

Observe that the multiplication on $N_{\max}(X)$ is not jointly continuous with respect to the locally convex GB*-topology on $N_{\max}(X)$ (since this topology is the weak-operator topology). In this regard, we note that if a commutative (locally convex) GB*-algebra A is complete and has jointly continuous multiplication, then A has no nonzero derivations [35, Theorem 3.3].

The above example motivates the problem of characterizing commutative GB*-algebras having at least one nonzero derivation $\delta : A[B_0] \to A$, and below, we give such a characterization. Let $A[\tau]$ be a commutative GB*-algebra. Recall that the Kähler module $E(A[B_0])$ of the commutative C*-algebra $A[B_0]$ is generated by the range of the Kähler derivation of $A[B_0]$. Furthermore, if $A[B_0]$ is $*$-isomorphic to $C(X)$ for some compact Hausdorff space X, where $C(X)$ has at least one element with an infinite range, we get from Theorem 3.2 that the Kähler derivation of $A[B_0]$ is nonzero. Therefore, by Theorem 3.1, the following result is now immediate.

Proposition 4.1. *Let $A[\tau]$ be a commutative GB*-algebra such that there is an element $a \in A[B_0]$ for which $\sigma_{A[B_0]}$ is infinite. Then there exists a nonzero derivation $\delta : A[B_0] \to A$ if and only if there exists a nonzero A-module map $h : E(A[B_0]) \to A$.*

5. Derivations of Fréchet GB*-algebras whose underlying C*-algebra is a W*-algebra

An open problem is whether or not every derivation of a Fréchet GB*-algebra is continuous. In this section, we positively answer this question for the case where $A[B_0]$ is a W*-algebra. In fact, we show in such a case that all derivations are inner (see Corollary 5.6 below). Recall that if $A[\tau]$ is a Fréchet GB*-algebra, then the multiplication on A is automatically jointly continuous.

Theorem 5.1. *Let M be a von Neumann algebra with a faithful finite normal trace, and let $X[\tau']$ be a complete locally convex M-bimodule contained in $LS(M)$ satisfying the following conditions.*

(i) *For every seminorm p on X defining the topology τ', we have for every $a \in M$ and $x \in X$ that $p(ax) \leq \|a\| p(x)$.*
(ii) *The topology τ' is stronger than τ_{lcm} on X.*

Then every derivation $\delta : LS(M) \to LS(M)$, with $\delta(M) \subseteq X$, is $\tau_{lcm} - \tau_{lcm}$ continuous.

Proof. If ϕ denotes the faithful finite normal trace on M, we let $\widetilde{M}$ denote the $*$-algebra of ϕ-measurable operators affiliated with M, which is a Fréchet topological $*$-algebra in the topology of convergence in measure [21]. Since the trace ϕ is also finite, $LS(M) = \widetilde{M}$, implying that $LS(M)[\tau_{lcm}]$ is a Fréchet algebra, since the topology τ_{lcm} also coincides with the topology of convergence in measure in this case. Now exactly the same proof of [9, Lemma 6.9] holds, except that we apply [34, Theorem 3.8], in place of the Ringrose theorem. □

If $X[\tau']$ in Theorem 5.1 is a Banach M-bimodule, then τ' is stronger than τ_{lcm} on X [9, Lemma 6.7].

If $A[\|\cdot\|]$ is a C*-algebra and $X[\tau]$ is a complete locally convex A-bimodule having $\|\cdot\| \times \tau - \tau$ jointly continuous module actions, then the topology τ on X can be defined by a family of seminorms Γ such that for every $q \in \Gamma$, $q(ax) \leq \|a\| q(x)$ and $q(xa) \leq \|a\| q(x)$ for all $a \in A$ and $x \in X$ (this is a special case of [29, Proposition 3.8]).

Corollary 5.2. *Let $A[\tau]$ be a Fréchet GB*-algebra such that $A[B_0]$ is a W*-algebra having a faithful finite normal trace. Then every derivation $\delta : A \to A$ is inner, and hence continuous.*

Proof. The algebra A can be regarded as an EW*-algebra over $M = A[B_0]$, and therefore as a $*$-subalgebra of $LS(M)$ [14, Corollary 2].

Also, A is a locally convex bimodule over M having $\tau \times \tau$-jointly continuous multiplication, and hence $\|\cdot\| \times \tau$-jointly continuous multiplication. By the previous paragraph, there is a family of seminorms defining the topology τ on A satisfying condition (i) of Theorem 5.1.

We now show that the topology τ on A is stronger than the topology τ_{lcm} when restricted to A. This is the same as showing that the identity map $id : A[\tau] \to A[\tau_{lcm}] \subseteq LS(A[B_0])[\tau_{lcm}]$ is continuous. Since $LS(A[B_0])[\tau_{lcm}]$ is a Fréchet topological $*$-algebra (due to the trace on $A[B_0]$ being finite), it is sufficient to use the closed graph theorem. Let (a_n) be a sequence in A with $a_n \to 0$ w.r.t τ, and such that $a_n \to a \in LS(A[B_0])$ w.r.t τ_{lcm}. Then $b_n = \mathrm{Re}(a_n) \to 0$ w.r.t τ and $b_n \to \mathrm{Re}(a) = b$ w.r.t τ_{lcm}. This implies that $b_n^* b_n \to 0$ w.r.t τ and $b_n^* b_n \to b^2$ w.r.t τ_{lcm}. Since $A[\tau]$ is a Fréchet GB*-algebra, the multiplication on A is jointly continuous, and hence hypocontinuous. Therefore, by [18, Theorem 6.5], the set of positive elements A^+ of A is τ-closed. Therefore, by a similar argument to the proof of [32, Lemma 3.6], applied to $b_n^* b_n \to 0$ w.r.t τ, there is a subsequence (c_n) of $(b_n^* b_n)$ and $c \in A^+$ such that $2^n c_n \leq c$ for all $n \in \mathbb{N}$. Hence, by the proof of [32, Lemma 3.7], there is a subsequence (c_{n_k}) of (c_n) such that $c_{n_k} \to 0$ w.r.t τ_{lcm}. Since $c_n \to b^2$ w.r.t τ_{lcm}, it follows that $c_{n_k} \to b^2$ w.r.t τ_{lcm}. Therefore, $b^2 = 0$, and hence $b = 0$. By a similar argument, $\mathrm{Im}(a) = 0$, implying that $a = 0$. Therefore the topology τ on A is stronger than the topology τ_{lcm} when restricted to A.

Let $\delta : A \to A$ be a derivation. By [8, Theorem 4.8], there is a derivation $\overline{\delta} : LS(M) \to LS(M)$ extending δ. It follows from Theorem 5.1 that $\overline{\delta}$ is $\tau_{lcm} - \tau_{lcm}$

continuous. Therefore, by [9, Theorem 4.1], $\overline{\delta}$ is inner. It is now immediate from [10, Proposition 5.19] that δ is inner, and hence continuous. □

Every derivation of a GB*-algebra, with $A[B_0]$ a properly infinite W*-algebra, is inner, and hence continuous [9, Theorem 5.1(ii)]. This is an extension of [35, Proposition 5.13].

From this observation, as well as Corollary 5.2, the following result is immediate. We still, however, give the (short) proof for sake of completeness.

Corollary 5.3. *If $A[\tau]$ is a Fréchet GB*-algebra such that $A[B_0]$ is a type II von Neumann algebra, then every derivation of A is inner, and hence continuous.*

Proof. We can write $A[B_0]$ as a direct sum of a type II_1 von Neumann algebra, which has a faithful finite normal trace, and a type II_∞ von Neumann algebra, which is properly infinite. Hence, from Corollary 5.2, the previous paragraph and [8, Proposition 2.1], the result follows. □

If $A[\tau]$ is a GB*-algebra with $M = A[B_0]$ a type III von Neumann algebra, then M is properly infinite, and therefore every derivation $\delta : A \to A$ is inner.

Therefore, by [8, Proposition 2.1], in order to show that every derivation of a Fréchet GB*-algebra, with $A[B_0]$ a (arbitrary) W*-algebra, is inner, we only have to prove that every derivation of a Fréchet GB*-algebra, with $A[B_0]$ a finite type I W*-algebra, is inner. This is what we do in what follows below. For this, we remark that if B is an EW*-algebra over the von Neumann algebra M, then, since M is τ_{lcm}-dense in B, it follows that the center of M is contained in the center of B.

We also recall that a derivation $\delta : A \to A$, where $A[\tau]$ is a locally convex $*$-algebra, is said to be *approximately inner* if there exists a net (a_α) in A such that $\delta(x) = \lim(a_\alpha x - x a_\alpha)$ for all $x \in A$.

Proposition 5.4. *Let $A[\tau]$ be a Fréchet GB*-algebra for which $A[B_0]$ is a finite W*-algebra. Then A is $*$-isomorphic to an EW*-algebra B over the von Neumann algebra $M \cong A[B_0]$ such that all derivations on B are approximately inner with respect to τ_{lcm}.*

Proof. By hypothesis, A is $*$-isomorphic to an EW*-algebra B over the von Neumann algebra $M \cong A[B_0]$ [14, Corollary 2]. Since $A[B_0]$ is finite, M is a finite von Neumann algebra. Due to the facts that $A[\tau]$ is a Fréchet GB*-algebra and that A is $*$-isomorphic to B, it easily follows that B can be equipped with a topology τ' such that $B[\tau']$ is a Fréchet GB*-algebra topologically $*$-isomorphic to $A[\tau]$.

Since M is a finite von Neumann algebra, it is semifinite, and therefore there is a faithful semifinite normal trace ϕ on M. If the trace ϕ is finite, then our result follows from Corollary 5.2. Therefore, we may assume that the trace ϕ is not finite. Since M is finite, $\phi|_{Z(M)}$ is a faithful semifinite normal trace on $Z(M)$, the center of M [17, Proposition 10, p. 12]. Therefore, there is an increasing net of central projections (p_α) with least upper bound being the identity element 1 of M, and such that $\phi(p_\alpha) < \infty$ for every α. Hence $(1 - p_\alpha)$ is a net of central projections decreasing toward zero. By [9, Corollary 2.4], $1-p_\alpha \to 0$ with respect to τ_{lcm}, and hence $p_\alpha \to 1$ with respect to τ_{lcm}.

Let $\delta : B \to B$ be derivation. If p is a central projection M, it follows that $\delta(px) = p\delta(x)$ for all $x \in B$, and hence $\delta|_{pB}$ is a derivation of pB. Since $p_\alpha B$ is a τ'-closed $*$-subalgebra of B for every α, it follows for every α that $p_\alpha B$ is a Fréchet GB*-algebra over the von Neumann algebra $p_\alpha M$ with identity element p_α [24]. Since, for every α, the trace of $p_\alpha M$ is finite, it follows from Corollary 5.2 that there exists, for every α, an element $a_\alpha \in p_\alpha B$ such that

$$\delta(p_\alpha x) = a_\alpha(p_\alpha x) - (p_\alpha x)a_\alpha$$

for all $x \in B$. Since $p_\alpha \in Z(B)$ for every α, it follows that

$$p_\alpha \delta(x) = (a_\alpha p_\alpha)x - x(a_\alpha p_\alpha)$$

for every α and $x \in B$. If we let $b_\alpha = a_\alpha p_\alpha$ for every α, it follows that

$$\delta(x) = \tau_{lcm} - \lim_\alpha(b_\alpha x - x b_\alpha),$$

for every $x \in B$, implying that δ is approximately inner with respect to τ_{lcm}. □

Proposition 5.5. *If $A[\tau]$ is a Fréchet GB*-algebra such that $A[B_0]$ is a finite type I W*-algebra, then every derivation of A is inner.*

Proof. By Proposition 5.4, A is $*$-isomorphic to an EW*-algebra B over a finite von Neumann algebra M, on which every derivation $\delta : B \to B$ is approximately inner with respect to τ_{lcm}, i.e., there exists a net (b_α) in B such that $\delta(x) = \tau_{lcm} - \lim_\alpha(b_\alpha x - x b_\alpha)$ for all $x \in B$. It is now immediate that $\delta(ax) = a\delta(x)$ for all $a \in Z(M)$, the center of M. Therefore, since M is of type I, we get from [2, Theorem 3.9] that δ is spatial in $LS(M)$. It follows from [10, Proposition 5.19] that δ is inner. □

All results given in this section culminate in the following result, which extends all results in [35, Section 5].

Corollary 5.6. *Every derivation of a Fréchet GB*-algebra $A[\tau]$, with $A[B_0]$ a W*-algebra, is inner, and hence continuous.*

Acknowledgement

The authors thank Professor H. France-Jackson of the Nelson Mandela Metropolitan University for bringing references [27] and [28] to their attention, which inspired the main results of section three of this manuscript.

The authors are indebted to Professor M. Fragoulopoulou for helpful comments and suggestions, which led to an improvement of this manuscript.

The authors express their gratitude to the referee, who made helpful suggestions for improvement of this manuscript, including a slight simplification of the example given in Section 4.

References

[1] S. Albeverio, Sh.A. Ayupov and K.K. Kudaybergenov, *Non commutative Arens algebras and their derivations*, J. Funct. Anal. **253**, 2007, 287–302.

[2] S. Albeverio, Sh.A. Ayupov and K.K. Kudaybergenov, *Derivations on the algebra of measurable operators affiliated with a type I von Neumann algebra*, Siberian. Adv. Math. **18**, 2008, 86–94.

[3] G.R. Allan, *A spectral theory for locally convex algebras*, Proc. London Math. Soc. **15**, 1965, 399–421.

[4] G.R. Allan, *On a class of locally convex algebras*, Proc. London Math. Soc. **17**, 1967, 91–114.

[5] J.-P. Antoine, A. Inoue and C. Trapani, *O*-dynamical systems and *-derivations of unbounded operator algebras*, Math. Nachr. **204**, 1999, 5–28.

[6] R. Becker, *Derivations on LMC*-Algebras*, Math. Nachr. **155**, 1992, 141–149.

[7] A.F. Ber, V.I. Chilin and F.A. Sukochev, *Derivations in commutative regular algebras*, Extracta Math. **21**, 2006, 107–147.

[8] A.F. Ber, V.I. Chilin and F.A. Sukochev, *Continuity of derivations in algebras of locally measurable operators*, ArXiv, 1 June 2012.

[9] A.F. Ber, V.I. Chilin and F.A. Sukochev, *Innerness of continuous derivations on algebras of locally measurable operators*, ArXiv, 8 April 2013.

[10] A.F. Ber, B. de Pagter and F.A. Sukochev, *Derivations in algebras of operator valued functions*, J. Operator Theory **66**, 2011, 261–300.

[11] S.J. Bhatt, *A note on generalized B*-algebras*, J. Indian Math. Soc. **43**, 1979, 253–257.

[12] C. Brödel and G. Lassner, *Derivationen auf gewissen Op*-algebren*, Math. Nachr. **67**, 1975, 53–58.

[13] V.I. Chilin and G.B. Levitina, *Derivations on ideals in commutative AW*-algebras*, arXiv, 28 May 2012.

[14] V.I. Chilin and B.S. Zakirov, *Abstract characterization of EW*-algebras*, Funct. Anal. Appl. **25**, 1991, 63–64.

[15] D.L. Cohn, *Measure Theory*, Birkhäuser, Boston, 1980.

[16] H.G. Dales, *Banach Algebras and Automatic Continuity*, Clarendon Press, Oxford, 2000.

[17] J. Dixmier, *Von Neumann Algebras*, North-Holland Mathematical Library, Amsterdam – New York – Oxford, 1981.

[18] P.G. Dixon, *Generalized B*-algebras*, Proc. London Math. Soc. **21**, 1970, 693–715.

[19] P.G. Dixon, *Unbounded operator algebras*, Proc. London Math. Soc. **23**, 1971, 53–69.

[20] D. Eisenbud, *Commutative Algebra with a View Toward Algebraic Geometry*, Springer-Verlag, New York, 1995.

[21] T. Fack and H. Kosaki, *Generalised s-numbers of τ-measurable operators*, Pacific J. Math. **123**, 1986, 269–300.

[22] M. Fragoulopoulou, *Topological Algebras with Involution*, North-Holland, 2005.

[23] M. Fragoulopoulou, A. Inoue and K.-D. Kürsten, *Old and new results on Allan's GB*-algebras*, Banach Center Publ. **91**, 2010, 169–178.

[24] M. Fragoulopoulou, A. Inoue and M. Weigt, *Tensor products of GB^*-algebras*, to appear in J. Math. Anal. Appl., available online at http://dx.doi.org/10.1016/j.jmaa.2014.06.046.

[25] M. Fragoulopoulou, M. Weigt and I. Zarakas, *Derivations of locally convex $*$-algebras*, Extracta Math. **26**, 2011, 45–60.

[26] A. Inoue and S. Ota, *Derivations on algebras of unbounded operators*, Trans. Amer. Math. Soc. **261**, 1980, 567–577.

[27] H. Osborn, *Modules of differentials I*, Math. Ann. **170**, 1967, 221–244.

[28] H. Osborn, *Derivations of commutative algebras*, Illinois J. Math. **13**, 1969, 137–144.

[29] A.Yu. Pirkovskii, *Arens–Michael envelopes, homological epimorphisms, and relatively quasi-free algebras*, Trans. Moscow. Math. Soc. **69**, 2008, 24–104.

[30] S. Sakai, *Operator Algebras in Dynamical Systems: The Theory of Unbounded Derivations in C^*-Algebras*, Cambridge University Press, Cambridge, 1991.

[31] G.F. Simmons, *Introduction to Topology and Modern Analysis*, McGraw-Hill, New York, 1963.

[32] M. Weigt, *Jordan homomorphisms between algebras of measurable operators*, Quaest. Math. **32**, 2009, 203–214.

[33] M. Weigt, *Derivations of τ-measurable operators*, Operator Theory: Advances and Applications, Birkhäuser Verlag, Basel, Switzerland **195**, 2009, 273–286.

[34] M. Weigt and I. Zarakas, *A note on derivations of pro-C^*-algebras into complete locally convex bimodules*, New Zealand J. Math. **41**, 2011, 143–152.

[35] M. Weigt and I. Zarakas, *Derivations of Generalized B^*-algebras*, Extracta Mathematicae **28**, 2013, 77–94.

[36] F.J. Yeadon, *Convergence of measurable operators*, Proc. Camb. Phil. Soc. **74**, 1973, 257–268.

[37] I. Zarakas, *Hilbert pro-C^*-bimodules and applications*, Revue Roumaine de Mathématiques Pures et Appliquées **17**, 2012, 289–310.

M. Weigt
Department of Mathematics and Applied Mathematics
Nelson Mandela Metropolitan University
Summerstrand Campus (South)
Port Elizabeth
6031 South Africa
e-mail: `weigt.martin@gmail.com`
`Martin.Weigt@nmmu.ac.za`

I. Zarakas
Department of Mathematics
University of Athens
Panepistimiopolis
Athens 15784, Greece
e-mail: `gzarak@hotmail.gr`

Operator Theory:
Advances and Applications, Vol. 247, 83–86
© Springer International Publishing Switzerland 2015

Generalized Sunder Inequality

Sangeeta Jhanjee and Alan Pryde

Abstract. V. Sunder proved that for $n \times n$ complex matrices A and B, with A being Hermitian and B being skew Hermitian with eigenvalues $\{\alpha_i\}_{i=1}^n$ and $\{\beta_i\}_{i=1}^n$ respectively (counting multiplicity) such that

$$|\alpha_1| \geq \cdots \geq |\alpha_n|,$$
$$|\beta_1| \leq \cdots \leq |\beta_n|,$$

then

$$|\alpha_i - \beta_i| \leq \|A - B\|,$$

where $\|.\|$ is the operator bound norm. We generalize Sunder's result to the case of an m-tuple of $n \times n$ complex matrices, using the Clifford operator.

Mathematics Subject Classification (2010). Primary: 15A42; secondary 15A18.

Keywords. Joint eigenvalues, joint spectra, Clifford algebra, commuting tuple of matrices, Sunder's inequality.

1. Introduction and Notation

Let $A = (A_1, \ldots, A_m)$ be an m-tuple of $n \times n$ complex matrices. A *joint eigenvalue* of A is an element $\alpha \in \mathbb{C}^m$, $\alpha = (\alpha_1, \ldots, \alpha_m)$ such that there exists a non-zero element $x \in \mathbb{C}^n$ with $A_j x = \alpha_j x$ for $1 \leq j \leq m$. Such an x is called a *joint eigenvector*. The set of all joint eigenvalues of A is called the *joint spectrum* of A and is denoted by $\mathrm{Sp}(A)$. This is a compact subset of $\mathbb{C}^m$. The *joint spectral radius* $r(A)$ is defined to be

$$r(A) = \max\{|\alpha| : \alpha \in \mathrm{Sp}(A)\}.$$

The *spectral set* $\gamma(A)$ was defined in [4] by

$$\gamma(A) = \left\{\alpha \in \mathbb{R}^m : 0 \in \sigma\left(\sum_{j=1}^{m}(A_j - \alpha_j I)^2\right)\right\},$$

where $\mathbb{R}$ denotes the field of real numbers. $\gamma(A)$ is always a non-empty finite subset of $\mathbb{R}^n$, if A_i are commuting [4]. Also, it is proved in [5] that for commuting matrices A_j with $\sigma(A_j) \subseteq \mathbb{R}$, $\gamma(A)$ coincides with $\mathrm{Sp}\, A$ and also with other well-known spectra [6], [9].

2. Clifford algebra

Alan McIntosh and Alan Pryde [5] introduced the use of Clifford algebra in the study of joint spectra. This approach has been pursued by other authors, for example [2], [3].

Let $\mathbb{R}^m$ be a real m-dimensional vector space with the basis $e_1, \ldots, e_m$. The Clifford algebra $\mathbb{R}_{(m)}$ is defined to be the real associative algebra generated by $e_1, \ldots, e_m$ with the relations $e_ie_j + e_je_i = 2\delta_{ij}$. The elements e_S, where S runs over the subsets of $\{1, \ldots, m\}$, form a basis of $\mathbb{R}_{(m)}$, if we define $e_\phi = 1$ and $e_S = e_{s_1} \ldots e_{s_k}$ when $S = \{s_1, \ldots, s_k\}$ and $1 \leq s_1 < s_2 < \cdots < s_k \leq m$. Elements of $\mathbb{R}_{(m)}$ are denoted by $\alpha = \sum_S \alpha_S e_S$, where $\alpha_S \in \mathbb{R}$. Under the inner product $\langle \alpha, \beta \rangle = \sum_S \alpha_S \beta_S$, $\mathbb{R}_{(m)}$ becomes a 2^m-dimensional Hilbert space with orthonormal basis $\{e_S\}$. The tensor product

$$\mathbb{C}^n \otimes \mathbb{R}_{(m)} = \left\{ \sum_S x_S \otimes e_S, x_S \in \mathbb{C}^n \right\},$$

is a Hilbert space, if we define an inner product

$$\langle x, y \rangle = \left\langle \sum_S x_S \otimes e_S, \sum_S y_S \otimes e_S \rangle = \sum_S \langle x_S, y_S \right\rangle$$

where $x_S, y_S \in \mathbb{C}^n$, and $\langle x_S, y_S \rangle$ is the usual inner product in $\mathbb{C}^n$, and the norm by

$$\left\| \sum_S x_S \otimes e_S \right\| = \left(\sum_S \|x_S\|^2 \right)^{1/2}.$$

The tensor product $M_n \otimes \mathbb{R}_{(m)}$ is a $n^2 \times 2^m$-dimensional linear space over $\mathbb{C}$. The linear space $M_n \otimes \mathbb{R}_{(m)}$ over $\mathbb{C}$ may be identified with a subalgebra of $L(\mathbb{C}^n \otimes \mathbb{R}_{(m)})$ by defining

$$\left(\sum_S A_S \otimes e_S \right) \left(\sum_T x_T \otimes e_T \right) = \sum_{S,T} A_S(x_T) \otimes e_S e_T.$$

So, $M_n \subset M_n \otimes \mathbb{R}_{(m)}$ by the identification $A \longrightarrow A \otimes e_\phi$ and $M_n \otimes \mathbb{R}_{(m)} \subset L(\mathbb{C}^n \otimes \mathbb{R}_{(m)})$. For $A = \sum_S A_S \otimes e_S \in M_n \otimes \mathbb{R}_{(m)}$, its adjoint is given by $A^* = \sum_S A_S^* \otimes \overline{e}_S$ where $\overline{e}_S = \pm e_S$, the sign being chosen so that $e_S \overline{e}_S = \overline{e}_S e_S = 1$.

Given an m-tuple $A = (A_1, \ldots, A_m)$ of $n \times n$ matrices, we define its Clifford operator $\mathrm{Cliff}(A) \in M_n \otimes \mathbb{R}_{(m)}$ by

$$\mathrm{Cliff}(A) = i \sum_{j=1}^m A_j \otimes e_j.$$

The symbol $\otimes$ is often omitted. The Clifford norm $\|\mathrm{Cliff}(A)\|$ is defined to be the operator bound norm as an element of $L(\mathbb{C}^n \otimes \mathbb{R}_{(m)})$. Pryde [7] introduced the Clifford operator to prove a generalization of the Bauer–Fike theorem.

3. Generalized Sunder inequality

Theorem 1. *Let* $A = (A_1, \dots, A_m)$ *and* $B = (B_1, \dots, B_m)$ *be* m*-tuples of commuting* $n \times n$ *matrices,* A_j *Hermitian and* B_j *skew Hermitian. Label the joint eigenvalues*

$$|\alpha_1| \geq \cdots \geq |\alpha_n|,$$
$$|\beta_1| \leq \cdots \leq |\beta_n|,$$

then

$$\max_k |\alpha_k - \beta_k| \leq \| \operatorname{Cliff}(A - B)\|.$$

Proof. As $A_1, \dots, A_m$ commute, there exists a common eigenvector $x_i \in \mathbb{C}^n$ associated with joint eigenvalues $\alpha_i = (\alpha_i^1 \dots \alpha_i^m) \in \mathbb{R}^m$ such that

$$A_j x_i = \alpha_i^j x_i, \qquad j = 1, \dots, m, \qquad i = 1, \dots, n.$$

As the A_i are Hermitian, so $\{x_1, \dots, x_n\}$ can be chosen to form an orthonormal basis of $\mathbb{C}$. Similarly, there exists joint eigenvectors $y_1, \dots, y_n$ in $\mathbb{C}^n$ corresponding to joint eigenvalues $\beta_1, \dots, \beta_n \in i\mathbb{R}^m$ of $B_1, \dots, B_n$ such that they form an orthonormal basis of $\mathbb{C}^n$ and

$$B_j y_i = \beta_i^j y_i \qquad i = 1, \dots, n \qquad j = 1, \dots, m.$$

Fix k. Let $M = \operatorname{Span}\{x_1, \dots, x_k\}$ and $N = \operatorname{Span}\{y_k, \dots, y_n\}$, then

$$\dim M \cap N \geq 1$$

so

$$M \cap N \neq \{0\}.$$

Let $z \in M \cap N$, $\|z\| = 1$, $a_i \in \mathbb{C}$, $b_i \in \mathbb{C}$ with

$$z = \sum_{i=1}^{k} a_i x_i = \sum_{i=k}^{n} b_i y_i.$$

Then

$$\begin{aligned}
\| \operatorname{Cliff} Az\|^2 &= \left\| i \sum_{j=1}^{m} A_j e_j \sum_{l=1}^{k} a_l x_l \right\|^2 = \sum_{j=1}^{m} \left\| \sum_{l=1}^{k} \alpha_l^j a_l x_l \right\|^2 = \sum_{j=1}^{m} \sum_{l=1}^{k} |\alpha_l^j a_l|^2 \\
&= \sum_{l=1}^{k} \left(\sum_{j=1}^{m} |\alpha_l^j|^2 \right) |a_l|^2 = \sum_{l=1}^{k} |\alpha_l|^2 |a_l|^2 \geq \sum_{l=1}^{k} |\alpha_k|^2 |a_l|^2 \\
&= |\alpha_k|^2 \sum_{l=1}^{k} |a_l|^2 = |\alpha_k|^2.
\end{aligned}$$

So

$$\| \operatorname{Cliff} Az\| \geq |\alpha_k|.$$

Similarly, we can show

$$\| \operatorname{Cliff} Az\| \geq |\beta_k|.$$

Also

$$\begin{aligned}|\alpha_k - \beta_k|^2 &= |\alpha_k|^2 + |\beta_k|^2 - 2Re\langle \alpha, \beta\rangle = |\alpha_k|^2 + |\beta_k|^2 - 2Re \sum \alpha_k^j \overline{\beta_k^j} \\ &= |\alpha_k|^2 + |\beta_k|^2 + 2Re \sum \alpha_k^j \beta_k^j = |\alpha_k|^2 + |\beta_k|^2 \\ &\le \| \operatorname{Cliff} Az\|^2 + \| \operatorname{Cliff} Bz\|^2 \\ &= \frac{1}{2}\| \operatorname{Cliff} Az + \operatorname{Cliff} Bz\|^2 + \frac{1}{2}\| \operatorname{Cliff} Az - \operatorname{Cliff} Bz\|^2 \\ &\le \| \operatorname{Cliff} A - \operatorname{Cliff} B\|^2 = \| \operatorname{Cliff}(A-B)\|^2\end{aligned}$$

as $\| \operatorname{Cliff} A - \operatorname{Cliff} B\| = \| \operatorname{Cliff} A + \operatorname{Cliff} B\|$, Cliff A being Hermitian and Cliff B skew Hermitian. So

$$\| \operatorname{Cliff}(A-B)\| \ge |\alpha_k - \beta_k|.$$

Since k was chosen arbitrarily, we have

$$\| \operatorname{Cliff}(A-B)\| \ge \max_k |\alpha_k - \beta_k|. \qquad \square$$

References

[1] R. Bhatia, *Perturbation Bounds for Matrix Eigenvalues*, Longman Scientific & Technical, Essex, England, 1987.

[2] R. Bhatia and T. Bhattacharyya, *A Henrici theorem for joint spectra of commuting matrices*, Proc. Amer. Math. Soc. 118(1993), 9–14.

[3] R. Bhatia and T. Bhattacharyya, *On the Joint Spectral Radius of Commuting Matrices*, Studia Math., 114 (1995), 29–38.

[4] G. Greiner and W.J. Ricker, *Joint spectra and commutativity of systems of* (2 × 2) *selfadjoint matrices*, Linear and Multi. Alg.36 (1993), 47–58.

[5] A. McIntosh and A. Pryde, *A functional calculus for several commuting operators*, Indiana U. Math. J. 36 (1987), 421–439.

[6] A. McIntosh, A. Pryde and W. Ricker, *Comparison of joint spectra for certain classes of commuting operators*, Studia Math. 88 (1987), 23–36.

[7] A.J. Pryde, *A Bauer–Fike Theorem for the joint spectrum of commuting matrices*, Linear Alg. Appl. 173 (1992), 221–230.

[8] V.S. Sunder, *Distance between Normal Operators*, Proc. Amer. Math. Soc. 84 (1982), 483–484.

[9] J.L. Taylor, *A joint spectrum for several commuting operators*, J. Funct. Anal. 6 (1970), 172–191.

Sangeeta Jhanjee and Alan Pryde
School of Math. Sciences
P.O. Box 28M
Monash University
Victoria 3800, Australia
e-mail: `sangeeta.jhanjee@monash.edu`
`alan.pryde@monash.edu`

Operator Theory:
Advances and Applications, Vol. 247, 87–100
© Springer International Publishing Switzerland 2015

The Berezin Number, Norm of a Hankel Operator and Related Topics

Mubariz T. Garayev

Abstract. We give, in terms of the Berezin number, necessary and sufficient conditions providing unitarity of an invertible operator. Also we obtain in terms of the Berezin number a new inequality for the norm of the Hankel operator H_φ which is better than the classical inequality $\|H_\varphi\| \leq \|\varphi\|_\infty$. The Berezin number is also used to generalize the Douglas lemma on zero Toeplitz products.

Mathematics Subject Classification (2010). Primary 47B32; Secondary 47A12, 47B35.

Keywords. Berezin number, numerical radius, unitary operator, Toeplitz operator, Hankel operator.

1. Introduction and notations

A Hilbert space $\mathcal{H} = \mathcal{H}(\Omega)$ consisting of functions defined on some set Ω is called a reproducing kernel Hilbert space (shortly, RKHS) if all point evaluations $f \to f(\lambda)$ $(\lambda \in \Omega)$ are continuous. Equivalently, there exists a function $k : \Omega \times \Omega \to \mathbb{C}$ such that all functions of the form $k(\cdot, \lambda) : \Omega \to \mathbb{C}$ belong to $\mathcal{H}$ and, moreover, satisfy the equality

$$\langle f, k(\cdot, \lambda) \rangle = f(\lambda) \quad (f \in \mathcal{H},\ \lambda \in \Omega).$$

The function k with these properties is easily seen to be unique and is usually called the reproducing kernel of $\mathcal{H}$.

Let $A : \mathcal{H} \to \mathcal{H}$ be a bounded linear operator (i.e., $A \in \mathcal{B}(\mathcal{H})$. The Berezin symbol of A is defined by

$$\widetilde{A}(\lambda) := \left\langle A \frac{k(\cdot, \lambda)}{\|k(., \lambda)\|}, \frac{k(\cdot, \lambda)}{\|k(\cdot, \lambda)\|} \right\rangle \ (\lambda \in \Omega),$$

where the function $\widehat{k_\lambda}(z) := \frac{k(z,\lambda)}{\|k(z,\lambda)\|}$ is called the normalized reproducing kernel of $\mathcal{H}$. A detailed presentation of the theory of functional Hilbert spaces, reproducing kernels and Berezin symbols is given, for instance, in Aronzajn [3], Saitoh [17, 18] and Nordgren and Rosenthal [14].

Let us denote

$$\operatorname{Ber}(A) = \left\{\widetilde{A}(\lambda) : \lambda \in \Omega\right\} \text{ and } \operatorname{ber}(A) := \sup_{\lambda\in\Omega}\left|\widetilde{A}(\lambda)\right|,$$

called the Berezin set and Berezin number of the operator A, respectively. We recall that $W(A) := \{\langle Af, f\rangle : \|f\|_{\mathcal{H}} = 1\}$ is the numerical range of the operator A and

$$w(A) := \sup\{|\langle Af, f\rangle| : \|f\|_{\mathcal{H}} = 1\}$$

is the numerical radius of A. Since $\operatorname{ber}(A) \le w(A)$ and $\operatorname{Ber}(A) \subset W(A)$, the investigation of these numerical characteristics of the bounded linear operators is of interest in the spectral theory of such operators.

We remark that the numerical range $W(A)$ is always convex. However, it is easy to see that the same property does not hold for the Berezin set of A. Indeed, let φ be a function in H^∞ such that Range (φ) is not a convex set, and let T_φ be a corresponding Toeplitz operator on the Hardy space H^2. Then since $\widetilde{T_\varphi} = \varphi$, we see that $\operatorname{Ber}(T_\varphi)$ is not a convex set.

We also remark that in [8], Engliš showed that there is no constant C such that either

$$\|T_f\|_e \le C \lim_{z\to\partial\mathbb{D}} \sup\left|\widetilde{f}(z)\right| \ \forall f \in L^\infty(\mathbb{D}, dm_2)$$

or

$$\|T_f\| \le C \sup_{z\in\mathbb{D}}\left|\widetilde{f}(z)\right| \ \ (= C\ ber(T_f)) \ \forall f \in L^\infty(\mathbb{D}, dm_2)$$

holds. Here $\|T_f\|_e$ denotes the essential norm (i.e., $\|T_f\|_e = \operatorname{dist}(T_f, \mathcal{K})$, where $\mathcal{K}$ is the set of compact operators on L_a^2) of the Toeplitz operator T_f on the Bergman space L_a^2 defined by $T_f g(z) = P(fg)(z) = \int_{\mathbb{D}} \frac{g(w)f(w)}{(1-z\overline{w})^2} dm_2(w)$.

This result implies that in general there is no universal constant $C > 0$ such that $\|A\| \le C \operatorname{ber}(A)$. More recently, Nazarov has shown that the inequality

$$\|T_f\| \le C \sup_{z\in\mathbb{D}}\left\|T_f\widehat{k}_{L_a^2,z}\right\| \ \forall f \in L^\infty(\mathbb{D}, dm_2)$$

can not hold for any constant $C > 0$ (see Miao and Zheng [13, Section 6]).

It is well known that unitary operators can be characterized as invertible contractions with contractive inverses, i.e., as operators A with $\|A\| \le 1$ and $\|A^{-1}\| \le 1$. For further results along this line, see for instance, Maeda [12], Singh and Mangla [20], Badea and Crouzeix [4].

Recently Sano and Uchiyama [19] proved that if A is an invertible operator on the abstract Hilbert space H such that $w(A) \le 1$ and $w(A^{-1}) \le 1$, then A is unitary (see also Stampfli [21, Corollary 1]). In [2, Theorem 1.1] Ando and Li generalized the latter by using the so-called ρ-radius of operator $A \in \mathcal{B}(H)$ defined by

$$w_\rho(A) := \inf\{\mu > 0 : \mu^{-1}A \in C_\rho\},$$

where C_ρ denotes the class of operators $T \in \mathcal{B}(H)$ which admits a unitary ρ-dilation; that is, there is a unitary operator U on a superspace $\mathcal{K} \supset H$ such that

$$T^n = \rho P_H U^n | H \text{ for } n = 1, 2, \ldots,$$

where $P_H : \mathcal{K} \to H$ is the orthoprojection. When $\rho = 1$ and $\rho = 2$, this definition reduces to the operator norm and the numerical radius, respectively.

Thus,the following question naturally arises.

Question. *Is an invertible operator A unitary if* $\operatorname{ber}(A) \leq 1$ *and* $\operatorname{ber}\left(A^{-1}\right) \leq 1$*?*

In this article, which is motivated mainly by this question, we obtain in terms of the Berezin numbers of the operators AA^* and $(AA^*)^{-1}$ necessary and sufficient conditions for unitarity of the invertible operator A on the reproducing kernel Hilbert space (Theorem 1 in Section 2). We also give in terms of the Berezin number a new estimate for the norm of the Hankel operator H_φ, which improves the classical inequality $\|H_\varphi\| \leq \|\varphi\|_\infty$ (Section 3). We also use the notion of Berezin number in generalizing the Douglas lemma for so-called zero Toeplitz products.

Before stating our results, let us give some more notations. The Hardy space $H^2 = H^2(\mathbb{D})$ is defined as the space of all analytic functions f in the unit disc $\mathbb{D} = \{z \in \mathbb{C} : |z| < 1\}$ for which the norm

$$\|f\|_2 = \sup_{0<r<1} \left(\frac{1}{2\pi} \int_0^{2\pi} \left| f\left(re^{it}\right) \right|^2 \right)^{1/2}$$

is finite. The space H^∞ consists of all bounded analytic functions f in the unit disc with the norm

$$\|f\|_\infty = \sup_{z \in \mathbb{D}} |f(z)|.$$

For functions in H^2 the radial limit

$$(bf)\left(e^{it}\right) := \lim_{r \to 1} f\left(re^{it}\right)$$

exists almost everywhere in t (Fatou's theorem; see, for instance, Hoffman [9]), and indeed $bf \in L^2(\mathbb{T})$, where $\mathbb{T}$ denotes the unit circle which is equipped with normalized Lebesgue measure. Moreover $\|f\|_{H^2} = \|bf\|_{L^2}$. We normally identify f with bf, and can thus regard H^2 as a closed subspace of $L^2(\mathbb{T})$.

The reproducing kernel of H^2 is the function

$$k_\lambda(z) = \frac{1}{1 - \overline{\lambda} z}.$$

For a function $f(z) = \sum_{n=0}^{\infty} a_n z^n \in H^2$ we have $\|f\|_2 = \left(\sum_{n=0}^{\infty} |a_n|^2 \right)^{\frac{1}{2}}$. We use P_+ to denote the orthogonal projection from $L^2(\mathbb{T})$ onto H^2 (called the Riesz projection), so that

$$P_+ : \sum_{n=-\infty}^{\infty} a_n e^{int} \to \sum_{n=0}^{\infty} a_n e^{int}.$$

Let dm_2 denote Lebesgue area measure on the unit disc $\mathbb{D}$, normalized so that the measure of $\mathbb{D}$ equals 1.

For a function $\varphi \in L^\infty(\mathbb{T})$ the corresponding Toeplitz operator T_φ on H^2 is defined by

$$T_\varphi f = P_+ \varphi f, \; f \in H^2.$$

The Hankel operator $H_\varphi : H^2 \to H^2_-$ is defined by

$$H_\varphi f = P_- \varphi f, \; f \in H^2,$$

where $P_- := I - P_+$ and $H^2_- = L^2(\mathbb{T}) \ominus H^2$.

If $\varphi \in H^\infty$ and $\varphi(z) \equiv z$, then T_z is the usual shift operator S, $(Sf)(z) = zf(z)$. For any function φ in $L^\infty(\mathbb{T})$ the harmonic extension into the unit disk $\mathbb{D}$ is denoted by the symbol $\widetilde{\varphi}$. It is well known that $\widetilde{\varphi} = \widetilde{T_\varphi}$ for any Toeplitz operator T_φ, $\varphi \in L^\infty(\mathbb{T})$, on the Hardy space H^2 (see, for example, Zhu [23], Engliš [7] and Karaev [10]).

2. The Berezin number and unitarity

In this section, we characterize unitary operators in terms of the Berezin number. Note that A is said to be a unitary operator, if $A^*A = AA^* = I$; that is, $A^* = A^{-1}$.

Definition 1 ([11]). Let $\mathcal{H} = \mathcal{H}(\Omega)$ be a reproducing kernel Hilbert space of complex-valued functions defined on some set Ω. We say that $\mathcal{H}$ possesses (Ber) *property*, if for any two operators $A_1, A_2 \in \mathcal{B}(\mathcal{H})$ $\widetilde{A_1}(\lambda) = \widetilde{A_2}(\lambda)$ for all $\lambda \in \Omega$ implies $A_1 = A_2$.

It is well known, for example, that any reproducing kernel Hilbert space of analytic functions in the unit disc $\mathbb{D}$ (including the Hardy and Bergman spaces) has the (Ber) property (see Zhu [23], Proposition 6.2).

The main result of this section is the following theorem, which essentially improves a result of the paper [11, Theorem 1].

Theorem 1. *Let $\mathcal{H} = \mathcal{H}(\Omega)$ be a reproducing kernel Hilbert space with the* (Ber) *property and $A \in \mathcal{B}(\mathcal{H})$ be an invertible operator. Then A is unitary if and only if* $\mathrm{ber}(AA^*) \leq 1$ *and* $\mathrm{ber}((AA^*)^{-1}) \leq 1$.

Proof. From the definition it is clear that

1) $\mathrm{ber}(AA^*) \leq 1$ if and only if $\left\| A^* \widehat{k}_\lambda \right\| \leq 1$ $(\forall \lambda \in \Omega)$;
2) $\mathrm{ber}\left(A^{-1*}A^{-1}\right) \leq 1$ if and only if $\left\| A^{-1} \widehat{k}_\lambda \right\| \leq 1$ $(\forall \lambda \in \Omega)$.

Then, by considering assertions 1) and 2), we have for all $\lambda \in \Omega$ that

$$\begin{aligned}\left\|\left(A^* - A^{-1}\right)\widehat{k}_\lambda\right\|^2 &= \left\langle \left(A^* - A^{-1}\right)\widehat{k}_\lambda, \left(A^* - A^{-1}\right)\widehat{k}_\lambda \right\rangle \\ &= \left\|A^*\widehat{k}_\lambda\right\|^2 + \left\|A^{-1}\widehat{k}_\lambda\right\|^2 - \left\langle A^*\widehat{k}_\lambda, A^{-1}\widehat{k}_\lambda\right\rangle - \left\langle A^{-1}\widehat{k}_\lambda, A^*\widehat{k}_\lambda\right\rangle \\ &= \left\|A^*\widehat{k}_\lambda\right\|^2 + \left\|A^{-1}\widehat{k}_\lambda\right\|^2 - \left\langle \widehat{k}_\lambda, AA^{-1}\widehat{k}_\lambda\right\rangle - \left\langle AA^{-1}\widehat{k}_\lambda, \widehat{k}_\lambda\right\rangle \\ &= \left\|A^*\widehat{k}_\lambda\right\|^2 + \left\|A^{-1}\widehat{k}_\lambda\right\|^2 - 2 \leq 0,\end{aligned}$$

and thus $\left(A^* - A^{-1}\right)k_\lambda = 0$ for all $\lambda \in \Omega$. Since $\{k_\lambda : \lambda \in \Omega\}$ is a total set, we deduce that A is unitary.

Conversely, it is obvious that $\operatorname{ber}(AA^*) = \operatorname{ber}\left(\left(AA^*\right)^{-1}\right) = 1$ if A is unitary. □

3. Equations with skew-symmetric operators

Recall that an operator $A \in \mathcal{B}(H)$ is a skew-symmetric operator if $A^* = -A$. For example, for any self-adjoint operator A on H, iA is a skew-symmetric operator; also the Volterra integral operator V_0, $V_0 f(x) := \int_{-x}^{x} f(t)\,dt$, is a skew-symmetric operator on the Lebesgue space $L^2(-1,1)$.

Following Zhu [24], note that many classical results in matrix theory deal with complex symmetric matrices (that is, $T = T^t$) and skew-symmetric matrices (that is, $T = -T^t$), which appear naturally in a variety of applications such as complex analysis, functional analysis, and even quantum mechanics.

Recently, there has been growing interest in skew-symmetric operators, which are closely related to the study of complex symmetric operators (that is, $CTC = T^*$ for some conjugation C on H).

There are several motivations for the study of skew-symmetric operators. For one thing, skew-symmetric operators have been extensively studied for many years in the finite-dimensional setting, and have many applications in pure mathematics, applied mathematics and even in engineering disciplines. In particular, real skew-symmetric matrices are very important in applications such as function theory, the solution of linear quadratic optimal control problems, robust control problems, crack following in anisotropic materials and others.

The second motivation for the study of skew-symmetric operators lies in the connections to complex symmetric operators. For example, it is known that if T is complex symmetric, then $T^*T - TT^*$ is skew-symmetric. In view of the description of skew-symmetric normal operators, this provides another approach to describing complex symmetric operators. On the other hand, each operator T on H can be written as the sum of a complex symmetric operator and a skew-symmetric operator. In fact, arbitrarily choose a conjugation C on H and set $A = \frac{1}{2}(T + CT^*C)$, $B = \frac{1}{2}(T - CT^*C)$. Then $CAC = A^*$, $CBC = -B^*$ and

$T = A+B$. This reflects a certain universality of complex symmetric operators and skew-symmetric operators. More informations about skew-symmetric operators can be found, for instance, in Zhu [24] and references therein.

The proof of Theorem 1 allows us to investigate solvability of the operator equations $T_1X = I + Y_1$ and $XT_2 = I + Y_2$, where Y_1, Y_2 are skew-symmetric operators, in terms of reproducing kernels. (For more information about these equations, see, for instance [5] and its references.)

Theorem 2. *Let $T : \mathcal{H}(\Omega) \to \mathcal{H}(\Omega)$ be a nonzero bounded linear operator on the RKHS $\mathcal{H} = \mathcal{H}(\Omega)$, and let $Y_1, Y_2 \in \mathcal{B}(H)$ be two nonzero skew-symmetric operators.*

(a) *If $X \in \mathcal{B}(\mathcal{H})$ satisfies the equation $TX = I + Y_1$, then there exists $\lambda_0 \in \Omega$ such that*
$$\left\|T^*\widehat{k}_{\mathcal{H},\lambda_0}\right\|^2 + \left\|X\widehat{k}_{\mathcal{H},\lambda_0}\right\|^2 > 2;$$

(b) *If $X \subset \mathcal{B}(\mathcal{H})$ satisfies the equation $XT = I + Y_2$, then there exists $\lambda_0 \in \Omega$ such that*
$$\left\|T\widehat{k}_{\mathcal{H},\lambda_0}\right\|^2 + \left\|X^*\widehat{k}_{\mathcal{H},\lambda_0}\right\|^2 > 2.$$

Proof. (a) Suppose to the contrary that
$$\left\|T^*\widehat{k}_{\mathcal{H},\lambda}\right\|^2 + \left\|X\widehat{k}_{\mathcal{H},\lambda}\right\|^2 \le 2$$
for all $\lambda \in \Omega$. Then by using this inequality and the identity
$$\left\langle \widehat{k}_{\mathcal{H},\lambda}, Y_1\widehat{k}_{\mathcal{H},\lambda}\right\rangle + \left\langle \widehat{k}_{\mathcal{H},\lambda}, Y_1^*\widehat{k}_{\mathcal{H},\lambda}\right\rangle = 0$$
we have for every $\lambda \in \Omega$ that (see the proof of Theorem 1)
$$\left\|(T^* - X)\,\widehat{k}_{\mathcal{H},\lambda}\right\|^2 = \left\|T^*\widehat{k}_{\mathcal{H},\lambda}\right\|^2 + \left\|X\widehat{k}_{\mathcal{H},\lambda}\right\|^2 - 2 \le 0,$$
which implies $(T^* - X)\,k_{\mathcal{H},\lambda} = 0$ for all $\lambda \in \Omega$, and consequently $X = T^*$. Then $TT^* = I + Y_1$, which implies that $I + Y_1$ is self adjoint. But, since $Y_1^* = -Y_1$ and $Y_1 \neq 0$, this gives a contradiction, which proves (a).

The proof of (b) is analogous. □

Corollary 1. *Let $T_1, T_2 \in \mathcal{B}(\mathcal{H})$ be two nonzero operators on the RKHS $\mathcal{H} = \mathcal{H}(\Omega)$, and let $Y_1, Y_2 \in \mathcal{B}(\mathcal{H})$ be two nonzero skew-symmetric operators. If the equations $T_1X = I + Y_1$ and $XT_2 = I + Y_2$ have a common solution X, then there exists $\lambda_1, \lambda_2 \in \Omega$ such that*
$$\min\left\{\left\|T_1^*\widehat{k}_{\mathcal{H},\lambda_1}\right\|^2 + \left\|X\widehat{k}_{\mathcal{H},\lambda_1}\right\|^2, \left\|T_2\widehat{k}_{\mathcal{H},\lambda_2}\right\|^2 + \left\|X^*\widehat{k}_{\mathcal{H},\lambda_2}\right\|^2\right\} > 2.$$

Corollary 2. *Let $\mathbb{X}_1$ denote the unit ball of $\mathcal{B}(\mathcal{H})$ in the operator norm and let Y be a nonzero skew-symmetric operator on $\mathcal{B}(\mathcal{H})$.*

(a) *If the equation $TX = I + Y$ is solvable on $\mathbb{X}_1$, then $\left\|T^*\widehat{k}_{\mathcal{H},\lambda_0}\right\| > 1$ for some $\lambda_0 \in \Omega$.*

(b) *If* $\sup_{\lambda\in\Omega}\left\|T^*\widehat{k}_{\mathcal{H},\lambda}\right\| \leq 1$ (*or, equivalently,* $\operatorname{ber}(TT^*) \leq 1$), *then the equation* $TX = I + Y$ *is not solvable on* $\mathbb{X}_1$.

(c) *If* $\sup_{\lambda\in\Omega}\left\|T\widehat{k}_{\mathcal{H},\lambda}\right\| \leq 1$ (*or, equivalently,* $\operatorname{ber}(T^*T) \leq 1$), *then the equation* $XT = I + Y$ *is not solvable on* $\mathbb{X}_1$.

(d) *If* $\|T_1\| \leq 1$ *and* $\|T_2\| \leq 1$, *then the equations* $XT_1 = I + Y_1$ *and* $T_2X = I + Y_2$ (*where* Y_1, Y_2 *are nonzero skew-symmetric operators*) *are not solvable on the set* $\mathbb{X}_1$.

4. The Berezin number and the norm of a Hankel operator

For any $\varphi \in L^\infty = L^\infty(\mathbb{T})$, let H_φ be a Hankel operator on the space H^2. It is well known that $H_{\varphi+\psi} = H_\varphi$ for any $\psi \in H^\infty$, and $\|H_\varphi\| = \operatorname{dist}(\varphi, H^\infty)$ (Nehari's theorem) and $\|H_\varphi\| \leq \|\varphi\|_\infty$ (see, for example, Peller [16]). In this section, we give in terms of the Berezin number a better estimate for the norm $\|H_\varphi\|$.

For $\varphi \in L^\infty$, $\psi \in H^\infty$, $\|\psi\|_\infty \leq 1$, and $A \in \mathcal{B}(H^2)$, let us denote

$$N_{\varphi,\psi,A} := T_\varphi\left(I - T_\psi A T_\psi^*\right),$$

where T_φ, T_ψ and $T_\psi^* = T_{\overline{\psi}}$ are Toeplitz operators on H^2.

For any operator $A \in \mathcal{B}(H^2)$, we will denote $\widetilde{A}^{\mathrm{rad}}(e^{it}) := \lim_{r\to 1^-} \widetilde{A}(re^{it})$ if these radial limits exist almost everywhere on the unit circle $\mathbb{T}$, and $\widetilde{A}^{\mathrm{rad}} \in L^\infty(\mathbb{T})$.

Let us set

$$\left(\mathcal{B}(H^2)\right)_\varphi := \left\{A \in \mathcal{B}(H^2) : \widetilde{A}^{\mathrm{rad}} \in L^\infty(\mathbb{T}) \text{ and } \operatorname{ber}(A) \leq \frac{\|H_\varphi\|}{\|\varphi\|_\infty}\right\}.$$

This set obviously contains every compact operator $A \in \mathcal{B}(H^2)$ with $\|A\| < \frac{\|H_\varphi\|}{\|\varphi\|_\infty}$, and therefore $\left(\mathcal{B}(H^2)\right)_\varphi$ is a nonempty set.

Let $(H^\infty)_1 := \{h \in H^\infty : \|h\|_\infty \leq 1\}$ denote the unit ball of H^∞.

The main result of this section is the following.

Theorem 3. *Let* $\varphi \in L^\infty$, $\psi \in (H^\infty)_1$ *and let* $A \in \mathcal{B}(H^2)$ *be any operator such that* $\widetilde{A}^{\mathrm{rad}} \in L^\infty(\mathbb{T})$. *Then* $\widetilde{N}^{\mathrm{rad}}_{\varphi,\psi,A}$ *exists and is finite, and*

$$\sup_{A\in(\mathcal{B}(H^2))_\varphi}\left(\operatorname{ber}(A)\|\varphi\|_\infty\right) \leq \|H_\varphi\|$$

$$\leq \inf_{A\in(\mathcal{B}(H^2))_\varphi}\left[\operatorname{ber}(A)\|\varphi\|_\infty + \inf_{\psi\in(H^\infty)_1}\left\|H_{\widetilde{N}^{\mathrm{rad}}_{\varphi,\psi,A}}\right\|\right] \leq \|\varphi\|_\infty.$$

Proof. It is clear from the definition of the class $\left(\mathcal{B}(H^2)\right)_\varphi$ that

$$\sup_{A\in(\mathcal{B}(H^2))_\varphi}\left(\operatorname{ber}(A)\|\varphi\|_\infty\right) \leq \|H_\varphi\|.$$

On the other hand, a simple calculation shows that

$$\widetilde{N}_{\varphi,\psi,A}(\lambda) = \widetilde{\varphi}(\lambda) - \widetilde{\varphi}(\lambda)|\psi(\lambda)|^2 \widetilde{A}(\lambda) + \left\langle \left(I - T_\psi A T_{\overline{\psi}}\right)\widehat{k}_\lambda, \left(T_{\overline{\varphi}} - \widetilde{T_{\overline{\varphi}}}(\lambda) I\right)\widehat{k}_\lambda \right\rangle \tag{4.1}$$

for all $\lambda \in \mathbb{D}$, where $\widetilde{\varphi} := \widetilde{T_\varphi}$ is the harmonic extension of the function $\varphi \in L^\infty(\mathbb{T})$ into the unit disk $\mathbb{D}$. Since $\widetilde{\varphi}$ is a harmonic function in $\mathbb{D}$, $\lim_{r\to 1^-} \widetilde{\varphi}(r\xi) = \varphi(\xi)$ for a.a. $\xi \in \mathbb{T}$. Then, by considering that $\sup_{\lambda\in\mathbb{D}} \left|\widetilde{A}(\lambda)\right| \le \|A\|$, $\widetilde{A}^{\mathrm{rad}} \in L^\infty(\mathbb{T})$ and $\|\psi\|_{H^\infty} \le 1$, and also the fact that (see [7])

$$\lim_{r\to 1^-} \left\| T_{\overline{\varphi}} \widehat{k}_{re^{it}} - \widetilde{T_{\overline{\varphi}}}\left(re^{it}\right) \widehat{k}_{re^{it}} \right\| = 0$$

for almost all $t \in [0, 2\pi)$, by applying the Cauchy–Schwarz inequality we assert from (1) that $\widetilde{N}^{\mathrm{rad}}_{\varphi,\psi,A}\left(e^{it}\right) := \lim_{r\to 1^-} \widetilde{N}_{\varphi,\psi,A}\left(re^{it}\right)$ exists and is finite for almost all $t \in [0, 2\pi)$. Also, it follows from (1) that

$$\left|\widetilde{\varphi}\left(re^{it}\right) - \widetilde{N}_{\varphi,\psi,A}\left(re^{it}\right)\right| \le \|\varphi\|_\infty \operatorname{ber}(A) + (1 + \|A\|) \left\| \left(T_{\overline{\varphi}} - \overline{\widetilde{\varphi}}\left(re^{it}\right) I\right) \widehat{k}_{re^{it}} \right\|.$$

From this inequality, by passing the radial limit, we have

$$\left|\varphi\left(e^{it}\right) - \widetilde{N}^{\mathrm{rad}}_{\varphi,\psi,A}\left(e^{it}\right)\right| = \lim_{r\to 1^-} \left|\widetilde{\varphi}\left(re^{it}\right) - \widetilde{N}_{\varphi,\psi,A}\left(re^{it}\right)\right| \le \operatorname{ber}(A) \|\varphi\|_\infty,$$

for almost all $t \in [0, 2\pi)$, for all $\psi \in (H^\infty)_1$ and $A \in \left(\mathcal{B}\left(H^2\right)\right)_\varphi$. This implies that

$$\left\|\varphi - \widetilde{N}^{\mathrm{rad}}_{\varphi,\psi,A}\right\|_\infty = \operatorname*{ess\,sup}_{t\in[0,2\pi)} \left|\varphi\left(e^{it}\right) - \widetilde{N}^{\mathrm{rad}}_{\varphi,\psi,A}\left(e^{it}\right)\right| \le \operatorname{ber}(A) \|\varphi\|_\infty$$

for all $\psi \in (H^\infty)_1$ and $A \in \left(\mathcal{B}\left(H^2\right)\right)_\varphi$. Then, we have for any $h \in H^\infty$ that

$$\|\varphi - h\|_\infty - \left\|\widetilde{N}^{\mathrm{rad}}_{\varphi,\psi,A} - h\right\|_\infty \le \operatorname{ber}(A) \|\varphi\|_\infty.$$

That is

$$\inf_{g\in H^\infty} \|\varphi - g\|_\infty \le \|\varphi - h\|_\infty \le \operatorname{ber}(A) \|\varphi\|_\infty + \left\|\widetilde{N}^{\mathrm{rad}}_{\varphi,\psi,A} - h\right\|_\infty$$

for every $h \in H^\infty$, $\psi \in (H^\infty)_1$ and $A \in \left(\mathcal{B}\left(H^2\right)\right)_\varphi$. Therefore we obtain that

$$\operatorname{dist}(\varphi, H^\infty) \le \|\varphi - h\|_\infty \le \operatorname{ber}(A) \|\varphi\|_\infty + \left\|\widetilde{N}^{\mathrm{rad}}_{\varphi,\psi,A} - h\right\|_\infty$$

for all $h \in H^\infty$, $\psi \in (H^\infty)_1$ and $A \in \left(\mathcal{B}\left(H^2\right)\right)_\varphi$. This implies by the Nehari formula $\|H_\varphi\| = \operatorname{dist}(\varphi, H^\infty)$ that

$$\|H_\varphi\| \le \operatorname{ber}(A) \|\varphi\|_\infty + \left\|\widetilde{N}^{\mathrm{rad}}_{\varphi,\psi,A} - h\right\|_\infty \tag{2}$$

for all $h \in H^\infty$, $\psi \in (H^\infty)_1$ and $A \in \left(\mathcal{B}\left(H^2\right)\right)_\varphi$. Clearly, since $A \in \left(\mathcal{B}\left(H^2\right)\right)_\varphi$, $\|H_\varphi\| - \operatorname{ber}(A)\|\varphi\|_\infty \geq 0$. Then, it follows from (2) that

$$\|H_\varphi\| - \operatorname{ber}(A)\|\varphi\|_\infty \leq \inf_{h\in H^\infty}\left\|\widetilde{N}^{\text{rad}}_{\varphi,\psi,A} - h\right\|_\infty = \operatorname{dist}\left(\widetilde{N}^{\text{rad}}_{\varphi,\psi,A}, H^\infty\right)$$
$$= \left\|H_{\widetilde{N}^{\text{rad}}_{\varphi,\psi,A}}\right\|,$$

which yields

$$\|H_\varphi\| \leq \operatorname{ber}(A)\|\varphi\|_\infty + \left\|H_{\widetilde{N}^{\text{rad}}_{\varphi,\psi,A}}\right\|$$

for all $\psi \in (H^\infty)_1$ and $A \in \left(\mathcal{B}\left(H^2\right)\right)_\varphi$. This means that

$$\sup_{A\in(\mathcal{B}(H^2))_\varphi} \operatorname{ber}(A)\|\varphi\|_\infty$$
$$\leq \|H_\varphi\| \leq \inf_{A\in(\mathcal{B}(H^2))_\varphi}\left[\operatorname{ber}(A)\|\varphi\|_\infty + \inf_{\psi\in(H^\infty)_1}\left\|H_{\widetilde{N}^{\text{rad}}_{\varphi,\psi,A}}\right\|\right].$$

Now it remains only to prove that

$$\inf_{A\in(\mathcal{B}(H^2))_\varphi}\left[\operatorname{ber}(A)\|\varphi\|_\infty + \inf_{\psi\in(H^\infty)_1}\left\|H_{\widetilde{N}^{\text{rad}}_{\varphi,\psi,A}}\right\|\right] \leq \|\varphi\|_\infty. \tag{3}$$

Indeed, we have

$$\inf_{A\in(\mathcal{B}(H^2))_\varphi}\left[\operatorname{ber}(A)\|\varphi\|_\infty + \inf_{\psi\in(H^\infty)_1}\left\|H_{\widetilde{N}^{\text{rad}}_{\varphi,\psi,A}}\right\|\right] \leq \operatorname{ber}(B)\|\varphi\|_\infty + \left\|H_{\widetilde{N}^{\text{rad}}_{\varphi,\mathbf{1},B}}\right\|$$

for $\psi = \mathbf{1}$, and $B := \delta I$, where $0 < \delta < 1$, and $\delta < \frac{\|H_\varphi\|}{\|\varphi\|_\infty}$. Then $\left\|H_{\widetilde{N}^{\text{rad}}_{\varphi,\mathbf{1},B}}\right\| \leq \left\|\widetilde{N}^{\text{rad}}_{\varphi,\mathbf{1},B}\right\|_\infty$, $\lim_{r\to 1^-}\widetilde{N}^{\text{rad}}_{\varphi,\mathbf{1},B}\left(re^{it}\right) = \lim_{r\to 1^-}\left(\widetilde{\varphi}\left(re^{it}\right)\left(1-\widetilde{B}\left(re^{it}\right)\right)\right)$ $= \lim_{r\to 1^-}\widetilde{\varphi}\left(re^{it}\right)(1-\delta) = \varphi\left(e^{it}\right)(1-\delta)$ for a.a. $t\in[0,2\pi)$ (see (1)) and $\operatorname{ber}(B) = \delta$. From the latter we obtain that

$$\inf_{A\in(\mathcal{B}(H^2))_\varphi}\left[\operatorname{ber}(A)\|\varphi\|_\infty + \inf_{\psi\in(H^\infty)_1}\left\|H_{\widetilde{N}^{\text{rad}}_{\varphi,\psi,A}}\right\|\right]$$
$$\leq \delta\|\varphi\|_\infty + \left\|\widetilde{N}^{\text{rad}}_{\varphi,\mathbf{1},B}\right\|_\infty$$
$$= \delta\|\varphi\|_\infty + \operatorname*{ess\,sup}_{t\in[0,2\pi)}\left|\lim_{r\to 1^-}\widetilde{N}^{\text{rad}}_{\varphi,\mathbf{1},B}\left(re^{it}\right)\right|$$
$$= \delta\|\varphi\|_\infty + \operatorname*{ess\,sup}_{t\in[0,2\pi)}\left|\lim_{r\to 1^-}\left(\widetilde{\varphi}\left(re^{it}\right)\left(1-\widetilde{B}\left(re^{it}\right)\right)\right)\right|$$
$$= \delta\|\varphi\|_\infty + \operatorname*{ess\,sup}_{t\in[0,2\pi)}\left|\lim_{r\to 1^-}\widetilde{\varphi}\left(re^{it}\right)(1-\delta)\right|$$
$$= \delta\|\varphi\|_\infty + (1-\delta)\|\varphi\|_\infty = \|\varphi\|_\infty,$$

which proves (3). □

Recall that for any compact operator $K \in \mathcal{B}(H^2)$, its Berezin symbol $\widetilde{K}$ vanishes on the boundary; i.e., $\widetilde{K}^{\mathrm{rad}}(\xi) = 0$ for almost all $\xi \in \mathbb{T}$.

Example 1. If $\varphi \in L^\infty(\mathbb{T})$ is a function such that the set $(\mathcal{B}(H^2))_\varphi$ contains an operator A such that $I - A$ is compact, then by putting $\psi = 1$ we obtain from Theorem 3 that $\|H_\varphi\| = \mathrm{ber}(A)\|\varphi\|_\infty$ which is a better estimate than $\|H_\varphi\| \le \|\varphi\|_\infty$.

5. The Berezin number and a Douglas type lemma

Note that the proof of Theorem 3 essentially used the known fact that (see Engliš [7]) for any $\varphi \in L^\infty$

$$\lim_{r \to 1^-} \left\| T_\varphi \widehat{k}_{re^{it}} - \widetilde{\varphi}(re^{it}) \widehat{k}_{re^{it}} \right\| = 0 \tag{4}$$

for almost all $t \in [0, 2\pi)$. Our next results show that assertion (4) can also be used for the proof of the well-known Douglas lemma in [6] concerning zero Toeplitz products. In fact, here we will prove a more general result which essentially generalizes and improves Douglas' lemma (see Proposition 2).

To start with, let us give some notations. For any operator $A \in \mathcal{B}(\mathcal{H})$, where $\mathcal{H} = \mathcal{H}(\Omega)$ is a standard RKHS on some set Ω (i.e., $\widehat{k}_{\mathcal{H},\lambda}$ weakly converges to zero when $\lambda \to \xi \in \partial\Omega$), let us define the following set (the "maximal Berezin set"):

$$\widetilde{W_0}(A) := \left\{ \lambda : \lambda = \lim_{\lambda_n \to \partial\Omega} \widetilde{A}(\lambda_n) \text{ and } \lim_{\lambda_n \to \partial\Omega} \left\| A\widehat{k}_{\mathcal{H},\lambda_n} \right\| = \|A\| \right\},$$

where $\widetilde{A}$ is the Berezin symbol of the operator A. Clearly,

$$\widetilde{W_0}(A) \subset W_0(A) := \{\lambda : \langle Tx_n, x_n \rangle \to \lambda, \text{ for } \|x_n\| = 1 \text{ and } \|Tx_n\| \to \|T\|\},$$

where $W_0(A)$ is the maximal numerical range of A (see Stampfli [22]). It is also obvious that $\widetilde{W_0}(A) = W_0(A)$ for any scalar operator $A = \lambda I$, and $\widetilde{W_0}(K) = \emptyset$ for every nonzero compact operator K on the standard RKHS $\mathcal{H}(\Omega)$. (However, the situation is not so trivial for other operators and RKHSs.)

Here we shall firstly be interested in the boundary behavior when $\lambda \to \xi \in \partial\Omega$ of

$$\left\| A\widehat{k}_{\mathcal{H},\lambda} - \widetilde{A}(\lambda) \widehat{k}_{\mathcal{H},\lambda} \right\|_{\mathcal{H}}.$$

Note that this problem is also related with the so-called stationary distance value of the generalized eigenvalue problem $Tf = \lambda Af$ for the Hilbert space operators T and A; see Paul [15]. More general questions are discussed in Engliš [7].

Indeed, by considering that $A\widehat{k}_{\mathcal{H},\lambda} - \widetilde{A}(\lambda)\widehat{k}_{\mathcal{H},\lambda} \perp \widetilde{A}(\lambda)\widehat{k}_{\mathcal{H},\lambda}$, we have

$$\left\| A\widehat{k}_{\mathcal{H},\lambda} \right\|_{\mathcal{H}}^2 = \left\| A\widehat{k}_{\mathcal{H},\lambda} - \widetilde{A}(\lambda)\widehat{k}_{\mathcal{H},\lambda} \right\|_{\mathcal{H}}^2 + \left| \widetilde{A}(\lambda) \right|^2$$

or

$$\left\| A\widehat{k}_{\mathcal{H},\lambda} - \widetilde{A}(\lambda)\widehat{k}_{\mathcal{H},\lambda} \right\|_{\mathcal{H}}^2 = \left\| A\widehat{k}_{\mathcal{H},\lambda} \right\|_{\mathcal{H}}^2 - \left| \widetilde{A}(\lambda) \right|^2, \tag{5}$$

for all $\lambda \in \Omega$. This implies that $\lim\limits_{\lambda \to \xi \in \partial\Omega} \left\| A\widehat{k}_{\mathcal{H},\lambda} - \widetilde{A}(\lambda)\widehat{k}_{\mathcal{H},\lambda} \right\|_{\mathcal{H}} = 0$ if and only if

$$\lim_{\lambda \to \xi \in \partial\Omega} \left\| A\widehat{k}_{\mathcal{H},\lambda} \right\|_{\mathcal{H}} = \lim_{\lambda \to \xi \in \partial\Omega} \left| \widetilde{A}(\lambda) \right|. \tag{6}$$

Thus, (5) and (6) allow us to prove the following.

Proposition 1. *Let $A : \mathcal{H}(\Omega) \to \mathcal{H}(\Omega)$ be a bounded linear operator on the RKHS $\mathcal{H} = \mathcal{H}(\Omega)$ such that* $\operatorname{ber}(A) \in \widetilde{W_0}(A)$. *Then there exists a sequence $\{\lambda_n\} \subset \Omega$ such that*

$$\lim_{\lambda_n \to \xi \in \partial\Omega} \left\| A\widehat{k}_{\mathcal{H},\lambda_n} - \widetilde{A}(\lambda)\widehat{k}_{\mathcal{H},\lambda_n} \right\|_{\mathcal{H}} = 0$$

if and only if $\|A\| = \operatorname{ber}(A)$.

Proof. Since $\operatorname{ber}(A) \in \widetilde{W_0}(A)$, there exists a sequence $\{\lambda_n\} \subset \mathbb{D}$ such that $\operatorname{ber}(A) = \lim\limits_{\lambda_n \to \xi \in \partial\Omega} \widetilde{A}(\lambda_n)$ and $\lim\limits_{\lambda_n \to \xi \in \partial\Omega} \left\| A\widehat{k}_{\mathcal{H},\lambda_n} \right\|_{\mathcal{H}} = \|A\|$. Then the assertion of the theorem is immediate from (5) and (6). □

Let $\widetilde{\mathcal{B}}^0(\mathcal{H}(\mathbb{D}))$ denote the set of all operators A such that:

1. $\widetilde{A}^{\mathrm{rad}}(e^{it})$ exists for almost all $t \in [0, 2\pi)$;
2. $\lim\limits_{r \to 1^-} \left\| \left(A^* - \widetilde{A^*}(re^{it}) I \right) \widehat{k}_{\mathcal{H},re^{it}} \right\| = 0$ for almost all $t \in [0, 2\pi)$.

Our next result generalizes and improves the well-known Douglas lemma for Toeplitz operators on H^2 (see Douglas [6]).

Proposition 2. *Let $A_1, A_2, \ldots, A_n \in \widetilde{\mathcal{B}}^0(\mathcal{H}(\mathbb{D}))$ be operators on the standard RKHS $\mathcal{H}(\mathbb{D})$ such that $A_1A_2 \cdots A_n$ is compact. Then*

$$\widetilde{A}_1^{\mathrm{rad}}(e^{it})\, \widetilde{A}_2^{\mathrm{rad}}(e^{it}) \cdots \widetilde{A}_n^{\mathrm{rad}}(e^{it}) = 0$$

for almost all $t \in [0, 2\pi)$.

Proof. Let $A_1A_2 \cdots A_n = \mathcal{K}$, where $\mathcal{K}$ is a compact operator on the standard RKHS $\mathcal{H}(\mathbb{D})$. Then, denoting

$$B_{i,\lambda} := \left\langle A_{i+1} \cdots A_n, A_i^*\widehat{k}_{\mathcal{H},\lambda} - \widetilde{A_i^*}(\lambda)\widehat{k}_{\mathcal{H},\lambda} \right\rangle \quad (i = 1, 2, \ldots, n-1)$$

and considering that $\widetilde{A_i^*} = \overline{\widetilde{A_i}}$, we have:

$$\begin{aligned}
\widetilde{\mathcal{K}}(\lambda) &= \left\langle A_1A_2 \cdots A_n\widehat{k}_{\mathcal{H},\lambda}, \widehat{k}_{\mathcal{H},\lambda} \right\rangle = \left\langle A_2 \cdots A_n\widehat{k}_{\mathcal{H},\lambda}, A_1^*\widehat{k}_{\mathcal{H},\lambda} \right\rangle \\
&= \left\langle A_2 \cdots A_n\widehat{k}_{\mathcal{H},\lambda}, \left(A_1^*\widehat{k}_{\mathcal{H},\lambda} - \widetilde{A_1^*}\widehat{k}_{\mathcal{H},\lambda} \right) + \widetilde{A_1^*}(\lambda)\widehat{k}_{\mathcal{H},\lambda} \right\rangle \\
&= B_{1,\lambda} + \widetilde{A_1}(\lambda) \left\langle A_2A_3 \cdots A_n\widehat{k}_{\mathcal{H},\lambda}, \widehat{k}_{\mathcal{H},\lambda} \right\rangle \\
&= B_{1,\lambda} + \widetilde{A_1}(\lambda) \left\langle A_3A_4 \cdots A_n\widehat{k}_{\mathcal{H},\lambda}, A_2^*\widehat{k}_{\mathcal{H},\lambda} \right\rangle
\end{aligned}$$

$$
\begin{aligned}
&= B_{1,\lambda} + \widetilde{A_1}(\lambda)\Big[\Big\langle A_3 \cdots A_n \widehat{k}_{\mathcal{H},\lambda}, A_2^* \widehat{k}_{\mathcal{H},\lambda} - \widetilde{A_2^*} \widehat{k}_{\mathcal{H},\lambda}\Big\rangle \\
&\qquad + \widetilde{A_2}(\lambda)\Big\langle A_3 \cdots A_n \widehat{k}_{\mathcal{H},\lambda}, \widehat{k}_{\mathcal{H},\lambda}\Big\rangle\Big] \\
&= B_{1,\lambda} + \widetilde{A_1}(\lambda) B_{2,\lambda} + \widetilde{A_1}(\lambda)\widetilde{A_2}(\lambda)\Big\langle A_3 \cdots A_n \widehat{k}_{\mathcal{H},\lambda}, \widehat{k}_{\mathcal{H},\lambda}\Big\rangle = \cdots \\
&= B_{1,\lambda} + \widetilde{A_1}(\lambda) B_{2,\lambda} + \widetilde{A_1}(\lambda)\widetilde{A_2}(\lambda) B_{3,\lambda} \\
&\qquad + \cdots + \widetilde{A_1}(\lambda)\widetilde{A_2}(\lambda)\cdots\widetilde{A_n}(\lambda).
\end{aligned}
$$

For $A_i \in \widetilde{\mathcal{B}}^0(\mathcal{H}(\mathbb{D}))$, $i = 1, 2, \ldots, n$,

$$\lim_{r \to 1^-} \widetilde{A}_i^{\mathrm{rad}}\left(re^{it}\right)$$

exists and for almost all $t \in [0, 2\pi)$, $\lim_{r \to 1^-} \left\| A_i^* \widehat{k}_{\mathcal{H},re^{it}} - \widetilde{A_i^*}\left(re^{it}\right) \widehat{k}_{\mathcal{H},re^{it}} \right\| = 0$. Also, $\widetilde{\mathcal{K}}(\lambda) \to 0$ as $\lambda \to \xi \in \mathbb{T}$. Hence by the Cauchy–Schwarz inequality we conclude that

$$\widetilde{A}_1^{\mathrm{rad}}\left(e^{it}\right)\widetilde{A}_2^{\mathrm{rad}}\left(e^{it}\right)\cdots\widetilde{A}_n^{\mathrm{rad}}\left(e^{it}\right) = 0$$

for almost all $t \in [0, 2\pi)$, as desired. □

The following are immediate from Propositions 1 and 2 (obviously, $\mathrm{ber}(A^*) = \mathrm{ber}(A)$).

Corollary 3. *Let $A_1, A_2, \ldots, A_n \in \mathcal{B}(\mathcal{H}(\mathbb{D}))$ be operators on the RKHS $\mathcal{H}(\mathbb{D})$ (not necessarily standard) such that $\mathrm{ber}(A_i) \in \widetilde{W_0}(A_i^*)$ $(i = 1, 2, \ldots, n)$ and $\widetilde{A}_i^{\mathrm{rad}}\left(e^{it}\right)$ exists for all $i \in \{1, 2, \ldots, n\}$ and for almost all $t \in [0, 2\pi)$. If $\mathrm{ber}(A_i) = \|A_i\|$ $(i = 1, 2, \ldots, n)$ and $A_1 A_2 \cdots A_n = 0$, then almost everywhere on the unit circle $\mathbb{T}$, $\widetilde{A}_1^{\mathrm{rad}}\left(e^{it}\right)\widetilde{A}_2^{\mathrm{rad}}\left(e^{it}\right)\cdots\widetilde{A}_n^{\mathrm{rad}}\left(e^{it}\right) = 0$.*

Corollary 4 (Douglas lemma [6]). *Let $\varphi_1, \varphi_2, \ldots, \varphi_n \in L^\infty(\mathbb{T})$, and T_{φ_i}, $i = 1, \ldots, n$, be Toeplitz operators on a Hardy space H^2. If $T_{\varphi_1} T_{\varphi_2} \cdots T_{\varphi_n} = 0$, then $\varphi_1 \varphi_2 \cdots \varphi_n = 0$.*

Note that the zero Toeplitz product problem in a Hardy space H^2 has been solved by Aleman and Vukotic [1]. However, it is still an open problem (for arbitrary n) whether n Toeplitz operators on the Bergman space $L_a^2(\mathbb{D})$, none of which is 0, can have a product that equals 0.

Acknowledgement

The author would like to extend his sincere appreciation to the Deanship of Scientific Research at King Saud University for funding of this research through the Research Group Project no RGP-VPP-323.

The author also wishes to thank the referee for many helpful remarks and suggestions, which has greatly improved the presentation of the manuscript.

References

[1] Aleman, A. and Vukotić, D., Zero products of Toeplitz operators, Duke Math. J., 148 (3) (2009), 373–403.

[2] Ando, T. and Li, C.-K., Operator Radii and Unitary Operators, Oper. Matrices, 4(2) (2010), 273–281.

[3] Aronzajn, N., Theory of reproducing kernels, Trans. Amer. Math. Soc., 68 (1950), 337–404.

[4] Badea, C. and Crouzeix, M., Numerical radius and distance from unitary operators, Operators and Matrices, 7 (2) (2013), 285–292.

[5] Dajič, A. and Koliha, J.J., Equations $ax = c$ and $xb = d$ in rings and rings with involution with applications to Hilbert space operators, Linear Algebra Appl., 429(7)(2008), 1779–1809.

[6] Douglas, R.G., Banach Algebra Techniques in the Theory of Toeplitz Operators, CBMS Reg. Couf. Ser. Math., vol. 15, Amer. Math. Soc., Providence, RI, 1973, p. 53.

[7] Engliš, M., Toeplitz operators and the Berezin transform on H^2, Linear algebra and its applications, 223/224(1995), 171–204.

[8] Engliš, M., Compact Toeplitz operators via the Berezin transform on bounded symmetric domains, Integr. Equat. Oper. Theory, 33 (1999), 426–455.

[9] Hoffman, K., Banach spaces of analytic functions, Prentice-Hall, Englewood Cliffs, 1965.

[10] Karaev, M.T., Berezin symbols and Schatten–von Neumann classes, Math. Notes, 72 (2) (2002), 185–192; translated from Matematicheskie Zametki, 72 (2) (2002), 207-215.

[11] Karaev, M.T. and Iskenderov, N.Sh., Berezin number of operators and related questions, Methods of Funct. Anal. Topology, (19) (1) (2013), 68–72.

[12] Maeda, F.-Y., Generalized unitary operators, Bull. Amer. Math. Soc., 71 (1965), 631–633.

[13] Miao, J. and Zheng, D., Compact operators on Bergman spaces, Integr. Equat. Oper. Theory, 48 (2004), 61–79.

[14] Nordgren, E. and Rosenthal, P., Boundary values of Berezin symbols, Oper. Theory: Advances and Applications, 73 (1994), 362–368.

[15] Paul, K., Translatable radii of an operator in the direction of another operator II, Mathematica Slovaca, 60(2010), 121–128.

[16] Peller, V.V., Hankel operators and their applications, Springer, 2003.

[17] Saitoh, S., Theory of reproducing kernels and its applications, Pitman Research Notes in Mathematics Series, v. 189, 1988.

[18] Saitoh, S., Integral transforms, reproducing kernels and their applications, Pitman Research Notes in Mathematics Series, v. 369, 1997.

[19] Sano, T. and Uchiyama, A., Numerical radius and unitarity, Acta Sci. Math. (Szeged), 76 (2010), 581–584.

[20] Singh, U.N. and Mangla, K., Operators with inverses similar to their adjoints, Proc. Amer. Math. Soc., 38 (2) (1973), 258–260.

[21] Stampfli, J.G., Minimal range theorems for operators with thin spectra, Pacif. J. Math., 23 (1967), 601–612.

[22] Stampfli, J.G., The norm of a derivation, Pacif. J. Math., 33(1970), 737–747.

[23] Zhu, K., Operator Theory in Function Spaces, Second edition. Mathematical Surveys and Monographs, 138. Amer. Math. Soc., Providence, RI, 2007.

[24] Zhu, S., Approximate unitary equivalence to skew-symmetric operators, Comp. Anal. Oper. Theory, 8 (2014), 1565–1580.

Mubariz T. Garayev
Department of Mathematics
College of Science
King Saud University
P.O. Box 2455
Riyadh 11451, Saudi Arabia
e-mail: mgarayev@ksu.edu.sa

Operator Theory:
Advances and Applications, Vol. 247, 101–121
© Springer International Publishing Switzerland 2015

Čebyšev Subspaces of C^*-Algebras – a Survey

M.N.N. Namboodiri, S. Pramod and A.K. Vijayarajan

Abstract. This article aims at a survey of the developments in the theory of Čebyšev subspaces of C^*-algebras. The classical theory was mainly due to the St. Petersburg school of mathematical analysis of P.L. Čebyšev and his collaborators, a detailed survey of which can be seen in Ivan Singer [87] and Karl-Georg Steffens [91]. Compared to the classical theory and its abstract formulation by Singer, the non-commutative theory initiated by A.G. Robertson is still in its infancy. In this survey, a detailed account of the non-commutative theory is furnished.

Mathematics Subject Classification (2010). Primary 46L05; Secondary 41A50, 46L52.

Keywords. Approximation, C^*-algebras, Čebyšev subspaces, Haar condition.

1. Introduction

This article is about developments in the theory of Čebyšev subspaces of C^*-algebras. Compared to the vast literature of the classical theory, its non-commutative counterpart is yet to grow as a reasonably complete one, though important discoveries were made by A.G. Robertson et al. [81], who initiated the study. During the last thirty to forty years nothing much has happened after the pioneering papers of Robertson, Pedersen and so on. Here we intend to give all major theorems and examples due to various authors.

The classical results are proved mainly using the lattice theoretic properties of scalar functions and the topology involved. But most of the pioneering results were proved using constructive hard analysis techniques. Excellent surveys are due to Ivan Singer [87], Karl-Georg Steffens [91], H. Berens and G.G. Lorentz [16] to cite a few important ones. During the 1930s P.P. Korovkin from the Russian school of analysis initiated the study of unifying many approximation processes such as Bernstein polynomial approximation, Fejër trigonometric polynomial approximation etc., using positivity of linear operators on spaces of continuous functions. His celebrated theorems, known as Korovkin type theorems, have attained a wide range of attention that also led to Korovkin type theorems and Korovkin sets in general Banach algebras and C^*-algebras in particular. One of the three major results of Korovkin connects Čebyšev systems and Korovkin's test sets in certain special cases. Extensive surveys of Korovkin type theorems have also appeared. The third section of this article gives a brief account of this. The first and second sections deal respectively with the commutative and non-commutative Čebyšev systems.

The list of references given, without any claim of exhaustiveness, contains most of the published work that we came across and which are relevant to this article.

2. Čebyšev systems and subspaces of function spaces

As mentioned in the well-known books by Samuel Karlin and William J. Studdens [46], George Steffens [91], the history of evolution of the theory of Čebyšev systems (sometimes known as T-systems) began with the work of P.L. Čebyšev and his collaborators from the St. Petersburg school who studied power functions and polynomials of minimal norm. Their study was motivated by problems in approximation theory, numerical analysis, oscillations of eigenfunctions of Sturm–Liouville differential equations and so on. But the modern theory of best approximation uses functional analytic techniques by considering the functions to be approximated and the approximating functions as elements in a normed linear space or more generally in a metric space. This approach unifies many approximation processes. Moreover, this modern approach essentially due to Singer laid a rigorous foundation for the classical theory and provided powerful tools for obtaining new results. The complete picture of these developments can be found in the monographs of Singer ([87], [88]). The classical concept of Čebyšev sets in normed linear spaces is closely related to the more general theory of best approximation. We recall below the notion of best approximation and a few basic results relevant to our discussion.

Definition 2.1. Let $(\mathcal{X}, \rho)$ be a metric space, $\mathcal{G}$ be a subset of it and x be a point in it. An element g_0 in $\mathcal{G}$ is called *a point of best approximation of x* if

$$\rho(x, g_0) = \inf\{\rho(x, g) : g \in \mathcal{G}\}.$$

For $\mathcal{X}$ and its subspace $\mathcal{G}$ as above, let

$$\mathcal{P}_{\mathcal{G}}(x) = \{g_0 \in \mathcal{G} : \rho(x, g_0) = \inf \rho(x, g); \quad g \in \mathcal{G}\}.$$

In what follows, most of the important theorems are stated as given in [88]. This will help in identifying future possible formulations in the non-commutative setting. The first main theorem that characterizes best approximation in linear subspaces of normed linear spaces is as follows [88].

Theorem 2.2 ([88], Theorem 1.1). *Let $\mathcal{X}$ be a normed linear space and $\mathcal{G}$ be a subspace of it, $x \in \mathcal{X}\backslash\overline{\mathcal{G}}$ and $g_0 \in \mathcal{G}$. Then $g_0 \in \mathcal{P}_{\mathcal{G}}(x)$ if and only if there exists $f \in \mathcal{X}^*$ such that*

(i) $\|f\| = 1$;
(ii) $f(g) = 0 \quad (g \in \mathcal{G})$ *and*
(iii) $f(x - g_0) = \|x - g_0\|$.

Remark 2.3. The functional f mentioned in the above theorem is 'maximal' in nature which can be determined for many function spaces. This is the importance of the above theorem. The following theorem illustrates this.

For a compact Hausdorff space Ω, $\mathcal{C}(\Omega)$ (respectively $\mathcal{C}_{\mathcal{R}}(\Omega)$) will denote the Banach Space of all complex continuous functions (respectively real continuous functions) on Ω, with supremum norm.

Theorem 2.4 ([88], Theorem 1.2). *Let $\mathcal{G}$ be a linear subspace of $C_R(\Omega)$, $x \in \mathcal{C}_{\mathcal{R}}(\Omega)\backslash\overline{\mathcal{G}}$. Then $g_0 \in \mathcal{P}_{\mathcal{G}}(x)$, if and only if there exist two disjoint closed sets $E_{g_{0+}}$ and $E_{g_{0-}}$ of Ω and a Radon measure μ on Ω such that*
(i) $\mid \mu \mid (\Omega) = 1$;
(ii) $\int_\Omega g(t)d\mu(t) = 0$, *for all g in $\mathcal{G}$;*
(iii) $\mu \geq 0$ *on* $E_{g_{0+}}$ *and* $\mu \leq 0$ *on* $E_{g_{0-}}$ *and support* $\mu \subseteq E_{g_{0+}} \cup E_{g_{0-}}$ *and*
(iv) $x(q) - g_0(q) = \begin{cases} \|x - g_0\| & \text{for } q \text{ in } E_{g_{0+}} \\ -\|x - g_0\| & \text{for } q \text{ in } E_{g_{0-}}. \end{cases}$

There are a few more interesting results in this setting but we restrict to the following one.

Theorem 2.5 ([88], Theorem 1.4). (a) *For a positive measure space (Ω, ν), $\mathcal{X} = \mathcal{L}^P(\Omega, \nu)$, $1 < p < \infty$, $\mathcal{G}$ be a linear subspace of $\mathcal{X}$, $x \in \mathcal{X}\backslash\overline{\mathcal{G}}$ and $g_0 \in \mathcal{G}$. We have $g_0 \in \mathcal{P}_{\mathcal{G}}(x)$ if and only if*

$$\int_\Omega g(t)|x(t) - g_0(t)|^{p-1} sign\ [x(t) - g_0(t)] d\nu(t) = 0, \quad (g \in \mathcal{G}).$$

(b) *Let $\mathcal{H}$ be an inner product space, $\mathcal{G}$ be a linear subspace of $\mathcal{H}$. Let $x \in \mathcal{H}\backslash\overline{\mathcal{G}}$ and $g_0 \in \mathcal{G}$. We have $g_0 \in \mathcal{P}_{\mathcal{G}}(x)$ if and only if*

$$\langle g, x - g_0 \rangle = 0, \qquad (g \in \mathcal{G})$$

and $\langle .,. \rangle$ denotes the inner product in $\mathcal{H}$.

Now we define the notion of Čebyšev subspace of a normed linear space.

Definition 2.6. A subspace $\mathcal{G}$ of a normed space $\mathcal{X}$ is called *a semi Čebyšev subspace* (respectively *proximinal subspace*) if each vector in $\mathcal{X}$ has at most one (respectively at least one) closest point in $\mathcal{G}$. A subspace $\mathcal{G}$ of $\mathcal{X}$ is called a *Čebyšev subspace* if it is both semi Čebyšev and proximinal: i.e., each vector in $\mathcal{X}$ admits a unique closest point in $\mathcal{G}$. If $\mathcal{G}$ is a subset of a metric space $\mathcal{X}$, there are corresponding notions of semi-Čebyšev sets, proximinal sets and Čebyšev sets.

Remark 2.7. Čebyšev sets were also called 'Haar sets' by some authors, e.g., by N. Efimov and S.B. Stečhkin [30].

Theorem 2.8 ([88], Theorem 3.1). *A linear subspace $\mathcal{G}$ of a normed linear space $\mathcal{X}$ is a semi-Čebyšev subspace if and only if there do not exist f in $\mathcal{X}^*$, x in $\mathcal{X}$ and g_0 in $\mathcal{G} \setminus \{0\}$ such that*

$$\|f\| = 1; \quad f(g) = 0, \quad (g \in \mathcal{G}); \quad f(x) = \|x\| = \|x - g_0\|.$$

We state a couple of other general theorems before considering concrete cases. If $\mathcal{G}$ is a set in a metric space $\mathcal{X}$, denote by $\pi_{\mathcal{G}}$ the multi-valued mapping $D(\pi_{\mathcal{G}}) \mapsto \mathcal{G}$ defined by $\pi_{\mathcal{G}}(x) \in \mathcal{P}_{\mathcal{G}}(x) \quad (x \in D(\pi_{\mathcal{G}}))$, where $D(\pi_{\mathcal{G}})$ denotes the domain of $\pi_{\mathcal{G}}$. In the particular case when $D(\pi_{\mathcal{G}}) = \mathcal{X}$ and $\pi_{\mathcal{G}}$ is one-valued (i.e., $\mathcal{G}$ is a Čebyšev set), $\pi_{\mathcal{G}}$ is called the metric projection of $\mathcal{X}$ onto $\mathcal{G}$. We use the following notation: For $\mathcal{X}$ and its subspace $\mathcal{G}$ as above,

$$\pi_{\mathcal{G}}^{-1}(0) = \{x \in \mathcal{X}; \quad 0 \in \mathcal{P}_{\mathcal{G}}(x)\}.$$

Theorem 2.9 ([88], Proposition 3.1). *For a closed linear subspace $\mathcal{G}$ of a normed linear space $\mathcal{X}$, the following statements are equivalent.*

(i) *$\mathcal{G}$ is a Čebyšev subspace;*
(ii) *$\mathcal{X} = \mathcal{G} \oplus \pi_{\mathcal{G}}^{-1}(0)$, where $\oplus$ means that the sum decomposition of each element $x \in \mathcal{X}$ is unique;*
(iii) *$\mathcal{G}$ is proximinal and $\pi_{\mathcal{G}}^{-1}(0) - \pi_{\mathcal{G}}^{-1}(0) \cap \mathcal{G} = \{0\}$;*
(iv) *$\mathcal{G}$ is proximinal and the restriction $\omega_{\mathcal{G}|\pi_{\mathcal{G}}{}^{-1}(0)}$ of the canonical mapping $\omega_{\mathcal{G}}$: $\mathcal{X} \longrightarrow \mathcal{X}/\mathcal{G}$ to the set $\pi_{\mathcal{G}}^{-1}(0)$ is one-to-one.*

The next theorem characterizes finite-dimensional Čebyšev subspaces of normed linear spaces.

Theorem 2.10 ([87], Theorem 2.1, pp. 210–211). *Let $\mathcal{X}$ be a normed linear space. An n-dimensional linear subspace $\mathcal{G} = \overline{\operatorname{span}}\{x_1, x_2, \dots, x_n\}; x_i \in \mathcal{X}$, $(i = 1, 2, \dots, n)$ of $\mathcal{X}$ is a Čebyšev subspace if and only if there do not exist h extremal points $f_1, f_2, \dots, f_h$ of $S_{\mathcal{X}^*}$ (unit sphere of $\mathcal{X}^*$), where $1 \le h \le n$ if the scalars are real and $1 \le h \le 2n-1$ if the scalars are complex, h numbers $\lambda_1, \lambda_2, \dots, \lambda_h \ge 0$ with $\sum_{j=1}^{h} \lambda_j = 1$ and $x \in \mathcal{X}, g_0 \in \mathcal{G} \setminus \{0\}$ such that we have $\sum_{j=1}^{h} \lambda_j f_j(x_k) = 0 (k = 1, 2, \dots, n)$ and $f_j(x) = \|x\| = \|x - g_0\|, (j = 1, 2, \dots, h)$.*

When $\mathcal{X} = C(\Omega)$ (Ω compact), we get the celebrated theorem due to Haar which characterizes the n-dimensional Čebyšev subspaces of $C(\Omega)$.

Theorem 2.11 ([39]). *Let $\mathcal{G}$ be an n-dimensional linear subspace of $C(\Omega)$ spanned by the elements $x_1, x_2, \dots, x_n$. Then $\mathcal{G}$ is a Čebyšev subspace of $C(\Omega)$ if and only if $x_1, x_2, \dots, x_n$ form a Čebyšev system of order n (i.e., every $\sum_{i=1}^{n} \alpha_i x_i = 0$ has at most $n-1$ zeros on Ω).*

We will refer to the above equivalence condition for a Čebyšev subspace as the classical Haar condition. Thus Haar condition connects the geometrical and algebraic properties of functions. The characterization below of finite-dimensional Čebyšev subspace of $C_{\mathbb{R}}[a,b]$ by Y. Ikebe [41] is also noteworthy.

Theorem 2.12 ([41]). *A finite-dimensional subspace $\mathcal{G}$ of $C_{\mathbb{R}}[a,b]$ is a Čebyšev subspace if and only if*

$$\|g_0\| < 2\|x\|, (x \in C_{\mathbb{R}}[a,b]\backslash\{0\}, g_0 \in P_{\mathcal{G}}(x)).$$

Remark 2.13. It may be interesting to know whether function spaces on arbitrary compact topological spaces admit Čebyšev systems or not. The following theorem gives the answer. In spite of the nice Haar condition for a given finite-dimensional subspace to be Čebyšev, it is not quite possible to find Čebyšev subspaces on arbitrary compact spaces. In fact the Mairhuber–Curtis theorem states that a compact space admits a Čebyšev system of order $n+1$ if and only if it is homeomorphic to a subset of the unit circle $\mathbb{T} = \{(x,y) \in \mathbb{R}^2/x^2+y^2=1\}$ in $\mathbb{R}^2$ and the compact space can be homeomorphic to the unit circle if and only if n is even (Mairhuber [61], Curtis [24]).

In the case of infinite-dimensional normed linear spaces, there always exists a linear functional which is not continuous, the kernel of which is not closed and hence dense (see [23] and [58]). If $\mathcal{G}$ is dense in a normed linear space $\mathcal{X}$, we have $\mathcal{P}_{\mathcal{G}}(x) = \emptyset$ for every $x \in \mathcal{X} \setminus \mathcal{G}$. This means that every infinite-dimensional normed space contains a semi-Čebyšev subspace. For Banach spaces, the problem of existence of closed semi-Čebyšev subspaces is not trivial but has got an affirmative answer, namely, every Banach space contains at least one semi-Čebyšev closed hyperplane [88]. In the case of Čebyšev subspaces of Banach spaces, the situation is different. In fact, Garkavi [35] gives an example of a Banach space for which there are no Čebyšev subspaces. Here we quote equivalence conditions for the existence of Čebyšev subspaces of Banach spaces.

Theorem 2.14 ([88], Theorem 3.11). *For a Banach space $\mathcal{X}$, the following statements are equivalent.*

(i) *All closed linear subspaces of $\mathcal{X}$ are Čebyšev subspaces;*
(ii) *All closed subspaces of $\mathcal{X}$ of a certain fixed finite co-dimension m where $1 \leq m \leq \dim E - 1$ are Čebyšev subspaces;*
(iii) *E is reflexive and strictly convex.*

2.1. Čebyšev subspaces of tensor products of Banach spaces

In the field of approximation of multi-variate functions by combinations of univariate ones, the setting is often a Banach space which is the tensor product of two or more simpler spaces. The usual questions in approximation theory that can be posed here are also relevant in the non-commutative case involving C^*-algebras which we will discuss later in the article. We briefly describe the notion of tensor product of Banach spaces and list a couple of results relevant to us.

In [57], William A. Light and Elliot W. Cheny consider proximinal subspaces of tensor products of Banach spaces, equipped with cross norms. Here we bypass the general theory of tensor products and related aspects which are given in [57].

Instead only those results that are connected to Čebyšev systems are quoted with a brief sketch of proof. If X and Y are two Banach spaces, there are many methods to define a norm on $X \otimes Y$ using the norms on X and Y. Here we will use the notation $X \otimes_\alpha Y$ to denote the fact that the norm we use on $X \otimes Y$ is α which is to assign to $\sum_{i=1}^{n} x_i \otimes y_i$, the norm it receives when regarded as an operator from X^* to Y, *viz.*,

$$\alpha\left(\sum_{i=1}^{n} x_i \otimes y_i\right) = \sup\{\|\sum_{i=1}^{n} \phi(x_i)y_i\|, \phi \in X^*, \|\phi\| = 1\}.$$

An element of the form $x \otimes y \in X \otimes Y$ is called a dyad. It is easy to see that, $\alpha(x \otimes y) = \|x\|\|y\|$, for all dyads $x \otimes y \in X \otimes Y$. Such a norm on $X \otimes Y$ for which the norm of a dyad equals the product of the norms of its components is called a cross norm. Given two Banach spaces X and Y, there is a rich supply of cross norms on $X \otimes Y$. Let X and Y be Banach spaces with tensor product $X \otimes_\lambda Y$ equipped with the cross norm λ which is the injective tensor norm, namely for $\sum_{i=1}^{n} x_i \otimes y_i \in X \otimes_\lambda Y$, $\lambda(x_i \otimes y_i) = \sup\limits_{\substack{x^* \in X^*, y^* \in Y^* \\ \|x^*\| \leq 1, \|y^*\| \leq 1}} |\sum_{i=1}^{n} x^*(x_i)y^*(y_i)|$. $X \otimes_\lambda Y$ is the completion of the algebraic tensor product with respect to the injective tensor norm. If S is a compact Hausdorff space and Y is a Banach space let $C(S, Y)$ denote the Banach space of all continuous maps from S into Y.

Theorem 2.15 ([57], Theorem 2.1). *Let S be a compact Hausdorff space and let H be a subspace of the Banach space Y. If there exists a continuous proximity map of Y onto H, then $C(S)\overline{\otimes}H$ (closure of $C(S) \otimes H$) is proximinal in $C(S) \otimes_\lambda Y$ and in fact it has a continuous proximity map.*

As a consequence, we get the following corollaries.

Corollary 2.16 ([57], Corollary 2.2). Let S be a compact Hausdorff space and let H be a subspace of a Banach space Y. If either

(i) H is finite dimensional and Čebyšev; or
(ii) Y is uniformly convex,
then $C(S, H)$ is proximinal in $C(S, Y)$.

Corollary 2.17 ([57], Corollary 2.3). Let S and T be two compact Hausdorff spaces. If H is a finite-dimensional Čebyšev subspace in $C(T)$, then $C(S) \otimes_\lambda H$ is proximinal in $C(S \times T)$.

3. Čebyšev subspaces of C^*-algebras

The study of Čebyšev subspaces in the general operator algebra setting was initiated by A.G. Robertson [81] followed by Robertson and Yost [82] and then Pedersen [77]. In [81], Robertson gives a characterization of one-dimensional Čebyšev subspaces of von Neumann algebras. The result is as follows:

Theorem 3.1 ([81], Theorem 1). *Let M be a von Neumann algebra. Let x be an operator in M. Then the one-dimensional subspace $\mathbb{C}x$ spanned by x is a Čebyšev subspace of M if and only if there exists a projection p in the centre of M such that px is left invertible in pM and $(1-p)x$ is right invertible in $(1-p)M$.*

The proof uses the existence of central projections in von Neumann algebras together with Hann–Banach and Krein–Milman theorems. Another important result of Robertson is regarding the non existence of higher-dimensional Čebyšev subspaces of infinite-dimensional von Neumann algebras which are also $*$-subalgebras.

Theorem 3.2 ([81], Theorem 6). *Let M be an infinite-dimensional von Neumann algebra. Let N be a finite-dimensional $*$-subalgebra of M with dimension greater than one. Then N is not a Čebyšev subspace of M.*

For the proof, Robertson uses the rich structural properties of von Neumann algebras. Attempts to prove the analogue of Haar's theorem [39] led to quite a few interesting results in the non-commutative C^*-algebra setting. A result in that direction by Robertson and Yost is the following.

Theorem 3.3 ([82], Theorem 2.3). *Let A be a norm-closed two-sided ideal in a von Neumann algebra, $x \in A$. Then $\mathbb{C}x$ is a Čebyšev subspace in A, if and only if there is no irreducible representation π of A for which 0 is an eigenvalue of both $\pi(x)$ and $\pi(x^*)$. When this happens, $x^*x + xx^*$ is strictly positive.*

A sketch of the proof is as follows. For the 'if' part, they use the existence of an extreme point of the unit ball of A^* satisfying certain properties, if $\mathbb{C}x$ is not a Čebyšev subspace of A. Assuming A to be acting on the Hilbert space H in its universal representation, we can write the above functional using a unit vector $\xi \in H$ and the corresponding vector state will be pure. Consequently, the representation π of A defined as the restriction of A to $A\xi$ will be irreducible and 0 will be an eigenvalue for both $\pi(x)$ and $\pi(x^*)$.
The 'only if' part is proved using central projections and Kadison's irreducibility theorem.
Robertson and Yost [82] also proved a remarkable result which says that every C^*-algebra in a certain large class contains an infinite-dimensional Hilbert subspace with the property that each of its closed subspaces is a Čebyšev subspace.

Theorem 3.4 ([82], Theorem 2.8). *Let M be a properly infinite von Neumann algebra, A a two-sided ideal in M. Suppose that A contains a strictly positive element (i.e., A has a one-dimensional Čebyšev subspace). Then A contains an infinite-dimensional Hilbert space V, which is Čebyšev in A. Moreover, each closed subspace of V is Čebyšev in A. So, A contains Čebyšev subspaces of all finite dimensions.*

The existence of a sequence of orthogonal projections each equivalent to identity adding up to identity together with the strictly positive element enables one to define an orthonormal basis, the span of which is the Hilbert space. Proximinality of reflexive subspaces together with best approximation property assured by compactness with respect to ultra weak topology implies that all closed subspaces of the Hilbert space so obtained are Čebyšev in A. It is to be noted that the above class of C^*-algebras includes $B(H)$ which means that it has got Čebyšev subspaces of all finite dimensions.

Remark 3.5. The work of Robertson and Yost established that there exists no Čebyšev subspace of finite dimension greater than one if the space under consideration is anyone of the following.

(i) An infinite-dimensional abelian von Neumann algebra.
(ii) An abelian non-separable C^*-algebra.

Theorem 3.4 tells us how different the situation is, in the non-commutative setting.

The following theorem [82] and the corollary establishes the dearth of Čebyšev subspaces of C^*-algebras which are $*$-subalgebras.

Theorem 3.6 ([82], Theorem 1.3). *Let A be a C^*-algebra, B, a C^*-subalgebra of A. Suppose that one of A, B is unital, and that B is a Čebyšev subspace of A. Then $1 \in A$ and $1 \in B$. If $B \neq \mathbb{C}1$, then every maximal abelian $*$-subalgebra of B is maximal abelian in A.*

Corollary 3.7 ([82], Corollary 1.4).

(1) Let A be an infinite-dimensional C^*-algebra, B a finite-dimensional $*$-subalgebra. If B is Čebyšev in A, then A is unital and $B = \mathbb{C}1$.
(2) Let A be a commutative C^*-algebra, B a finite-dimensional subalgebra of A. If B is Čebyšev in A, then A is unital and $B = \mathbb{C}1$.

Theorem 3.8 ([82], Theorem 1.5). *Let M be a von Neumann algebra, A a proper C^*-subalgebra of M with $A \neq \mathbb{C}1$. Suppose that M is not a factor of type II or III. If A is Čebyšev in M, then M is $M_2(\mathbb{C})$, with A being the algebra of diagonal matrices.*

Remark 3.9. The above result establishes the fact that the only exception of a C^*-algebra A having non-trivial Čebyšev subalgebra $B(B \neq A, B \neq \mathbb{C}1)$ is $A = M_2(\mathbb{C})$ for which the algebra of diagonal matrices is a Čebyšev subalgebra.

Now we turn to the results of G.K. Pedersen [77] who studied the finite-dimensional Čebyšev subspaces of C^*-algebras quite extensively. Pedersen, in his attempt to extend Haar's theorem to the non-commutative case, succeeds partially by giving a characterization of one-dimensional and two-dimensional Čebyšev subspaces of C^*-algebras. Another result of his further extends the work initiated by Robertson and Yost to the case of C^*-algebras.

Theorem 3.10 ([77], Theorem 1). *Let V be an n-dimensional subspace of a C^*-algebra $\mathcal{A}$ and assume that there is a unitary u in $M(\mathcal{A})$ (the C^*-algebra of multipliers of $\mathcal{A}$ in $\mathcal{A}''$) and a non-zero element x_0 in V such that $\phi_i(x_0^* x_0) = \phi_i(u x_0 x_0^* u^*) = 0$ for at least n orthogonal pure states $\phi_1, \phi_2, \dots, \phi_n$ of $\mathcal{A}$. Then V is not a Čebyšev subspace of $\mathcal{A}$.*

Theorem 3.11 ([77], Theorem 2). *Let V be an n-dimensional subspace of a C^*-algebra $\mathcal{A}$. The following conditions are equivalent.*

(i) *V is not a Čebyšev subspace;*

(ii) *There is a unitary operator u in $\tilde{\mathcal{A}}$, a non-zero element x_0 in V and an atomic state ϕ, which is a convex combination of m orthogonal pure states, such that $\phi(x_0^* x_0) = \phi(u x_0 x_0^* u^*) = 0$. If $m < n$, $\phi(uV) = 0$.*

In the following two theorems Pedersen characterizes the one-dimensional and two-dimensional Čebyšev subspaces of C^*-algebras in terms of irreducible representations, their eigenvalues and eigen vectors. These results can also be seen as the generalization of Haar's theorem to the first two dimensions. Pedersen remarks in the context of the theorem above that it seems to be the best one can do in generalizing Haar's theorem (Theorem 2.11). However a recent work [71] generalises Pedersen's result for all finite dimensions.

Let $\mathcal{A}$ be a C^*-algebra with unit 1 and let $x_0 \in \mathcal{A}$ is not a multiple of 1. In this setting Pedersen [77] obtained the following results.

Theorem 3.12 ([77], Theorem 3). *Let x_0 be a non-zero element in a C^*-algebra $\mathcal{A}$. The following conditions are equivalent.*

(i) *$\mathbb{C}x_0$ is a Čebyšev subspace of $\mathcal{A}$;*

(ii) *$x_0^* x_0 + u x_0 x_0^* u^*$ is strictly positive in $\mathcal{A}$ for every unitary u in $M(\mathcal{A})$ (the C^*-algebra of multipliers of $\mathcal{A}$ in $\mathcal{A}''$);*

(iii) *In no irreducible representation $(\pi, \mathcal{H})$ of $\mathcal{A}$ do the operators $\pi(x_0)$ and $\pi(x_0^*)$ both have zero as an eigenvalue.*

Proposition 3.13 ([77], Proposition 1). *Let $x_0 \in \mathcal{A}$ be as above. Then the following conditions are equivalent.*

(i) *The 2-dimensional subspace $G = \mathrm{span}\,(1, x_0)$ is a Čebyšev subspace of $\mathcal{A}$;*

(ii) *For a given $\lambda \in \mathbb{C}$, there exists at most one irreducible representation $(\pi, \mathcal{H})$ of $\mathcal{A}$ (up-to equivalence) in which x_0 and x_0^* have the eigenvalues λ and $\overline{\lambda}$ respectively. Moreover, none of the multiplicities of λ and $\overline{\lambda}$ in $\mathcal{H}$ exceed 1 and the corresponding eigenvectors are not orthogonal.*

Theorem 3.14 ([77], Theorem 4). *If A is a C^*-algebra without unit and B a Čebyšev C^*-subalgebra of A, then $B = A$.*

Theorem 3.15 ([77], Theorem 5). *If A is a C^*-algebra with unit, B, a Čebyšev C^*-subalgebra of A, then either $B = A, B = \mathbb{C}1$, or else $A = M_2$ and B is isomorphic to the algebra of diagonal matrices.*

Legg, Scranton and Waed [55] obtained some important results characterizing the semi-Čebyšev and Čebyšev subspaces of $K(\mathcal{H})$, the space of all compact operators on some Hilbert space $\mathcal{H}$. We quote a few of them:

Theorem 3.16 ([55], Theorem 3). *Let $\mathcal{H}$ be a separable Hilbert space. Then $K(\mathcal{H})$ has an N-dimensional Čebyšev subspace for each positive integer N.*

Theorem 3.17 ([55], Theorem 5). *An N-dimensional subspace $\mathcal{V} \subset K(\mathcal{H})$ is Čebyšev if and only if there does not exist a non-zero $C \in \mathcal{V}$, $C_j \in \mathcal{V}, j = 1, 2, \ldots, N-1$ and two sets A and B each consisting of m orthonormal elements so that*

(1) $\operatorname{span}(C, C_1, \ldots, C_{N-1}) = \mathcal{V}$;
(2) $0 \neq A = \{v_1, v_2, \ldots, v_m\} \in \ker C$. $B = \{u_1, u_2, \ldots, u_m\} \in \ker C^*$;
(3) *the* $(N-1) \times m$ *matrix* $M = (\langle C_i v_j, v_j \rangle) i = 1, 2, \ldots, N-1, j = 1, 2, \ldots, m$ *has linearly dependent columns.*

As a consequence of the above theorem we get the following corollary:

Corollary 3.18 ([55], Corollary 2). If $\mathcal{H}$ is not separable, $K(\mathcal{H})$ has no finite-dimensional Čebyšev subspaces.

If H is separable, $K(\mathcal{H})$ belongs to the class mentioned in Theorem 3.4. In particular, $K(\mathcal{H})$ has an infinite-dimensional Čebyšev subspace. This differs from the commutative theory, for c_0 has no infinite-dimensional Čebyšev subspace [88].

3.1. Čebyšev subalgebras of JB-algebras

Mirmostafaee and Niknam [64] looked at the Čebyšev subalgebras of JB-algebras. A Jordan algebra is an associative algebra $\mathcal{A}$, satisfying the conditions $a.b = b.a$ and $a^2.(a.b) = a.(a^2.b)$ for every $a, b \in \mathcal{A}$. Any associative algebra $\mathcal{A}$, with respect to the Jordan product $(a \circ b = (a.b + b.a)/2, a, b \in \mathcal{A})$ is a Jordan algebra. A JB-algebra is a Jordan algebra $\mathcal{A}$ with a complete norm $\|.\|$ satisfying the properties $\|ab\| \leq \|a\|\|b\|$ and $\|a\|^2 \leq \|a^2 + b^2\|; \forall a, b \in \mathcal{A}$.

Theorem 3.19 ([64], Theorem 3.1). *If $\mathcal{A}$ is a JB-algebra with unit and if $\mathcal{B}$ is a Čebyšev JB-subalgebra of $\mathcal{A}$, then either $\mathcal{B}$ is a trivial subalgebra of $\mathcal{A}$ or $\mathcal{A} = \mathcal{H} \oplus \mathbb{R}.1$ for some Hilbert space $\mathcal{H}$ with* $\dim(\mathcal{H}) \geq 2$.

Sketch of the Proof: If neither $\mathcal{A} \neq \mathcal{B}$ nor $\mathcal{A} = \mathbb{R}.1$, there is an element $e \in B$ such that $\sigma(e)$ contains two points. Then using the Shirshov–Cohn theorem ([64], Theorem 2.1) and spectral theory, the result is proved. It is also proved in [64] that a particular class of JB-algebras do not admit non-trivial Čebyšev subalgebras.

4. Čebyšev subspaces and boundary representations

In 1977, Pedersen in his work [77] tried to extend the classical Haar condition to the non-commutative case and he succeeded in the cases of dimensions one and two. Nothing much has happened in that direction for the last thirty to forty years. Recently a work by Namboodiri, Pramod and Vijayarajan [71] emerged extending the result of Pedersen to all finite dimensions. This work crucially involves the notion of non-commutative Haar condition introduced in [71]. This work also establishes that there is much to be explored in the relationship of Čebyšev subspaces with Arveson's notion of boundary representation.

The *non-commutative Haar condition* is as follows.

Definition 4.1. Let $\mathcal{A}$ be a C^*-algebra with unit $1_{\mathcal{A}}$. For $x_1, x_2, \ldots, x_{n-1} \in \mathcal{A}$, let $\mathcal{V} = \mathbb{C}1_{\mathcal{A}} + \mathbb{C}x_1 + \cdots + \mathbb{C}x_{n-1}$ be n-dimensional. Then $\{1_{\mathcal{A}}, x_1, \ldots, x_{n-1}\}$ is said to satisfy the non-commutative Haar condition if the following conditions are satisfied:
For a given $\lambda \in \mathbb{C}$,

(a) there are at most $n-1$ irreducible representations $(\pi_i, \mathcal{H}_i)$ (up to equivalence) and a non-zero vector $z_0 \in \operatorname{span}(x_1, x_2, \ldots, x_{n-1})$ such that λ and $\overline{\lambda}$ are eigenvalues of $\pi_i(z_0)$ and $\pi_i(z_0^*)$ respectively $(i = 1, 2, \ldots, n-1)$;
(b) Assume that there are $m \leq n-1$ irreducible representations $(\pi_i, \mathcal{H}_i)$ (up to equivalence) and a non-zero vector $z_0 \in \operatorname{span}(x_1, x_2, \ldots, x_{n-1})$ such that λ and $\overline{\lambda}$ are eigenvalues of $\pi_i(z_0)$ and $\pi_i(z_0^*)$ respectively $(i = 1, 2, \ldots, n-1)$. If n_i (respectively $\overline{n_i}$) $i = 1, 2, \ldots, m$ are the multiplicities of λ (respectively $\overline{\lambda}$) in $\mathcal{H}_i$, then $\sum_{i=1}^{m} n_i \leq n-1 \left(\text{respectively } \sum_{i=1}^{m} \overline{n_i} \leq n-1\right)$. Moreover, at least one eigenvector of $\tilde{\pi}_i(z_0^*)$ of the form $\tilde{\pi}_i(u)$ for some unitary $u \in \mathcal{A}$ which is not in $\mathcal{V}$ corresponding to $\overline{\lambda}$ is not orthogonal to $\tilde{\pi}_i(\mathcal{V})\tilde{\mathcal{H}}_i$ where $\left(\tilde{\pi}_i, \tilde{\mathcal{H}}_i\right)$ is the G.N.S representation corresponding to $(\mathcal{A}, \phi_i)$, ϕ_i is the pure state defined by $\phi_i(a) = \langle \pi_i(a)\xi_i, \xi_i \rangle; i = 1, 2, \ldots, m, a \in \mathcal{A}$ and ξ_i is an eigenvector of $\pi_i(z_0)$ corresponding to λ.

The following result shows that the non-commutative Haar condition is equivalent to the Haar condition in the classical case.

Proposition 4.2 ([71], Proposition 2.4). *Let $\mathcal{A} = C(X)$ be a C^*-algebra of all complex-valued continuous functions on a compact Hausdorff space X. Let $\mathcal{B} = \{1_{\mathcal{A}}, f_1, \ldots, f_{n-1}\} \subset \mathcal{A}$ be a linearly independent set and let $\mathcal{V} = \operatorname{span}\mathcal{B}$. Then $\mathcal{B}$ satisfies the non-commutative Haar condition if and only if it satisfies the classical Haar condition.*

Now we state a general version of Proposition 3.13 for finite-dimensional Čebyšev subspaces of C^*-algebras.

Theorem 4.3 ([71], Theorem 2.8). *Let $\mathcal{A}$ be a unital C^*-algebra. Consider a linearly independent set $\mathcal{B} = \{1_{\mathcal{A}}, x_1, x_2, \ldots, x_{n-1}\} \subseteq \mathcal{A}$ and define $\mathcal{V} = \operatorname{span}\mathcal{B}$. Then the following are equivalent:*

(i) *The subspace $\mathcal{V}$ is an n-dimensional Čebyšev subspace of $\mathcal{A}$;*
(ii) *$\mathcal{B}$ satisfies the non-commutative Haar condition.*

The following theorem in [71] establishes that the representation mentioned in Proposition 3.13 is indeed a boundary representation for the subspace, provided it generates the C^*-algebra. In this connection, we need the following definition introduced by Arveson in [7].

Definition 4.4. Let G be a linear subspace of a C^*-algebra $\mathcal{A}$ such that G contains the identity of $\mathcal{A}$ and $\mathcal{A} = C^*(G)$, the C^*-algebra generated by G. A *boundary* *representation* for G is an irreducible representation π of $\mathcal{A}$ on a Hilbert space such that $\pi_{|G}$ has a unique completely positive extension, namely π itself to $\mathcal{A}$.

Definition 4.5. A map $\phi \in UCP(\mathcal{A}, \mathcal{H})$ is called pure if whenever $\phi - \xi$ is completely positive for some $\xi \in CP(\mathcal{A}, \mathcal{H})$, $\exists\ 0 \leq t \leq 1$ such that $\xi = t\phi$.

Theorem 4.6 ([71], Theorem 2.13). *Let $\mathcal{A}$ be a unital C^*-algebra and let $x_0 \in \mathcal{A}$ such that $G = \operatorname{span}(1_{\mathcal{A}}, x_0)$ is a two-dimensional Čebyšev subspace of $\mathcal{A}$, with $\mathcal{A} = C^*(G)$. Given $\lambda \in \mathbb{C}$, let π_0 be an irreducible representation of $\mathcal{A}$ on $\mathcal{H}_{\pi_0}$ such that $\pi_{0|G}$ is pure with $\pi_0(x_0)(u_0) = \lambda u_0$ and $\pi_0(x_0^*)(v_0) = \overline{\lambda} v_0$ for some unit vectors $u_0, v_0 \in \mathcal{H}_{\pi_0}$. Also assume that every pure $\Psi \in \{\Phi \in CP(\mathcal{A}, \mathcal{H}_{\pi_0}) : \Phi_{|G} = \pi_{0|G}\}$ satisfies the condition $\|\Psi(u)(\xi_0)\| = \|\xi_0\|$ for every unitary $u \in \mathcal{A}$ and some $\xi_0 \in \mathcal{H}_{\pi_0}$. Then π_0 is a boundary representation for G.*

Remark 4.7. Theorem 4.3 can be applied wherever the irreducible representations of the C^*-algebra under consideration are completely known. One such special case of interest is $C(X) \otimes M_N$ where X is a compact Hausdorff space and M_N is the set of all $N \times N$ matrices over $\mathbb{C}$, which is nothing but the C^*-algebra of all M_N-valued continuous functions on X. Let $\mathcal{A}$ be the C^*-algebra $C(X) \otimes M_N$ with identity $1_X \otimes I_N$ where 1_X and I_N are the constant function 1 on X and the $N \times N$ identity matrix respectively.

Let $G = \operatorname{span}(f_0 \otimes a_0, f_1 \otimes a_1, \ldots, f_{n-1} \otimes a_{n-1})$ be an n-dimensional subspace of $\mathcal{A}$ where $f_k \in C(X)$ and $a_k \in M_N$ for $k = 1, 2, \ldots, n-1$, and $f_0 \otimes a_0 = 1_X \otimes I_N$. By Theorem 4.3, G is a Čebyšev subspace of $\mathcal{A}$ if and only if the spanning set satisfies the non-commutative Haar condition. In the following results the conditions (a) and (b) in the non-commutative Haar condition are made more explicit in comparison with the classical case by proving equivalent conditions for (a) and (b) for $C(X) \otimes M_N$.

Proposition 4.8 ([71], Proposition 2.15). *Let $\omega_0 \in \operatorname{span}\{f_j \otimes a_j; j = 0, 1, \ldots, n-1\}$. Then there exist at most $n-1$ irreducible representations $\pi_k, k = 1, 2, \ldots, n-1$ such that 0 is an eigenvalue of $\pi_k(\omega_0)$ and $\pi_k(\omega_0^*)$ if and only if given n distinct points $x_1, x_2, \ldots, x_n$ in X and non-zero vectors $\xi_1, \xi_2, \ldots, \xi_n$ in $\mathbb{C}^N$, the $Nn \times n$ matrix $(f_j(x_k) a_j(\xi_k))$ is of rank n.*

Proposition 4.9 ([71], Proposition 2.16). *A vector $\omega_0 = \sum_{j=0}^{n-1} \beta_j (f_j \otimes a_j)$ in G satisfies condition* (b) *of the non-commutative Haar condition if and only if there exist at most m cyclic vectors $\xi_1, \xi_2, \ldots, \xi_m$ $(m \leq n-1)$ in $\mathbb{C}^N$, distinct points $x_1, x_2, \ldots, x_m$ in X for the identity representation on C^N and unitary matrices $u_1, u_2, \ldots, u_m$ in M_N such that*

$$AB = \bar{\lambda} B \tag{4.1}$$

where

$$A = \begin{pmatrix} \bar{\beta}_1 \bar{f}_1(x_1) a_1^* & \bar{\beta}_2 \bar{f}_2(x_1) a_2^* & \dots \bar{\beta}_{n-1} \bar{f}_{n-1}(x_1) a_{n-1}^* \\ \bar{\beta}_1 \bar{f}_1(x_2) a_1^* & \bar{\beta}_2 \bar{f}_2(x_2) a_2^* & \dots \bar{\beta}_{n-1} \bar{f}_{n-1}(x_2) a_{n-1}^* \\ \cdot & & \\ \cdot & & \\ \cdot & & \\ \bar{\beta}_1 \bar{f}_1(x_m) a_1^* & \bar{\beta}_2 \bar{f}_2(x_m) a_2^* & \dots \bar{\beta}_{n-1} \bar{f}_{n-1}(x_m) a_{n-1}^* \end{pmatrix},$$

and

$$B = \text{diagonal}\,(u_1(\xi_1), \dots, u_m(\xi_m)).$$

Further, the diagonal matrix on the right-hand side of (4.1) *with non-zero diagonal entries is non-singular. Also the multiplicities n_i (respectively $\overline{n}_i$) of λ_0 (respectively $\bar{\lambda}_0$) satisfy the inequality* $\sum_{i=1}^{m} n_i \leq n-1 \left(\text{respectively } \sum_{i=1}^{m} \overline{n}_i \leq n-1 \right)$.

Remark 4.10. The above two propositions bring clarity to the obscure nature in the definition of non-commutative Haar condition. Note that condition (b) comes from mainly non-commutativity while condition (a) is shared by commutative case. When $N = 1$, the $nN \times n$ matrix $(f_j(x_k)a_j(\xi_k))$ of rank n becomes a non-singular matrix of order n. Thus in this case, condition (a) becomes the classical Haar condition. Further, when N=1, condition (b) holds trivially because each of the diagonal elements $u_i \xi_i$ on the right-hand side of equation 4.1 becomes a product of two non-zero scalars.

5. Korovkin type theorems

This section is meant to survey the relation between the Korvkin subspaces and Čebyšev subspaces of function spaces initially recognized by Korovkin and later by Saskin, Wulbert and Singer ([85], [101], [87]). These considerations lead to a number of interesting problems in the C^*-algebra frame work which are yet to be investigated. The classical approximation theorem due to P.P. Korovkin [50] in 1953, unified many existing approximation processes such as Bernstein polynomial approximation of continuous functions. Korovkin's discovery inspired many researchers that lead to Korovkin-type theorems and Korovkin sets in various settings such as more general functions, Banach algebras, Banach lattices and operator algebras. Another major advancement was the discovery of geometric aspects of Korovkin sets by Y.A. Šaškin [85] and D.E. Wulbert in 1968 [101]. A detailed survey of most of these developments can be found in the article of Berens and Lorentz in 1975 [16], the monograph of Altomare and Campiti [5] and the most recent survey by Altomare [4] which contains several new results also. Here we quote three major theorems due to Korovkin following the article of Berens and Lorentz [16]. We designate these as Korovkin's type I, type II and type III theorems. This section will be about 'quantization' of these three theorems!

Type I Korovkin's theorem

Let $\{\Phi_n : n = 1, 2, 3, \ldots\}$ be a sequence of positive linear maps on $C[a, b]$ and for each of the functions $g_k(x) = x^k$, $x \in [a, b]$, $k = 0, 1, 2$, let

$$\lim_{n\to\infty} \Phi_n(g_k) = g_k \text{ uniformly on } [a, b],\ k = 0, 1, 2.$$

Then

$$\lim_{n\to\infty} \Phi_n(f) = f \text{ uniformly on } [a, b] \text{ for all } f \text{ in } C[a, b].$$

Definition 5.1. A set S in $C[a, b]$ is called a test set or Korovkin set for positive linear operators on $C[a, b]$ if for every sequence $\{\Phi_n\}$ of positive linear operators on $C[a, b]$, $\lim_{n\to\infty} \Phi_n(s) = s$ uniformly on $[a, b]$ for every s in S implies that $\lim_{n\to\infty} \Phi_n(f) = f$ uniformly of $[a, b]$ for all $f \in C[a, b]$.

Thus, the Type I theorem says that $\{1, x, x^2\}$ is a test set.

Type II Korovkin's theorem

There is no test for $C[a, b]$ consisting only of two functions. Thus the cardinality of test sets is at least 3.

5.1. Type III Theorem for more general cases

This section gives a few known results that relate Korovkin sets and Čebyšev subspaces. However the study of above considerations in the C^*-algebra settings is yet to be carried out. In what follows a few important results in this area are given. Now we present the concepts of $\mathcal{D}_+$-subset for positive, bounded Radon measures, the notion of $\mathcal{K}_+$-subspaces of order n, the relation between Čebyšev subspaces and Korovkin sets in function spaces.

Definition 5.2 ([5]). Let X be a locally convex Hausdorff space and μ be a positive, bounded Radon measure on X. A subset $\mathcal{G}$ of $\mathcal{C}_0(X)$ is called a *determining set* for μ or more simply, a $\mathcal{D}_+$-subset for μ, if it satisfies the following properties: If $(\mu_i)_{i\epsilon I}$ is an arbitrary net of positive bounded Radon measures such that $\sup_{i\epsilon I} \parallel \mu_i \parallel < \infty$ and if $\lim_{i\epsilon I} \mu_i(h) = \mu(h)$ for all h in $\mathcal{G}$, then $\lim_{i\epsilon I} \mu_i(h) = \mu(h)$ for all h in $\mathcal{C}_0(X)$.

Definition 5.3 ([5]). Let X and Y be locally compact Hausdorff spaces, $\mathcal{G}$ be a subset of $\mathcal{C}_0(X)$ and $\mathcal{T}$ a positive linear operator from $\mathcal{C}_0(X)$ to $\mathcal{C}_0(Y)$. Then $\mathcal{G}$ is called a Korovkin subset for $\mathcal{T}$ or briefly $\mathcal{K}_+$ subset for $\mathcal{T}$ if it satisfies the following property:
if $(\mathcal{L}_i); i \in I$ is a net of positive linear operators from $\mathcal{C}_0(X)$ to $\mathcal{C}_0(Y)$ such that $\|\mathcal{L}_i\| < \infty$ for every i and if

$$\lim_{i\in I} \mathcal{L}_i(g) = \mathcal{T}(g)$$

for all $g \in \mathcal{G}$, then

$$\lim_{i\in I} \mathcal{L}_i(g) = \mathcal{T}(g)$$

for all $g \in \mathcal{C}_0(X)$. Here the convergence is in the strong sense and $\mathcal{T}$ is a positive linear transformation from $\mathcal{C}_0(X)$ to $\mathcal{C}_0(Y)$.

Definition 5.4 ([5]). Let X and Y be locally compact Hausdorff spaces and let $\mathcal{T} : \mathcal{C}_0(X) \to \mathcal{C}_0(Y)$ be a positive linear operator.

We shall say that $\mathcal{T}$ is finitely defined of order n if there exist n mappings $\phi_1, \phi_2, \dots, \phi_n : Y \to X$ and n real functions $\psi_1, \psi_2, \dots, \psi_n : \mathcal{Y} \to R$ such that

$$\mathcal{T}(f) = \sum_{i=1}^{n} \psi_i(f \circ \phi_i),$$

for each f in $\mathcal{C}_0(X)$.

Definition 5.5 ([5]). Let X be a locally compact, Hausdorff space and G be a subset of it. Then G is called a $\mathcal{K}_+$-subset of order n in $\mathcal{C}_0(X)$ if for every locally compact Hausdorff space Y and for every finitely defined operator $\mathcal{T}$ in $F_n(X, Y)$, $\mathcal{G}$ is a $\mathcal{K}_+$-subset for $\mathcal{T}$.
If in addition G is a subspace, we shall call it a $\mathcal{K}_+$-subspace of order n in $\mathcal{C}_0(X)$.

Theorem 5.6 ([5], Theorem 3.4.7). *Let $\mathcal{G}$ be a subspace of $\mathcal{C}_0(X)$. Then the following statements are equivalent:*

(1) *$\mathcal{G}$ is a $\mathcal{K}_+$-subspace of order n in $\mathcal{C}_0(X)$.*
(2) *For every choice of different points $x_1, x_2, \dots, x_n$ in X and for every positive, bounded Radon measure μ supported by $\{x_1, x_2, \dots, x_n\}$, $\mathcal{G}$ is a $\mathcal{D}_+$-subspace for μ.*
(3) *For every set of distinct points $x_1, x_2, \dots, x_n$ in X, for every compact subset $\mathcal{K}$ of X which does not intersect $\{x_1, x_2, \dots, x_n\}$, and for every ϵ positive nonzero, there exists h in $\mathcal{G}$ and u in $\mathcal{C}_0^+(X)$ such that $\| u \| < \epsilon$, $0 \le h + u$ on $\mathcal{K}$ and $h(x_i) + u(x_i) < \epsilon$ for every $i = 1, 2, \dots, n$.*

Theorem 5.7 ([5], Corollary 3.4.10). *Let X be a real interval $[a, b]$ or the unit circle Γ. If $\mathcal{G}$ is a subspace of $C(X)$ of dimension $n+1$, then the following statements hold:*

(1) *If n is even, $\mathcal{G}$ is a Čebyšev subspace of order $n+1$ in $C(X)$ if and only if $\mathcal{G}$ is $\mathcal{K}_+$-subspace of order $n/2$ in $C(X)$.*
(2) *If n is odd and $\mathcal{G}$ is a Čebyšev subspace of order $n+1$ in $C(X)$, then $\mathcal{G}$ is a $\mathcal{K}_+$-subspace of order $(n-1)/2$ in $C(X)$.*

Remark 5.8. The geometric formulation of Korovkin sets/spaces by Saskin[85] has its impact on Čebyšev systems/spaces too. The following theorem reveals it.

Theorem 5.9 ([16]). *Let $\mathcal{S}_o$ be a subset of $C(X)$ that separates points of X and contains the function 1_X. Then $\mathcal{S}_o$ is a Korovkin set with respect to F in $C(X)$ exactly when $\partial\mathcal{G} = X$, where $\mathcal{G} = \mathrm{span}(\mathcal{S}_o)$, and $\partial\mathcal{G}$ is the Choquet boundary of $\mathcal{G}$,where F is either the set of all positive linear maps or the set of all linear contractions on $C(X)$.*

According to a theorem of Carathéodory [16], a convex combination of points of X^* in $\mathbb{R}^{m+1}$ is also a convex combination of some $m+2$ points of X^*. According to a theorem of Fenchel [16] the number $m+2$ can be replaced by $m+1$ if X^* is connected. The following definition is due to K. Borsuk [16].

Definition 5.10.

(i) A subset X^* of $\mathbb{R}^{m+1}$ is k-independent if no $k+1$ points of X^* lie in a k-dimensional subspace of $\mathbb{R}^{m+1}$ or equivalently, if no point of X^* is a nontrivial linear combination of k other points of X^*.

(ii) A subset X^* of $\mathbb{R}^{m+1}$ is k-regular if no $k+1$ points of X^* lie in a $(k-1)$-dimensional plane of $\mathbb{R}^{m+1}$; in other words, if no point of X^* is a nontrivial linear combination, with sum of coefficients equal to 1, of some k points of X_+.

Remark 5.11. These notions can be used in the study of systems of functions $\mathcal{S} = \{g_0, g_1, \dots, g_n\}$ in $C(X)$. We consider the map $\Theta : x \longrightarrow (g_0(x), g_1(x), \dots, g_n(x))$ of the set X into the Euclidean space R^{n+1}. We call the set $\mathcal{S}$ (and the map Θ) k-independent if Θ is a homeomorphism and if the image X^* of X under Θ is k-independent. In particular, for a set $\mathcal{S}$, let $g_0 = 1_X$, and let X^* be the image of X, under $\Theta : x \longrightarrow (g_0(x), \dots, g_n(x))$. One can see that $\mathcal{S}$ is k-independent if and only if $\mathcal{S}$ separates points and if X^* is k-regular in $\mathcal{R}^n$.

As an application to Čebyšev systems, we have the following proposition [16].

Proposition 5.12 ([16], Proposition 2). *Let $\mathcal{S} = \{g_0, g_1, \dots, g_n\} \subset \mathcal{C}(X)$ be a set of functions that separates points of X. Then $\mathcal{S}$ is a Čebyšev system on X if and only if it is n-independent.*

Remark 5.13. We conclude this rather short survey of Čebyšev, semi-Čebyšev systems and spaces etc., for function spaces/C^*-algebras/W^*-algebras hoping that many celebrated results in classical analysis have their non-commutative counterparts called 'quantisation', for instance, of Korovkin sets/Čebyšev sets as Arveson called it. However, it is now clear that this is a very fertile at the same time challenging area for researchers. Another aspect is that for the special case $C(X)\otimes\mathcal{M}_n$, we have explicit results as in the classical case. So we may address these analogous results in this special case (respectively for the general case) as '**semi classical versions**' (respectively '**quantised versions**': see Arveson [10]) of '**classical**' versions.

Acknowledgment

The authors are very thankful to the referee for pointing out some errors and giving suggestions for improvement of the presentation of this article. They also wish to thank the organizers of the wonderful conference IWOTA-2013 held at Indian Institute of Science, Bangalore India for the invitation.

References

[1] A.R. Alimov and H. Berens, *Examples of Čebyšev sets in matrix spaces*, J. of Approximation Theory, vol. 99, issue 1, July 1999, 44–53.

[2] A.R. Alimov, *Convexity of Čebyšev sets contained in a subspace*, Mathematical notes, vol. 78, no. 1, 2005, 3–13.

[3] A.R. Alimov, *The geometric structure of Čebyšev sets in $l^\infty(n)^*$*, Functional Analysis and its applications, vol. 39, no. 1, 1–8, 2005.

[4] F. Altomare, *Korovkin type theorems and approximation by positive linear operators*, Surveys in approximation theory, vol. 5, 2010; 92–164.

[5] F. Altomare, M. Campiti, *Korovkin type approximation theory and its applications*, de Gruyter Studies in Mathematics, Berlin, 1994.

[6] H. Ardakani and H. Mazaheri, *Some results on f-weakly Čebyšev subspaces in Appx. theory*, J. of pure and applied mathematics: advances and applications, vol. 2, Number 2, 2009, 239–245.

[7] W.B. Arveson, *Subalgebras of C^*-algebras*, Acta. Math. 123 (1969), 141–224.

[8] W.B. Arveson, *Subalgebras of C^*-algebras II*, Acta. Math. 128 (1972), 271–308.

[9] W.B. Arveson, *The noncommutative Choquet boundary*, J. of Amer. Math. Soc, 21:1065–1084, 2008.

[10] W.B. Arveson, *The noncommutative Choquet boundary II: Hyperrigidity*, Israel J. Math., 10, 2011:349–385.

[11] W.B. Arveson, *The noncommutative Choquet boundary III: Operator systems in matrix algebras*, Math Scand 106 (2010), no. 2, 196–210.

[12] E. Asplund, *Čebyšev sets in Hilbert space*, Trans. of Amer. Math. Soc. 144 (1969), 236–240.

[13] A. Assadi, H. Haghshenas, T.D. Narang, *A look at proximinal and Čebyšev sets in Banach spaces*, Le Matematische, vol. LXIX (2014)-Fasc. I, pp. 77–87.

[14] Pradipta Bandopadhyay, Yongjins Li, Bor-Luh Lin, Darapaneni Narayana, *Proximinality in Banach spaces*, J. of Mathematical Analysis and Applications, vol. 341, issue 1, May 2008, 309–317.

[15] H. Berens, H.J. Schmidt, *An example of Čebyšev set, the complex case*, Rocky Mountain J. Math 19 (1988), 67–71.

[16] H. Berens and G.G. Lorentz, *Geometric theory of Korovkin sets*, J. Approx. Theory, 15, 161–189,1975.

[17] Petr A. Borodin, *On the convexity of N-Čebyšev sets*, Izvestiya. Math., 75, (2011) 889.

[18] P.A. Borodian, *Convexity of 2-Čebyšev sets in Hilbert spaces*, Moscow Univ. mathematics bull. June 2008, vol. 63, issue 3, 96–98.

[19] P.A. Borodin, *2-Čebyšev subspaces in the spaces L_1 and $\mathbb{C}$*, Mathematical Notes, May 2012, vol. 91, issue 5-6, 770–781.

[20] A.L. Brown, *Čebyšev subspaces of finite codimension in spaces of continuous functions*, J. Australian Mathematical Soc., Series A, vol. 26, issue 1, 1978, 99–109.

[21] F. Centrone and A. Martellotti, *Proximinal subspaces of $C(Q)$ of finite codimension*, J. of Approximation theory, vol. 101, issue 1, Nov. 1999, 78–91.

[22] E.W. Cheney and D.E. Wulbert, *Existence and unicity of best approximations*, Math. Scad. 24 (1969), 113–140.

[23] J.B. Conway, *A course in functional analysis*, Springer-Verlag, New York, 1990.

[24] P.C. Curtis, *n-parameter families and best approximation*, Pacific J. Math., 9 (1959), 1013–1027.

[25] I.K. Daugavet, *Some classes of Čebyšev systems*, J. of Mathematical sciences, April 2007, vol. 42, issue1, 1763–1768.

[26] Charles B. Dunham, *Families satisfying the Haar condition*, J. of Approximation Theory, 12, 291–298 (1974).

[27] Charles B. Dunham, *Uniqueness of best Čebyšev Appx. to nearby functions*, Periodica Mathematica Hungerica vol. 5 (3) (1974), 223–226.

[28] F. Deutsch, *Existence of best approximations*, J. of Approximation Theory, 28 (1980), 132–154.

[29] Frank Deutsch, Larry L. Shamaker and Zvi Ziegler, *Approximation in weak Čebyšev subspaces by Čebyšev subspaces*, J. of Approximation Theory, vol. 123, issue 2, August 2003, 188–213.

[30] N. Efimov and S.B. Stechkin, *Some properties of Čebyšev sets*, Dokl. Akad. Aauk. SSSR (1958), 17–19 (Russian).

[31] Efimov and Stechkin, *Approximative compactness and Čebyšev sets*, Soviet Math. Dokl., 2, 1961, 1226–1228.

[32] A.L. Garkavi, *On Čebyšev and almost Čebyšev subspaces*, Izv. Akad. Nauk SSSR Ser. Mat. 28:4(1964), 799–818.

[33] A.L. Garkavi, *Almost Čebyšev systems of continuous functions*, AMS Translations 96 (1970), 177–187.

[34] A.L. Garkavi, *The theory of best approximation in normed linear spaces*, Mathematical analysis, Progress in mathematics, vol. 8, 1970, 83–150.

[35] A.L. Garkavi, *Compact admitting Čebyšev systems of measures*, Matem. Sbornik 74 (1967), 641–656 (Russian).

[36] J.G. Glimm and R.V. Kadison, *Unitary operators on C^*-algebras*, Pacific Jour. Math. 10 (1960), 547–556.

[37] Günther Nürnberger and R. Shaback, *A local version of Haar's theorem in approximation theory*, Numerical Functional Analysis and Optimization Theory, vol. 5, issue 1, 1982, 21–46.

[38] A. Guichardet, *Tensor Products of C^*-algebras*, Aarhus University Lecture Note Series, no. 12, Aarhus, 1969.

[39] A. Haar, *Minkowskische Geometrie und die Annäherung an stetige Funktionen*, Math. Ann. 8 (1918), 294–311.

[40] M.R. Haddadi and H. Mazaheri, *Best appx. in vector spaces*, Mathematical Notes, June 2011, vol. 89, issue 5-6, 741–745.

[41] Y. Ikebe, *A characterization of Haar subspace in $C(a,b)$*, Proc. Japan Acad. 44, (1968), 219–220.

[42] Y. Ikebe, *A characterizantion of best Čebyšev appx. in function spaces*, Proc. Japan Acad. 44, (1968), 485–488.

[43] M. Iramanesh and H. Mohebi, *Best simultaneous appx. in tensor product spaces*, Applied mathematical sciences, vol. 1, no. 18, 2007, 883–894.

[44] A. Kamal, *On weak-Čebyšev subspaces and the property* AWC1 – *a survey*, Appx. theory and its applications, 17, 4, 2001, 77–89.

[45] D. Kamuntavicius, *Test for the existence of finite-dimensional Čebyšev subspaces in L_ϕ-spaces*, Lithuanian Mathematical J., Jan.–March 1990, vol. 30, issue 1, 27–35.

[46] Samuel Karlin and William J.Studden, *Tcebycheff system with applications in analysis and Statistics*, Interscience Publishers, New York, (1966).

[47] Kazuaki Kitahara, *A characterization of Čebyšev subspaces*, Proc. Japan Acad. Ser. A. Math. Sci. vol. 62, no. 10 (1986), 375–378.

[48] Victor Klee, *Convexity of Čebyšev sets*, Mathematische Annalen, 1961, vol. 142, issue 3, 292–304.

[49] C. Kleski, *Boundary representations and pure completely positive maps*, J. Operator Theory, to appear.

[50] P.P. Korovkin, *Linear operators and approximation theory*, Hindustan Publ. Corp., Delhi, India, 1960.

[51] V.A. Koshcheev, *Čebyšev systems of fns and vector fields on spheres*, Mathematical notes of the Academy of sciences of the USSR, Sept. 1989, vol. 46, issue 3, 695–700.

[52] V.A. Koshcheev, *Čebyšev systems of cont. fns with values in two-dimensional strictly convex normed space*, Mathematical notes of the Academy of sciences of the USSR, March 1983, vol. 33, issue 3, 186–195.

[53] Joanna Kowynia, *Best Appx. in Čebyšev subspaces of $L(l_1{}^n, l_1{}^n)$*, Opuscula Mathematica, vol. 29, no. 1, 2009, 57–67.

[54] K. Kumar, M.N.N. Namboodiri and S. Serra-Capizano, *Preconditioners and Korovkin-type theorems for infinite dimensional bounded linear operators via completely positive maps*, Studia Mathematica, 218(2), 2013, 95–118.

[55] D.A. Legg, B.E. Scranton and J.D. Ward, *Chebyshev subspaces in the space of compact operators*, J. of Approximation Theory 15, 1975, 326–334.

[56] Ka-Sing Lau, *On almost Čebyšev subspaces*, J. of Approximation Theory vol. 21, issue 4, Dec. 1977, 319–327.

[57] W.A. Light and E.W. Cheney, *Theory of best approximation of tensor product spaces*, Springer Verlag, New York (1985).

[58] B.V. Limaye, *Functional Analysis*, Wiley Eastern Limited, New Delhi, 1981.

[59] B.V. Limaye and S.D. Shirali, *Korovkin's theorem for positive functionals on *normed algebras*, J. Indian Math. Soc. (India), 40 (1976), 163–172.

[60] B.V. Limaye and M.N.N. Namboodiri, *Weak Korovkin approximation by completely positive linear maps on $\beta(H)$*, J. of Approximation Theory (U.S.A.), 42 (1984), pp. 202–211.

[61] J.C. Mairhuber, *On Haar's theorem concerning Čebesev approximation problem having unique solutions*, Proc. Amer. Math. Soc. 7(4) (1956), 609–615.

[62] H. Mazaheri and A. Norouzi, *Best appx. in reflexive subspaces of $C(Q)$*, Int. Mathematical Forum, 2, 2007, no. 15, 705–711.

[63] H. Mazaheri, *τ-Čebyšev and τ-coČebyšev subspaces of Banach spaces-coČebyšev subspaces of Banach spaces*, Analysis in theory and applications. June 2006, vol. 22, issue 2, 141–145.

[64] A.K. Mirmostafaee and A. Niknam, *Čebyšev subalgebras of JB-algebras*, J. of Sciences, I. R. Iran, vol. 9, no. 4, Autumn, 1998.

[65] H. Mohebi and H. Mazaheri, *Best approximation in reflexive subspaces of $L(X, Y)$* Soochew J. of Mathematics, vol. 27, no. 3, 255–266, July 2001.

[66] H. Mohebi, *On quasi-Čebyšev subspaces of Banach spaces*, J. Approximation Theory, volume 107, issue 1, Nov. 2000, 87–95.

[67] H. Mohebi and Sh. Rezpour, *On sum and quotient of quasi-Čebyšev subspaces in Banach spaces*, Analysis in theory and applications. Sept. 2003, vol. 19, issue 3, 266–272.

[68] H. Mohebi, *Pseudo- Čebyšev subspaces in* L1, Korean J. of computational and applied mathematics, May 2000, vol. 7, issue 2, 465–475.

[69] E.N. Morozov, *On Čebyšev subspaces of vector valued functions*, Mathematical notes of the academy of sciences, 1976.

[70] Oleg R. Musin, *Čebyšev systems and zeros of a function on a convex curve*, Proc. Steklov. Inst. of Math. 221, (1998), 236–246.

[71] M.N.N. Namboodiri, S. Pramod and A.K. Vijayarajan, *Finite dimensional Čebyšev subspaces of C^*-algebras*, J. of Ramanujan Math. Soc. 29, no. 1 (2014), 233–236.

[72] M.N.N. Namboodiri, *Developments in non-commutative Korovkin-type theorems*, RIMS Kokyuroku Series 737, 2011, 91–104.

[73] M.N.N. Namboodiri, *Geometric theory of weak Korovkin sets*, Operators and Matrices (2011).

[74] Darapaneni Narayana, T.S.S.R. K Rao, *Some remarks on quasi-Čebyšev subspaces*, J. of Mathematical Analysis and Applications, vol. 321, issue 1, Sept. 2006, 193–197.

[75] D.J. Newman and HS Shapiro, *Some theorems on Čebyšev approximation*, Duke Math. J. 30 (1963), 673–684.

[76] E.V. Oshman, *Continuity of metric proj. onto a Čebyšev subspace of finite codimension in a space of cont. fns*, Mathematical notes of the academy of sciences of the USSR, April 1980, vol. 27, issue 4, 293–298.

[77] G.K. Pedersen, *Čebyšev subspaces of C^*-algebras*, Math. Scand. 45 (1979), 147–156.

[78] N.K. Rakhmetov *On finite dimensional Čebyšev subspaces of spaces with an integral metric*, Mathematicheskii sbornik, 1991, vol. 182, no. 11, 1613–1614.

[79] S.H. Rezapour, *Best approximation in cone metric spaces*, Mathematica Moravica, vol. 11 (2007), 85–88.

[80] A.G. Robertson, *Strongly Čebyšev subspaces of matrices*, J. of Approximation theory, 55, 1988, 264–269.

[81] A.G. Robertson, *Best approximation in von Neumann algebras*, Math. Proc. Cambridge Phil. Soc. 81 (1977), 233–236.

[82] A.G. Roberston and D. Yost, *Chebyshev subspaces of operator algebras*, J. Lond. Math. Soc. (2) 19 (1979), 523–531.

[83] E. Rozema, *Almost Čebyšev subspaces of $L1(M, E)$*, Pacific J. of Math. 53 (1974), 585–603.

[84] S. Sakai, *C^*-Algebras and W^*-Algebras*, Springer-Verlag,Berlin, (1971).

[85] Y.A. Saskin, *Korovkin systems in spaces of cont.functions*, A.M.S translations (2) 54, 1966, 125–144.

[86] Maram Shams, Hamid Mazaheri and Sayed Manzsour Vaezpour, *On w-Čebyšev subspaces in Banach spaces*, Bull. of the Korean Math. Soc. 45 (2008), no. 3, 601–606.

[87] I. Singer, *Best approximation in normed linear spaces by elements of linear subspaces*, Springer-Verlag, 1970.

[88] I. Singer, *Best approximation in normed linear spaces*, Constructive aspects of functional analysis, G. Geymonat (ed.), CIME summer schools, Erice Italy, Springer, 1970, 683–792. Springer-Verlag, 1971.

[89] I. Singer, *On best approximation of continuous functions*, Math.Ann. 140 (1960), 165–168.

[90] I. Singer, *On normed linear spaces which are proximinal in every subspace*, J. of appx. theory, vol. 7, Issue 4, 1973, 399–402.

[91] Karl George Steffens, *History of approximation theory, from Euler to Bernstein*, Birkhäuser, 2006.

[92] Simon Stevin, *The unicity of best appx. in a space of compact operators*, Bull. of Belgian Mat. Soc., vol. 17, no. 5 (2010), 807–820.

[93] I.G. Tsarikov, *Bounded Čebyšev sets in finite dimensional Banach spaces*, Mathematical notes of the academy of sciences of the USSR, July 1984, vol. 36, issue 1, 530–537.

[94] G.M. Ustinov, *On existence of subspaces in $C(Q)$*, Mathematical notes, May 1999, vol. 65, issue 5, 610–619.

[95] M. Uchiyama, *Korovkin type theorems for Schwartz maps and operator monotone functions in C^*-algebras*, Mathematische Zeitschrift 230, 785–797, 1999.

[96] A.J. Vasilev, *Čebyšev sets and strong convexity of metric linear spaces.* Mathematical notes of the Academy of sciences of the USSR, May 1979, vol. 25, issue 5, 335–340.

[97] L.P. Vlasov, *Subspaces of finite co-dimension: The existence of elements of best approximation*, Mathematical notes of the Academy of sciences of the USSR, Jan. 1985, vol. 37, issue 1, 45–49.

[98] L.P. Vlasov, *Čebyšev sets and appx. convex sets*, Mathematical notes of the Academy of sciences of the USSR, Aug. 1967, vol. 2, issue 2, 600–605.

[99] L.P. Vlasov, *On Čebyšev sets*, Doklady Akad. Nauk. SSSR 173 (1967), 491–494.

[100] L.P. Vlasov, *Almost convex and Čebyšev sets*, Mathematical notes of the Academy of sciences of the USSR, Nov. 1970, vol. 8, issue 5, 776–779.

[101] D.E. Wulbert, *Convergence of operators and Korovkin's theorem*, J. Approximation theory 1 (1968), 381–390.

[102] David Yost, *Best approximation and intersections of the balls in Banach spaces*, Bulletin of the Australian mathematical soc. vol. 20, issue 2, April 1979, 285–300.

M.N.N. Namboodiri
Department of Mathematics
Cochin University of Science & Technology
Kochi, Kerala, India-682022
e-mail: `mnnadri@gmail.com`

S. Pramod
Department of Agriculture
Government of Kerala, India
e-mail: `pramodsuryan@yahoo.com`

A.K. Vijayarajan
Kerala School of Mathematics
Kozhikode, Kerala, India-673571
e-mail: `vijay@ksom.res.in`

Operator Theory:
Advances and Applications, Vol. 247, 123–156
© Springer International Publishing Switzerland 2015

Coherent States and Wavelets, a Contemporary Panorama

Jean-Pierre Antoine

Abstract. In a first part, we review the general theory of coherent states (CS). Starting from the canonical CS introduced by Schrödinger in 1926 and rediscovered by Glauber, Klauder and Sudarshan in the 1960s, we proceed to the derivation of general CS from square integrable group representations and some of their generalizations: Perelomov CS, general square integrable covariant CS, nonlinear CS, Gazeau–Klauder CS, with a hint to their application in quantization.

Next, we turn to signal processing and note that two of the most familiar tools, namely, the Gabor transform and the wavelet transform, are special cases of CS, associated to the Weyl–Heisenberg group (which yields the canonical CS) and the affine group of the line, respectively. Then we review the properties of the wavelet transform, both in its continuous and its discrete versions, in one or two dimensions, emphasizing mostly the mathematical properties. We also consider its extension to higher dimensions, to more general manifolds (sphere, hyperboloid,...) and to the space-time context, for the analysis of moving objects.

Mathematics Subject Classification (2010). 22D10, 42C40, 62H35, 68U10, 81R30.

Keywords. Coherent states, wavelets, square integrable group representations, higher-dimensional wavelets, wavelets on manifolds, signal processing, image processing.

1. Motivation

Coherent states (CS) were introduced in 1926 by Schrödinger for studying the quantum-to-classical transition, but quickly forgotten. They were rediscovered in the 1960s by Glauber, Klauder and Sudarshan in the context of quantum optics (coherent light from lasers) and it became readily clear that these so-called canonical CS, associated to a one-dimensional quantum harmonic oscillator, were in fact related to the Weyl–Heisenberg group. In 1972, Gilmore and Perelomov, independently, constructed CS associated to other groups, thus turning them into a problem in group representation theory. Since then, CS were considerably generalized, both mathematically and for applications into almost all domains of physics.

As for wavelets, they were invented around 1985 by Jean Morlet in the context of oil prospecting. Then it was soon realized, in collaboration with Alex Grossmann [21], that these continuous 1-D wavelets are in fact a particular case of CS, associated with the affine group of the line ('$ax + b$' group). There followed generalizations to higher dimensions and to various manifolds (two-sphere, two-sheeted hyperboloid, two-torus,...), then to more sophisticated mathematical objects, such as ridgelets, curvelets, shearlets, etc. In parallel, the theory exploded in a different direction, namely discrete wavelets, derived from multiresolution analysis. Nowadays, both the continuous and the discrete wavelet transforms have led to an immense number of applications, and they have become standard tools in signal/image processing. The aim of this chapter is to give an up-to-date panorama of the whole field, emphasizing its mathematical coherence. We mainly follow our textbooks [3, 10] and that of Gazeau [19], where plenty of applications are described in detail. In addition, all 'historical' papers on wavelets and their precursors have been collected in the compendium [22].

2. The pioneers: Canonical coherent states

2.1. Definitions

As we said above, the canonical CS are associated to a 1-D quantum harmonic oscillator. Starting from the canonical position and momentum operators Q, P, we introduce the operators

$$a = \frac{1}{\sqrt{2}}(Q + iP), \quad a^\dagger = \frac{1}{\sqrt{2}}(Q - iP), \quad [a, a^\dagger] = I. \tag{2.1}$$

Then the Hamiltonian of the system reads as $H = \frac{1}{2}(P^2 + Q^2) = a^\dagger a + \frac{1}{2} = N + \frac{1}{2}$ (we put $\hbar = 1$), where $N = a^\dagger a$ is the number operator, acting in the abstract Hilbert space $\mathcal{H}$. The spectrum of H is $\sigma(H) = \{n + \frac{1}{2},\ n = 0, 1, 2 \dots\}$, the eigenstates $\{|n\rangle\}$ are nondegenerate and normalized, $\langle n|m\rangle = \delta_{mn}$, thus an orthonormal basis in $\mathcal{H}$.[1] In particular, the ground state satisfies the relation $a|\,0\rangle = 0$ and $a, a^\dagger$ are ladder operators:

$$a|n\rangle = \sqrt{n}\,|\,n-1\rangle, \qquad a^\dagger|n\rangle = \sqrt{n+1}\,|\,n+1\rangle.$$

Thus a is called an *annihilation* operator and $a^\dagger$ a *creation* operator. Indeed one can recover all states by acting successively on the ground state with the creation operator $a^\dagger$:

$$|n\rangle = \frac{(a^\dagger)^n}{\sqrt{n!}}\,|\,0\,\rangle, \quad n = 1, 2, 3 \dots$$

[1] We use here, and throughout the paper, the Dirac bra-ket notation familiar in quantum physics. In particular, our inner product $\langle\cdot|\cdot\rangle$ is linear in the *second* factor, so that $|\psi\rangle\langle\phi| := \psi \otimes \overline{\phi}$.

The problem is that these energy eigenstates are inadequate for describing the classical limit of quantum mechanics, $n \gg 1$, because they don't have a definite phase. Thus one needs other states and this is the reason why, in 1926, Schrödinger introduced the following ones, nowadays called *canonical CS*:

$$| z \rangle = e^{-\frac{1}{2}|z|^2} \sum_{n=0}^{\infty} \frac{z^n}{\sqrt{n!}} | n \rangle, \quad z \in \mathbb{C}. \tag{2.2}$$

Unfortunately, they were quickly forgotten, but were rediscovered in 1963 by Glauber, Klauder and Sudarshan in the context of quantum optics, for describing a coherent light beam, namely, a laser beam.

2.2. Basic properties

The canonical CS have four basic properties, that we now list.

(P1) The states $|z\rangle$ saturate the Heisenberg inequality:

$$\langle \Delta Q \rangle_z \, \langle \Delta P \rangle_z = \frac{1}{2}\hbar, \tag{2.3}$$

where, for $A = Q, P$, $\langle \Delta A \rangle_z := [\langle z|A^2|z\rangle - \langle z|A|z\rangle^2]^{1/2}$ is the variance (uncertainty) of the observable A in the state $|z\rangle$.

(P2) The states $|z\rangle$ are eigenvectors of the annihilation operator, with eigenvalue $z \in \mathbb{C}$ (thus they are uncountable):

$$a|z\rangle = z|z\rangle, \quad z \in \mathbb{C}. \tag{2.4}$$

(P3) The states $|z\rangle$ are obtained from the ground state $|0\rangle$ by a displacement operator

$$|z\rangle = D(z)|0\rangle, \text{ where } D(z) := e^{za^\dagger - \overline{z}a}, \; z \in \mathbb{C}.$$

The operator $D(z)$ satisfies the relation $D(z)\,D(z') = e^{i\mathrm{Im}(z\overline{z'})}\, D(z+z')$.

(P4) The family $\{|z\rangle, z \in \mathbb{C}\}$ constitute an *overcomplete* (i.e., total) family of vectors in the Hilbert space $\mathcal{H}$. This property is encoded in the following *resolution of the identity*:

$$\int_{\mathbb{C}} |z\rangle\langle z| \, \frac{\mathrm{d}^2 z}{\pi} = I = \sum_{n=0}^{\infty} |n\rangle\langle n|, \tag{2.5}$$

where $\mathrm{d}^2 z := \mathrm{d}\,\mathrm{Re}\, z \;\mathrm{d}\,\mathrm{Im}\, z$ is the Lebesgue measure on the plane $\mathbb{C}$ and the integral is understood in the weak sense.

The properties (P1)–(P4) are equivalent for the canonical CS, in the sense that every one may be used as definition, but they lead to different generalizations.

Before going into these, let us discuss in more detail some consequences of the properties (P3) and (P4).

2.2.1. (P3) Connection with the Weyl–Heisenberg group. The three operators $\{a, a^\dagger, I\}$, with the commutation relation (2.1), generate a nilpotent Lie algebra. The corresponding Lie group, the *Weyl–Heisenberg* group $G_{\rm WH}$, plays a key role in nonrelativistic quantum mechanics. It consists of elements

$$\xi \equiv (\theta, z) := e^{\theta I + za^\dagger - \overline{z}a} = e^{\theta I} D(z),\ \theta \in [0, 2\pi),\ z = 2^{-1/2}(q + ip) \in \mathbb{C}.$$

The center of $G_{\rm WH}$ is $Z = \{(\theta, 0)\} \simeq \mathbb{S}^1$. The group $G_{\rm WH}$ is unimodular, with (left and right) invariant measure

$$\mathrm{d}\mu(\xi) = (2\pi)^{-1}\, \mathrm{d}\theta\, \mathrm{d}q\, \mathrm{d}p = (2\pi)^{-1}\, \mathrm{d}\theta\, \mathrm{d}^2 z.$$

Given a unitary irreducible representation (UIR) U of $G_{\rm WH}$, its restriction to Z is $U(\theta, 0) = \exp(i\lambda\theta), \lambda \in \mathbb{Z}$. By von Neumann's uniqueness theorem, all UIRs of $G_{\rm WH}$ are unitarily equivalent and of the form $U^\lambda(\theta, z) = e^{i\lambda\theta}\, D^\lambda(z)$, where $D^\lambda(z) = D(z) = e^{za^\dagger - \overline{z}a}$, acting in a Hilbert space $\mathcal{H}^\lambda$.

In addition, every representation U^λ $(\lambda \neq 0)$ is *square integrable*:

$$\int_{G_{\rm WH}} |\langle U^\lambda(\xi)\phi|\phi\rangle|^2\, \mathrm{d}\mu(\xi) = \int_{G_{\rm WH}/Z} |\langle D^\lambda(z)\phi|\phi\rangle|^2\, \mathrm{d}^2 z < \infty,\ \forall\, \phi \in \mathcal{H}^\lambda.$$

For U^λ and $|0\rangle \in \mathcal{H}^\lambda$, the orbit $\{|z\rangle = D^\lambda(z)|0\rangle, z \in \mathbb{C} \simeq G_{\rm WH}/Z\}$ is a *family of canonical CS*.

More generally, given any nonzero vector $\eta \in \mathcal{H}^\lambda$, the orbit $\{\eta_z = D^\lambda(z)\eta,\ z \in G_{\rm WH}/Z\}$ is a family of CS such that

$$\int_{G_{\rm WH}/Z} |\eta_z\rangle\langle\eta_z|\, \frac{\mathrm{d}^2 z}{\pi} := \int_{G_{\rm WH}/Z} \eta_z \otimes \overline{\eta}_z\, \frac{\mathrm{d}^2 z}{\pi} = I \quad \text{(weakly)}.$$

In that sense, one realizes that the construction of canonical CS is a problem in the theory of group representations!

In particular, in the Schrödinger realization of quantum mechanics, one has $\mathcal{H} = L^2(\mathbb{R}, \mathrm{d}x)$ and

$$D(q,p)f)(x) = e^{iqp/2}\, e^{ipx}\, f(x - q),\ x \in \mathbb{R},$$
$$U^\lambda(\theta, q, p) = e^{i\lambda\theta} D(q,p),$$

thus U^λ is a UIR of $G_{\rm WH}$ in $L^2(\mathbb{R}, \mathrm{d}x)$.

2.2.2. (P4) Functional analysis of canonical CS. Condition (P4) implies that

$$|\psi\rangle = \int_{\mathbb{C}} |\overline{z}\rangle\langle\overline{z}|\psi\rangle\, \mathrm{d}\mu(\overline{z}),\ \forall\, \psi \in \mathcal{H},\quad \text{where } \mathrm{d}\mu(z) := \frac{\mathrm{d}^2 z}{\pi}.$$

The function $\psi(z) := \langle\overline{z}|\psi\rangle$ is called the *symbol* of ψ. Using (P4) again, we get

$$\psi(z') = \int_{\mathbb{C}} K(\overline{z'}, z)\, \psi(z)\, \mathrm{d}\mu(z),$$

where $K(\overline{z'}, z) := \langle z'|z\rangle = \exp\left(-\frac{1}{2}|z'|^2 + \overline{z'}z - \frac{1}{2}|z|^2\right)$ is a *reproducing kernel*:

$$\int_{\mathbb{C}} K(\overline{z'}, z'')\, K(\overline{z''}, z)\, \mathrm{d}\mu(z'') = K(\overline{z'}, z).$$

From the resolution of the identity (2.5), one gets

$$\langle\psi_1|\psi_2\rangle = \int_{\mathbb{C}} \overline{\psi_1(z)}\,\psi_2(z)\,\mathrm{d}\mu(z)\,,\ \forall\,\psi_1,\,\psi_2\in\mathcal{H},$$

that is, the map $W:\mathcal{H}\to L^2(\mathbb{C},\,\mathrm{d}\mu)$ defined by $(W\psi)(z)=\psi(z)$ is an *isometry*. Hence $\mathcal{H}_{\rm FB}:=\operatorname{Ran}W$ is a closed subspace of $L^2(\mathbb{C},\,\mathrm{d}\mu)$. The corresponding projection operator $P_W : L^2(\mathbb{C},\,\mathrm{d}\mu)\to\mathcal{H}_{\rm FB}$ is an *integral* operator with kernel $K(\overline{z'},z)$. This approach leads to the so-called *Fock–Bargmann representation* of quantum mechanics.

Write

$$\psi(z)=\langle\overline{z}|\psi\rangle=e^{-\frac{1}{2}|z|^2}\,f(z),$$

where f is an *entire* holomorphic function. Then one has the unitary equivalence

$$\mathcal{H}\simeq\mathcal{H}_{\rm FB}:=\left\{f\text{ entire},\ \int_{\mathbb{C}}|\,f\,(z)|^2\,e^{-|z|^2}\,\mathrm{d}\mu(z)\,<\infty\right\}.$$

In this Fock–Bargmann realization, one has

$$a=\frac{\mathrm{d}}{\mathrm{d}z},\qquad a^\dagger = \text{ multiplication operator by } z,$$

and the orthonormal basis

$$u_n(z)=\frac{z^n}{\sqrt{n!}},\qquad n=0,1,2\ldots.$$

Since $z=2^{-1/2}\,(q+ip)$, the Fock–Bargmann representation is in fact a *phase space* representation. This fact will prove important for quantization!

2.3. Generalizations

As said above, all four conditions (P1)–(P4) lead to different generalizations of the canonical CS. We list them very briefly.

(i) *Minimal uncertainty states* (P1):

A state that verifies (P1) is either a CS or a *squeezed state*, that is, a state of the form

$$|\alpha,z\rangle=S(\alpha)|z\rangle,\ S(\alpha)=\exp\left[\tfrac{1}{2}(\alpha {a^\dagger}^2-\overline{\alpha}a^2)\right],\ \alpha,z\in\mathbb{C}.$$

Such states have been constructed experimentally in quantum optics.

(ii) *Eigenvalue property* (P2):

This property has led to the construction of *Barut–Girardello CS* for the discrete series representations of $\mathrm{SL}(2,\mathbb{R})\simeq\mathrm{SU}(1,1)\simeq\mathrm{SO}_o(1,2)$, or, more generally, for any Lie algebra with ladder operators $A, A^\dagger$, for instance, algebras of q-deformed oscillators or supersymmetric systems. In that case the Barut–Girardello CS are defined as eigenvectors of the lowering operator, i.e., solutions of the eigenvalue equation

$$A|\xi\rangle=\xi|\xi\rangle,\quad \xi\in\mathbb{C}.$$

(iii) *Group theory* (P3):

This approach has led, in successive generality, to

- *Covariant CS* derived from square integrable representations of locally compact groups: $\eta_g = U(g)\eta$.
  Example: 1-D wavelets corresponding to the $ax + b$ group.
- *Gilmore–Perelomov CS* on homogeneous spaces $X_\eta := G/H_\eta$, where H_η is the isotropy subgroup of η up to a phase.
  Examples: the n-D wavelets associated to the similitude group SIM(n), the canonical CS corresponding to $G_{\rm WH}$ or the spin CS stemming from SU(2).
- *Covariant CS* on arbitrary homogeneous spaces.
  Examples: the relativistic CS, corresponding to the Poincaré and Galilei groups.

We will discuss these three types of (generalized) CS in the following sections.

(iv) *Reproducing kernel* (P4):

This is the essential property and the guiding principle, for instance in the context of quantization. We will touch briefly on that point of view in Section 5.

3. CS derived from group representations

In this section, we shall discuss the three types of CS derived from group representations, in increasing generality. We follow [2] and [3].

3.1. CS from square integrable representations

The framework consists of a locally compact group G, with left Haar measure dg and a strongly continuous UIR U of G on a Hilbert space $\mathcal{H}$. Assume U is *square integrable* (equivalently, U belongs to the discrete series), i.e., there exists a nonzero vector $\eta \in \mathcal{H}$, called an *admissible vector*, such that

$$I(\eta) := \int_G |\langle U(g)\eta|\eta\rangle|^2 \, {\rm d}g < \infty. \tag{3.1}$$

Equivalently, η is admissible if

$$\int_G |\langle U(g)\eta|\phi\rangle|^2 \; {\rm d}g < \infty, \; \forall \phi \in \mathcal{H}. \tag{3.2}$$

Since η is admissible, the vector $\eta_g = U(g)\eta$ is admissible for all $g \in G$, so that the set $\mathcal{A}$ of all admissible vectors is invariant under U. Then, since U is irreducible, either $\mathcal{A} = \{0\}$ or $\mathcal{A}$ is dense in $\mathcal{H}$. In particular, $\mathcal{A} = \mathcal{H}$, i.e., every vector in $\mathcal{H}$ is admissible, if and only if G is unimodular.

Given a fixed admissible vector $\eta \in \mathcal{A}$, the elements of the orbit of η under U are called *covariant CS* and their set is denoted by $\mathcal{S} = \{\eta_g = U(g)\eta, \, g \in G\}$.

Then the map $W_\eta : \mathcal{H} \to L^2(G, {\rm d}g)$, defined as

$$(W_\eta\phi)(g) = [c(\eta)]^{-1/2}\langle\eta_g|\phi\rangle, \; \phi \in \mathcal{H}, \; g \in G, \quad \text{where } c(\eta) := I(\eta)/\|\eta\|^2,$$

is called the *coherent state or CS map*. This is the crucial object of the theory, as discovered in the pioneering work of Grossmann *et al.* [21].

The CS map W_η has the following properties.

(1) W_η is an isometry onto a closed subspace $\mathcal{H}_\eta$ of $L^2(G, \mathrm{d}g) : W_\eta^* W_\eta = I$. Thus $\mathcal{S}$ is a total set: $\mathcal{S}^\perp = \{0\}$. Equivalently, η_g generates a *resolution of the identity*

$$\frac{1}{c(\eta)} \int_G |\eta_g\rangle\langle\eta_g| \,\mathrm{d}g = I. \tag{3.3}$$

(2) The subspace $\mathcal{H}_\eta = W_\eta \mathcal{H} \subset L^2(G, \mathrm{d}g)$ is a *reproducing kernel Hilbert space.* The projection operator $P_\eta = W_\eta W_\eta^*$, $P_\eta L^2(G, \mathrm{d}g) = \mathcal{H}_\eta$, is an integral operator with reproducing kernel $K_\eta(g, g') = c(\eta)^{-1} \langle \eta_g | \eta_{g'} \rangle$:

$$(P_\eta \Phi)(g) = \int_G K_\eta(g, g') \Phi(g') \,\mathrm{d}g', \quad \Phi \in L^2(G, \mathrm{d}g).$$

(3) W_η may be inverted on its range by the adjoint operator, which yields a *reconstruction formula*:

$$\phi = W_\eta^* \Phi = [c(\eta)]^{-1/2} \int_G (W_\eta \phi)(g) \eta_g \,\mathrm{d}g, \quad \Phi = W_\eta \phi \in \mathcal{H}_\eta.$$

(4) W_η intertwines U and the left regular representation U_ℓ:

$$W_\eta U(g) = U_\ell(g) W_\eta, \quad \forall\, g \in G.$$

Therefore, U is equivalent to a subrepresentation of U_ℓ or, equivalently, U belongs to the discrete series, or W_η is *covariant* under the action of G.

As a matter of fact, square integrable representations behave in many respect like UIRs of compact groups. In particular, they lead to nice orthogonality relations. If G is compact and U is a UIR of G (necessarily finite-dimensional), the matrix elements $\langle U(g)\psi|\phi\rangle$ of U satisfy orthogonality relations and, by the Peter–Weyl theorem, they generate an orthonormal basis of $L^2(G, \mathrm{d}g)$. This fact underlies many properties of well-known special functions, such as spherical harmonics or Bessel functions. The same result holds for any locally compact group, if U is square integrable. Indeed, if G is a locally compact group and U a square integrable UIR of G, then there exists a unique positive, self-adjoint, invertible operator C in $\mathcal{H}$, with dense domain $D(C)$ equal to $\mathcal{A}$, such that, for η and η' admissible and ϕ, ϕ' arbitrary in $\mathcal{H}$, one has

$$\int_G \overline{\langle \eta'_g | \phi' \rangle} \langle \eta_g | \phi \rangle \,\mathrm{d}g = \langle C\eta | C\eta' \rangle \; \langle \phi' | \phi \rangle.$$

The operator C is called the *Duflo–Moore operator* and is often denoted by $C = K^{-1/2}$. Then, $C = \lambda I$, $\lambda > 0$, if and only if G is unimodular. If G is compact, $C = [\dim \mathcal{H}]^{-1/2} I$. If G is not compact, but unimodular, then $C = [c(\eta)]^{1/2} I$. In that case, $d_U := c(\eta)^{-1}$ is called the formal dimension of U.

Remark: instead of (3.3), one has also the following generalized resolution of the identity, for η, η' admissible such that $\langle C\eta | C\eta' \rangle \neq 0$,

$$\frac{1}{\langle C\eta | C\eta' \rangle} \int_G |\eta'_g\rangle\langle\eta_g| dg = I.$$

3.2. Gilmore–Perelomov CS

It is a fact that the usual admissibility conditions (3.1) or (3.1) are often too strong. This motivates the generalization due to Gilmore [20] and Perelomov [26] (independently).

Given an admissible vector η, let H_η be a subgroup of G that leaves it invariant up to a phase (this condition obviously comes from quantum mechanics, where a state is represented, not by a vector, but by a ray in the Hilbert space):

$$U(h)\eta = \exp i\alpha(h)\eta, \ \forall\, h \in H_\eta$$

Therefore, the integrand of the admissibility condition (3.1) depends only on the coset gH_η, not on η itself. Hence we can introduce a weaker admissibility condition on $X_\eta = G/H_\eta$ (we assume that X_η carries an invariant measure ν, a mild restriction):

$$\begin{aligned} c(\eta,\phi) &= \int_{X_\eta} |\langle U(g)\eta|\phi\rangle|^2 \, \mathrm{d}\nu(x) < \infty, \ \ \forall\, \phi \in \mathcal{H} \quad (x := gH_\eta) \\ &= \int_{X_\eta} |\langle U(\sigma(x))\eta|\phi\rangle|^2 \, \mathrm{d}\nu(x) < \infty, \ \ \forall\, \phi \in \mathcal{H}, \end{aligned} \tag{3.4}$$

with an *arbitrary* section $\sigma : X_\eta \to G$ in the principal fibre bundle $\pi : G \to X_\eta$. When these conditions are satisfied, we say that U is *square integrable mod* H_η.

In that case, we get a new set of CS: $\mathcal{S}_\sigma = \{\eta_{\sigma(x)} = U(\sigma(x))\eta, \, x \in X_\eta\}$, called *Gilmore–Perelomov CS* (η has been normalized by $c(\eta,\eta) = 1$). Note that the admissibility condition (3.4) does *not* depend on σ, but $\mathcal{S}_\sigma$ does !

The properties of Gilmore–Perelomov CS are exactly the same as those of the previous CS. First, $\mathcal{S}_\sigma$ is a total set in $\mathcal{H} : \mathcal{S}_\sigma^\perp = \{0\}$. Then, the CS map $W_\eta : \mathcal{H} \to L^2(X_\eta, \mathrm{d}\nu)$, which reads now as $(W_\eta\phi)(x) = \langle \eta_{\sigma(x)}|\phi\rangle$, is an isometry onto a closed subspace $\mathcal{H}_\eta$ of $L^2(X_\eta, \mathrm{d}\nu)$, which leads to the resolution of the identity

$$\int_{X_\eta} |\eta_{\sigma(x)}\rangle\langle\eta_{\sigma(x)}| \, \mathrm{d}\nu(x) = I.$$

Finally, the projection operator $P_\eta = W_\eta W_\eta^* : L^2(X_\eta, \mathrm{d}\nu) \to \mathcal{H}_\eta$ is an integral operator with reproducing kernel $K(x',x) = \langle \eta_{\sigma(x')}|\eta_{\sigma(x)}\rangle$.

As examples of Gilmore–Perelomov CS, we may give:

- If G is the Weyl–Heisenberg group G_{WH}, leading to canonical CS, the CS map W_η is the windowed Fourier or Gabor transform.
- If G is compact, any UIR is square integrable. For instance, if η is a highest weight state, then H_η is the maximal compact subgroup. A typical example is SU(2) with the spin CS.
- For G noncompact semisimple and U square integrable, typical examples are the discrete series representations of SU(1,1) (useful in the description of path integrals) or $\mathrm{Sp}(3,\mathbb{R})$ (nuclear structure theory).

- If G is the similitude group of $\mathbb{R}^n$, $\mathrm{SIM}(n) = \mathbb{R}^n \rtimes (\mathbb{R}_*^+ \times \mathrm{SO}(n))$, then the corresponding CS are the n-dimensional wavelets that we shall discuss in Section 9.1.

3.3. General square integrable covariant CS

In practice, it is often too restrictive to require that the subgroup H in a quotient $X = G/H$ be the invariance subgroup of a given admissible vector, even up to a phase, as Gilmore and Perelomov do. Thus a natural extension of the previous analysis consists in constructing CS over an arbitrary homogeneous space.

Thus we start with a locally compact group G, a strongly continuous UIR U in $\mathcal{H}$ and a closed subgroup H of G. Then we consider the homogeneous space $X := G/H$, equipped with an invariant measure ν (this is a weak restriction). Let $\sigma : X = G/H \to G$ be a Borel section (in the principal bundle $G \to G/H$). Then we say that U is *square integrable mod* (H, σ) for the vector $\eta \in \mathcal{H}$ (or that η is admissible for (U, σ)) if

$$0 < c_X(\eta, \phi) := \int_X |\langle U(\sigma(x))\eta|\phi\rangle|^2 \, \mathrm{d}\nu(x) = \langle \phi | A_\sigma \phi \rangle \ < \infty, \ \ \forall\, \phi \in \mathcal{H},$$

with A_σ a bounded positive *invertible* operator on $\mathcal{H}$, called the frame operator. Note that A_σ^{-1} may be unbounded.

Then coherent states based on X read as $\mathcal{S}_\sigma = \{\eta_{\sigma(x)} := U(\sigma(x))\eta,\ x \in X\}$ (here again we put $c_X(\eta, \eta) = 1$). Clearly these new CS depend on the choice of the section σ. Yet they have essentially the same properties as the previous ones.

1. $\mathcal{S}_\sigma$ is total (overcomplete) in $\mathcal{H} : \mathcal{S}_\sigma^\perp = \{0\}$.
2. The range $\mathcal{H}_\eta$ of the linear map (CS map) $W_\eta : \mathcal{H} \to L^2(X, \mathrm{d}\nu)$, given by $(W_\eta \phi)(x) = \langle \eta_{\sigma(x)} | \phi \rangle$, is complete with respect to the new scalar product $\langle \Phi | \Psi \rangle_\eta := \langle \Phi | W_\eta\, A_\sigma^{-1}\, W_\eta^{-1} \Psi \rangle$ and W_η is *unitary* from $\mathcal{H}$ onto $\mathcal{H}_\eta$. The new fact here is the occurrence of the operator A_σ^{-1} in the scalar product. Thus we get a resolution of the identity (with weak convergence, as usual):
$$\int_X |\eta_{\sigma(x)}\rangle\langle \eta_{\sigma(x)}| \, \mathrm{d}\nu(x) = A_\sigma .$$
If A_σ^{-1} is bounded, $\mathcal{S}_\sigma$ is a (possibly continuous) *frame*; if $A_\sigma = \lambda I$, $\mathcal{S}_\sigma$ is a *tight frame.*
3. The orthogonal projection $P_\eta : L^2(X, \mathrm{d}\nu) \to \mathcal{H}_\eta$ is an integral operator K_σ and $\mathcal{H}_\eta$ is a reproducing kernel Hilbert space of functions. The kernel is $K(x', x) = \langle \eta_{\sigma(x')} | A_\sigma^{-1}\, \eta_{\sigma(x)} \rangle$ if $\eta_{\sigma(x)} \in D(A_\sigma^{-1})$, $\forall\, x \in X$; otherwise, the relation must be understood in a distributional sense.
4. As a consequence, W_η may be inverted on its range by the adjoint operator, $W_\eta^{-1} = W_\eta^*$ on $\mathcal{H}_\eta$, to obtain (again for $\eta_{\sigma(x)} \in D(A_\sigma^{-1}), \forall\, x \in X$)
$$W_\eta^{-1} \Phi = \int_X \Phi(x)\, A_\sigma^{-1}\, \eta_{\sigma(x)} \, \mathrm{d}\nu(x), \quad \Phi \in \mathcal{H}_\eta .$$

Examples of this construction are, of course, Gilmore–Perelomov CS, for which H is the stability subgroup of η and $A_\sigma = I$, but also CS for the relativity groups (Euclidean, Galilei, Poincaré), which are *not* of the Gilmore–Perelomov type.

4. Algebraic CS

With retrospect, we realize that the crucial ingredient for the construction of CS is the reproducing kernel, not group theory. Thus we are led to an algebraic formulation, directly generalizing the original definition of Schrödinger (2.2).

4.1. Nonlinear CS

A new class of CS was introduced under the name of *Nonlinear CS*, simply by modifying the standard formula (2.2) for the canonical CS, namely,

$$|\, z\,\rangle = e^{-\frac{1}{2}|z|^2} \sum_{n=0}^{\infty} \frac{z^n}{\sqrt{n!}} \,|\, e_n\rangle,\ z \in \mathbb{C},$$

where $\{e_n\}$ is an orthonormal basis in $\mathcal{H}$. Given an increasing sequence of positive numbers $0 < \varepsilon_1 \leqslant \varepsilon_2 \leqslant \cdots \leqslant \varepsilon_n \leqslant \cdots$, one defines the new CS as

$$|z\rangle = \mathcal{N}(|z|^2)^{-1/2} \sum_{n=0}^{\infty} \frac{z^n}{\sqrt{\varepsilon_n!}}\, |e_n\rangle, \tag{4.1}$$

where $\varepsilon_n! := \varepsilon_1\varepsilon_2 \dots \varepsilon_n$ is a generalized factorial. In (4.1), the normalization factor $\mathcal{N}(|z|^2)$ is chosen so that $\langle z|z\rangle = 1$.

These CS are overcomplete and satisfy a resolution of the identity

$$\int_{\mathcal{D}} |z\rangle\langle z|\, \mathcal{N}(|z|^2)\ \mathrm{d}\nu(z,\overline{z}) = I\,,$$

where $\mathcal{D} \subset \mathbb{C}$ is an open disc of radius L, the radius of convergence of $\sum_{n=0}^{\infty} z^n \Big/ \sqrt{\varepsilon_n!}$. The measure ν is defined as $\mathrm{d}\nu = \mathrm{d}\theta\, \mathrm{d}\lambda(r)$ (for $z = re^{i\theta}$), where $\mathrm{d}\lambda$ is related to the ε_n through the moment condition:

$$\frac{\varepsilon_n!}{2\pi} = \int_0^L r^{2n}\, \mathrm{d}\lambda(r), \quad n = 0, 1, 2, \dots .$$

In addition, these CS possess a reproducing kernel, namely,

$$K(\overline{z}, z') = \langle z|z'\rangle = \left[\mathcal{N}(|z|^2)\mathcal{N}(|z'|^2)\right]^{-1/2} \sum_{n=0}^{\infty} \frac{(\overline{z}z')^n}{\varepsilon_n!}.$$

4.2. Gazeau–Klauder CS

The so-called Gazeau–Klauder CS are a special case of the general nonlinear CS just described. We consider three cases, depending whether the spectrum of the Hamiltonian is discrete, continuous or mixed.

(1) *Discrete spectrum dynamics*

Let H be a positive Hamiltonian with purely discrete nondegenerate spectrum $H|e_n\rangle = \omega\varepsilon_n|e_n\rangle$, where $0 = \varepsilon_0 < \varepsilon_1 < \varepsilon_2 < \cdots < \varepsilon_n < \cdots$, is a finite or infinite sequence. The corresponding Gazeau–Klauder or action-angle CS read as

$$|J,\gamma\rangle = \frac{1}{\sqrt{\mathcal{N}(J)}}\sum_{n\geqslant 0}\frac{J^{n/2}}{\sqrt{\varepsilon_n!}}e^{-i\varepsilon_n\gamma}|e_n\rangle, \quad J \geqslant 0,\ \gamma \in \mathbb{R}. \tag{4.2}$$

Of course, we recover the canonical CS if we put $\varepsilon_n = n,\ z = \sqrt{J}\,e^{-in\gamma}$. The CS (4.2) enjoy all the standard properties:

- Continuity of the map $[0,R)\times\mathbb{R} \ni (J,\gamma) \mapsto |J,\gamma\rangle \in \mathcal{H}$, where the convergence radius $R = \overline{\lim}_{n\to\infty}\sqrt[n]{\varepsilon_n!}$ is nonzero and $0 \leqslant J < R$.
- Temporal stability: $e^{-iHt}|J,\gamma\rangle =|\ J,\gamma+\omega t\rangle$.
- Action identity: $\langle J,\gamma|H|J,\gamma\rangle = \omega J$.
- Resolution of the identity:

$$I = \int |J,\gamma\rangle\langle J,\gamma|\ \mathrm{d}\mu_\rho(J,\gamma) := \lim_{\Gamma\to\infty}\frac{1}{2\Gamma}\int_{-\Gamma}^{\Gamma}\mathrm{d}\gamma\int_0^L |J,\gamma\rangle\langle J,\gamma|\ \mathcal{N}(J)w_\rho(J)\ \mathrm{d}J,$$

provided the following moment problem is satisfied:

$$\varepsilon_n! = \int_0^L J^n\, w_\rho(J)\ \mathrm{d}J, \quad w_\rho(J) \geqslant 0.$$

Next we exhibit two explicit examples of Gazeau–Klauder CS.

1. Coulomb-like discrete spectrum:

$$\varepsilon_n = 1 - \frac{1}{(n+1)^2}, \text{ for which } \varepsilon_n! = \frac{1}{2}\frac{n+2}{n+1}.$$

The distribution probability $w(J)$ is given by

$$\varepsilon_n! = \int_0^1 J^n w(J)\ \mathrm{d}J \text{ where } w(J) = \frac{1}{J}(1+\delta(J-1^-)).$$

2. Infinite square well and Pöschl–Teller potentials:

Consider the following Hamiltonian for a particle on the line:

$$H = -\frac{1}{2m}\frac{\mathrm{d}^2}{\mathrm{d}x^2} + V(x) - \frac{\lambda^2}{2m},$$

where

$$V(x) = \begin{cases} \dfrac{4\lambda(\lambda-1)}{\sin^2 x}, & 0 < x < \pi, \\ \infty, & x \leqslant 0,\ x \geqslant \pi, \end{cases}$$

For $\lambda = 1$, this yields the infinite square well and for $\lambda > 1$ the symmetric Pöschl–Teller potential. For these two cases, everything is computable explicitly: $\varepsilon_n = n(n + 2\lambda)$; $\omega = \frac{1}{2m}$; $R = \infty$; $w(J)$; the resolution of the identity.

(2) *CS for continuum dynamics*

Take now a Hamiltonian with positive, nondegenerate, purely continuous spectrum:

$$H|\mathsf{E}\rangle = \mathsf{E}|\mathsf{E}\rangle, \quad 0 \leqslant \mathsf{E} < \overline{\mathsf{E}}.$$

The corresponding normalized CS are

$$|J,\gamma\rangle = \frac{1}{\sqrt{\mathcal{N}_\rho(J)}} \int_0^{\overline{\mathsf{E}}} \frac{J^{\mathsf{E}/2}}{\sqrt{\rho(\mathsf{E})}} e^{-i\gamma\mathsf{E}} |\mathsf{E}\rangle \, \mathrm{d}\mathsf{E},$$

with the normalization condition

$$\mathcal{N}_\rho(J) = \int_0^{\overline{\mathsf{E}}} \frac{J^{\mathsf{E}}}{\rho(\mathsf{E})} \, \mathrm{d}\mathsf{E} < \infty, \text{ for } 0 \leqslant J < R.$$

For these CS, the continuity and temporal stability properties are satisfied and there is a nonnegative probability density $w_\rho(J)$ such that $\int_0^R J^{\mathsf{E}} w_\rho(J) \, \mathrm{d}J = \rho(\mathsf{E})$, from which follows the resolution of the identity

$$I = \int |J,\gamma\rangle\langle J,\gamma| \, \mathrm{d}\mu_\rho(J,\gamma) := \int_{-\infty}^{\infty} \mathrm{d}\gamma \int_0^R |J,\gamma\rangle\langle J,\gamma| \, \mathcal{N}_\rho(J) w_\rho(J) \, \mathrm{d}J.$$

(3) *CS for discrete and continuum dynamics*

For a Hamiltonian having both discrete and continuous spectra, $H = H_D \oplus H_C$, one simply combines the two previous cases.

4.3. More exotic CS

All the coherent states we have encountered so far live in complex Hilbert spaces. However the concept has been extended to a number of more exotic situations. We may mention:

(1) *Vector CS over matrix domains*

The algebraic definition (4.1) may be generalized to the case where the complex variable z is replaced by an $n \times n$ matrix $\mathcal{Z}$, taking its value in an appropriate domain, chosen in such a way that a resolution of the identity holds true [28]. In addition, the Hilbert space is taken of the form $\mathbb{C}^n \otimes \mathcal{H}$, which leads to the so-called Vector CS (VCS). This approach generalizes canonical CS, including their quaternionic version.

(2) *CS over quaternionic Hilbert spaces*

A quaternionic Hilbert space is a Hilbert space where the scalars are quaternions instead of complex numbers. Since quaternions are noncommutative, one has to distinguish between left and right quaternionic Hilbert spaces. In such a framework, the whole CS machinery may be developed, following essentially the same procedure as in the case of matrix-valued VCS described above [29].

(3) *CS on Hilbert modules*

First we may rewrite (4.1) in the following abstract form (this anticipates (5.1) below):

$$|z\rangle = \sum_{n=0}^{\infty} \phi_n(z)\, e_n\,,$$

where $\phi_n \in L^2(\mathcal{D})$ and $\{e_n\}$ is an orthonormal basis in the Hilbert space $\mathcal{K}$. Take now two unital C*-algebras $\mathcal{A}, \mathcal{B}$ and a Hilbert C*-module $\mathcal{E}$ over $\mathcal{B}$, with left action from $\mathcal{A}$ [23]. Next one defines the space $\mathcal{H} = \mathcal{E} \otimes \mathcal{K}$, which is again a Hilbert C*-module under the right action of $\mathcal{B}$. Then one chooses an orthonormal basis $\{F_n\}$ in $L^2(X,\, \mathrm{d}\mu; \mathcal{E})$, the space of square integrable, $\mathcal{E}$-valued functions on the measure space (X, μ) and a basis $\{\phi_n\}$ in the Hilbert space $\mathcal{K}$. Putting all together, one defines the so-called *module-valued CS* as

$$|x, a\rangle = \sum_n aF_n(x) \otimes \phi_n, \text{ where } a \in \mathcal{A} \text{ satisfies } aa^* = I_{\mathcal{A}}.$$

These CS $|x, a\rangle \in \mathcal{H}$ then verify all the expected properties, including the resolution of the identity [4]. They also generalize the VCS. A variant of this construction also leads to the Cuntz algebras.

5. Probabilistic aspects, quantization

CS may be defined for a general system in the following way. Take as "observable" space a measure space (X, μ), on which observables are defined as (generalized) functions $f \in L^2(X,\, \mathrm{d}\mu)$. Assume there is an orthonormal basis $\{\phi_n\}$ in $L^2(X,\, \mathrm{d}\mu)$ such that $0 < \mathcal{N}(x) := \sum_n |\phi_n(x)|^2 < \infty$ (a.e.). Then consider an abstract Hilbert space $\mathcal{H}$ (the space of quantum states) with orthonormal basis $\{|e_n\rangle\}$ in one-to-one correspondence $|e_n\rangle \leftrightarrow \phi_n$ with the previous basis. In this setup, CS may be defined as

$$X \ni x \mapsto |x\rangle = \frac{1}{\sqrt{\mathcal{N}(x)}} \sum_n \overline{\phi_n(x)} |e_n\rangle \in \mathcal{H}, \tag{5.1}$$

where one imposes

$$\langle x|x\rangle = 1\,, \qquad \int_X |x\rangle\langle x|\, \mathcal{N}(x)\, \mathrm{d}\mu(x) = I_{\mathcal{H}}.$$

Then the set $\{|x\rangle,\, x \in X\}$ may be interpreted as a *frame* on X, that selects a certain point of view on X. Next, the CS (5.1) determine two probability distributions, in Bayesian duality:

- A *prior* distribution on the set of indices, $n \mapsto |\phi_n(x)|^2/\mathcal{N}(x)$.
- A *posterior* distribution on the set of parameters, i.e., the classical measure space (X, μ), $x \mapsto |\phi_n(x)|^2$, parameterized by n.

These probabilistic aspects will play a key role in the quantization procedure, that is, the construction of a quantum system from a classical one. Take again for classical observable a (generalized) function $f(x)$. Typically, X is a phase space and $x = (q,p)$. Then, by *quantization*, we mean a map $f(x) \mapsto A_f$, where A_f is an operator on some (quantum) Hilbert space, satisfying a number of consistency conditions:

- $1 \mapsto I$,
- Poisson bracket $\{f,g\} \mapsto$ commutator $[A_f, A_g]$.

However, this procedure is plagued by a number of problems, such as noncommutativity of operator algebras, domain problems, consistency, ...

A consistent procedure, that avoids these problems, is the so-called *CS or Berezin quantization*, which is defined as follows:

$$f(x) \mapsto A_f := \int_X |x\rangle\langle x|\, f(x) \mathcal{N}(x)\, \mathrm{d}\mu(x),$$

where f may be a smooth function, $f \in L^2$ or a distribution. With this definition, $\widehat{A_f} := f(x)$ is called the *contravariant* or the *upper symbol* of A_f and $\check{f}(x) := \langle x|A_f|x\rangle$ the *covariant* or the *lower symbol* of f. For testing the validity of the method, one has to compare $\check{f}(x) \simeq f(x)$, at least in some limit, for instance $\hbar \to 0$, corresponding to the transition quantum $\to$ classical.

For a thorough study of this and other methods of quantization, we refer to the review [1], as well as [19] or [3, Chap. 11]. In particular, the connection between quantization, complex Hermite polynomials and reproducing kernel Hilbert spaces is explored in [5].

6. Remarks on signal processing: Time-frequency representation

The sequel of this paper will be devoted to wavelets, which are in fact a special case of CS, namely those associated to the affine group of the line. However, instead of proceeding along the same lines as in the preceding sections, we will start from scratch, with some general remarks on signal processing. Actually this means that we follow the chronological order in the development of wavelet theory. For more details, we refer to the classical treatises of Daubechies [17] or Mallat [24], and also to our own textbooks [3, 10].

The aim of signal processing is to transform raw signals, as given by receivers of some sort, in such a way that they can be analyzed, compressed and retransmitted. The traditional tool to that effect is simply the Fourier transform

$$s(x) \quad \longleftrightarrow \quad \widehat{s}(\xi) = \frac{1}{\sqrt{2\pi}} \int_{-\infty}^{\infty} e^{-i\xi x} s(x)\, \mathrm{d}x. \tag{6.1}$$

However, Fourier analysis is not sufficient. On one hand, time localization is totally lost: when does the $\widehat{s}(\xi)$ component occur? On the other hand, it is very *unstable*: a tiny perturbation of the signal, such as a Dirac δ, perturbs completely the Fourier spectrum. The reason, of course, is that the Fourier transform is global.

FIGURE 1. A traditional time-frequency representation of a signal (from Mozart's Don Giovanni, Act 1).

Therefore, signal analysts turn to *time-frequency* (TF) representations. The idea is that one needs *two* parameters. One, called a, characterizes the frequency, the other one, b, indicates the position in the signal. This concept of a TF representation is in fact quite old and familiar. The most obvious example is simply a musical score (Fig. 1)!

If one requires, in addition, the transform to be *linear*, a general TF transform will take the form:

$$s(x) \mapsto S(b,a) = \int_{-\infty}^{\infty} \overline{\psi_{b,a}(x)}\, s(x)\, dx, \tag{6.2}$$

where s is the signal and $\psi_{b,a}$ the analyzing function (we denote the time variable by x, in view of the extension to higher dimensions). The assumption of linearity is actually nontrivial, for there exists a whole class of quadratic or, more properly, sesquilinear time-frequency representations. The prototype is the so-called Wigner–Ville transform, introduced originally by E.P. Wigner [31] in quantum mechanics (in 1932!) and extended by J. Ville [30] to signal analysis.

Within the class of linear TF transforms, two stand out as particularly simple and efficient, the Windowed or Short-Time Fourier Transform (STFT)[2] and the wavelet transform (WT). For both of them, the analyzing function $\psi_{b,a}$ is obtained by a group action on a basic (or mother) function ψ, only the group differs. The essential difference between the two is in the way the frequency parameter a is introduced. In both cases, b is simply a time translation. The kernels of the two transforms can be written as follows.

- For the *Windowed Fourier Transform*, one chooses

$$\psi_{b,a}(t) = e^{it/a}\, \psi(t-b). \tag{6.3}$$

  Here ψ is a window function and the a-dependence is a modulation ($1/a \sim$ frequency); the window has constant width, but the lower a, the larger the number of oscillations in the window. The associated group is the Weyl–Heisenberg group G_{WH} Thus, Gabor analysis amounts to an expansion in terms of *canonical CS*!

- For the *wavelet transform*, instead, one takes

$$\psi_{b,a}(t) = \frac{1}{\sqrt{a}}\, \psi\Big(\frac{t-b}{a}\Big). \tag{6.4}$$

[2]Often called the *Gabor transform*, although Gabor considered only a discretized version [18].

The action of a on the function ψ is a dilation ($a > 1$) or a contraction ($a < 1$). The shape of the function is unchanged, it is simply spread out or squeezed. In particular, the effective support of $\psi_{b,a}$ varies as a function of a. Here the associated group is the $ax + b$ group, the affine group of the line. Thus wavelet analysis follows the pattern of general CS theory.

7. The continuous wavelet transform in 1-D

7.1. Basic formulas, interpretation

As announced above, the basic formulas read, in time domain and in frequency domain, respectively,

$$S(b,a) := \langle \psi_{b,a} | s \rangle = |a|^{-1/2} \int_{-\infty}^{\infty} \overline{\psi\left(a^{-1}(x-b)\right)}\, s(x)\, \mathrm{d}x \tag{7.1}$$

$$= |a|^{1/2} \int_{-\infty}^{\infty} \overline{\widehat{\psi}(a\xi)}\, \widehat{s}(\xi)\, e^{i\xi b}\, \mathrm{d}\xi, \tag{7.2}$$

where $a \neq 0$ is a scale parameter, $b \in \mathbb{R}$ is a translation parameter and the hat denotes a Fourier transform. Thus the pair (b, a) runs over the time-scale half-plane $\mathbb{R}^2_+$. In practice one generally uses the restriction $a > 0$, as in (6.4), which is physically more natural.

In these relations, s is a square integrable function and the analyzing wavelet ψ is assumed to be well localized *both* in the space (or time) domain and in the frequency domain. Here the minimal requirement is that $\psi \in L^1 \cap L^2$, but in practice stronger conditions are usually imposed (like $\psi \in \mathcal{S}$, Schwartz' space of fast decreasing functions). In consequence, the CWT provides good bandpass filtering both in x *and* in ξ.

Moreover, ψ must satisfy the following admissibility condition, which guarantees the invertibility of the WT (see (7.14) below):

$$c_\psi := 2\pi \int_{-\infty}^{\infty} \frac{|\widehat{\psi}(\xi)|^2}{|\xi|}\, \mathrm{d}\xi < \infty. \tag{7.3}$$

In most cases, this condition may be reduced to the requirement that ψ has zero mean (the condition is necessary, and becomes sufficient upon a slight restriction on ψ):

$$\widehat{\psi}(0) = 0 \iff \int_{-\infty}^{\infty} \psi(x)\, \mathrm{d}x = 0. \tag{7.4}$$

In addition, ψ is often required to have a certain number of *vanishing moments:*

$$\int_{-\infty}^{\infty} x^n\, \psi(x)\, \mathrm{d}x = 0,\ n = 0, 1, \ldots, N. \tag{7.5}$$

This property improves the efficiency of ψ at detecting singularities in the signal, since it is blind to polynomials up to order N, which constitute the smoother, and less informative, part of a signal (think of a Taylor expansion).

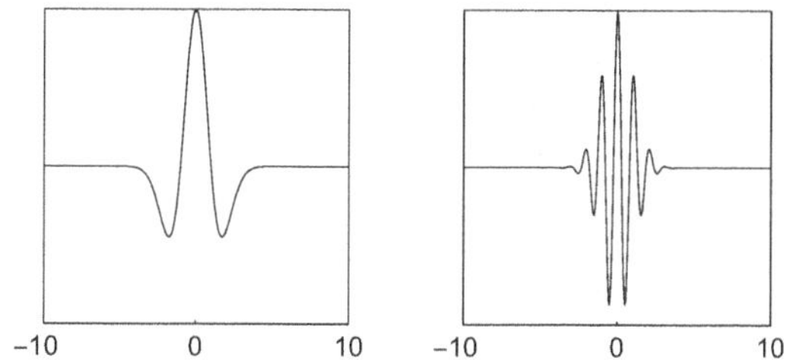

FIGURE 2. (left) The Mexican hat or Marr wavelet; (right) The real part of the Morlet wavelet, for $\xi_o = 5.6$.

In order to fix ideas, we exhibit here two commonly used, and well-documented, wavelets (see Fig. 2).

1. *The Mexican hat or Marr wavelet*
   This wavelet is simply the second derivative of a Gaussian:
   $$\psi_{\rm H}(x) = (1 - x^2)\, e^{-x^2/2}, \quad \widehat{\psi}_{\rm H}(\xi) = \xi^2\, e^{-\xi^2/2}. \tag{7.6}$$
   It is real and admissible, it has 2 vanishing moments $n = 0, 1$.
2. *The Morlet wavelet*
   This wavelet is essentially a plane wave within a Gaussian window:
   $$\psi_{\rm M}(x) = e^{ik_0 x}\, e^{-x^2/2} + h(x), \quad \widehat{\psi}_{\rm M}(\xi) = e^{-(\xi-\xi_0)^2/2} + \widehat{h}(\xi). \tag{7.7}$$
   Here the correction term h must be added in order to satisfy the admissibility condition (7.4), but in practice one will arrange that this term be numerically negligible ($\leqslant 10^{-4}$) and thus can be omitted (it suffices to choose the basic frequency $|\xi_0|$ large enough, namely, one has to take $|\xi_0| > 5.5$).

These two wavelets have very different properties and, naturally, they will be used in quite different situations. Typically, the Mexican hat is sensitive to singularities in the signal, and it yields a genuine time-scale analysis. On the other hand, since the Morlet wavelet is complex, it will catch the phase of the signal, hence it will be sensitive to frequencies, and will lead to a time-frequency analysis, somewhat closer to a Gabor analysis. In both cases, additional flexibility is obtained by adding a width parameter to the Gaussian.

We must now make more precise the localization conditions on the wavelet ψ. It turns out that, for large scales ($a \gg 1$), the CWT is sensitive to *low frequencies*, and thus yields a rough analysis. On the contrary, for very small scales ($a \ll 1$), the CWT is sensitive to *high frequencies*, that is, small details.

Combining now these localization properties with the fact that the CWT is a convolution with a zero mean filter, we conclude that the CWT provides a local filtering in time (b) and scale (a):
$$S(b,a) \not\approx 0 \quad \Longleftrightarrow \quad \psi_{b,a}(x) \approx s(x).$$

Thus it may be interpreted as a mathematical microscope, with optics ψ, position b, and magnification $1/a$, so that one may consider the CWT as a *singularity detector*. In addition, thanks to the covariance property under scaling (see (7.11) below), the CWT can measure the strength of a singularity, hence it is also a *singularity analyzer*.

7.2. The group-theoretical background

Combining dilations and translations, one gets *affine* transformations of the line

$$y = (b, a)x \equiv ax + b, \quad a \neq 0,\ b \in \mathbb{R},\ x \in \mathbb{R}.$$

Thus $\{(b, a)\} =: G_{\text{aff}} = \mathbb{R} \rtimes \mathbb{R}_*$, the *affine group* of the line.

The action of (b, a) on a signal ψ reads as

$$(U(b, a)\psi)(x) = |a|^{-1/2}\psi\left(\frac{x-b}{a}\right), \tag{7.8}$$

and U is a unitary irreducible representation of G_{aff} in $L^2(\mathbb{R}, \mathrm{d}x)$ (unique up to unitary equivalence). Moreover, U is *square integrable* and a function ψ is admissible if and only if

$$\iint_{G_{\text{aff}}} |\langle U(b, a)\psi|\psi\rangle|^2 \, \frac{\mathrm{d}b\, \mathrm{d}a}{a^2} = c_\psi \, \|\psi\|^2 < \infty, \tag{7.9}$$

where c_ψ is given in (7.3). Restricting to $a > 0$, one gets the *connected* affine group G^+_{aff} (or $ax + b$ group) and U is a UIR of it in $L^2(\mathbb{R}^+, \mathrm{d}x)$.

7.3. Mathematical properties

Given an admissible wavelet ψ, the CWT $W_\psi : s(x) \mapsto S(b, a)$ is the CS map associated with the representation U of the $ax+b$ group given in (7.8) (with $a > 0$). Hence, it enjoys all properties of CS maps described in the previous sections.

(1) *Covariance* under translation and dilation:

$$W_\psi : s(x - x_0) \mapsto S(b - x_0, a) \tag{7.10}$$

$$W_\psi : \frac{1}{\sqrt{a_0}}\, s\Big(\frac{x}{a_0}\Big) \mapsto S\Big(\frac{b}{a_0}, \frac{a}{a_0}\Big). \tag{7.11}$$

(2) *Energy conservation*:

$$\int_{-\infty}^{\infty} |s(x)|^2 \, \mathrm{d}x = c_\psi^{-1} \iint_{\mathbb{R}^2_+} |S(b, a)|^2 \frac{\mathrm{d}b\, \mathrm{d}a}{a^2}. \tag{7.12}$$

Thus, $|S(b, a)|^2$ may be interpreted as the energy density in the time-scale half-plane $\mathbb{R}^2_+$.

The relation (7.12) means that W_ψ is an *isometry* from the space of signals $L^2(\mathbb{R}, \mathrm{d}x)$ onto a *closed* subspace $\mathcal{H}_\psi$ of $L^2(\mathbb{R}^2_+, \mathrm{d}b\, \mathrm{d}a/a^2)$, namely, the space of wavelet transforms. An equivalent statement is that the wavelet ψ generates a *resolution of the identity* (weak integral, as usual):

$$c_\psi^{-1} \iint_{\mathbb{R}^2_+} |\psi_{b,a}\rangle \langle \psi_{b,a}| \frac{db\, da}{a^2} = I. \tag{7.13}$$

(3) *Reconstruction formula*: As a consequence, W_ψ is invertible on its range $\mathcal{H}_\psi$ by the adjoint map, thus we obtain an exact reconstruction formula:

$$s(x) = c_\psi^{-1} \iint_{\mathbb{R}^2_+} \psi_{b,a}(x)\, S(b,a)\, \frac{db\, da}{a^2}. \tag{7.14}$$

In other words, we have a representation of the signal $s(x)$ as a linear superposition of wavelets $\psi_{b,a}$ with coefficients $S(b,a)$.

(4) The projection $P_\psi : L^2(\mathbb{R}^2_+, db\, da/a^2) \to \mathcal{H}_\psi$ is an integral operator, with kernel

$$K(b', a'; b, a) = c_\psi^{-1} \langle \psi_{b',a'} | \psi_{b,a} \rangle. \tag{7.15}$$

The kernel K is the autocorrelation function of ψ and it is a *reproducing kernel.* Indeed, the function $f \in L^2(\mathbb{R}^2_+, db\, da/a^2)$ is the WT of a certain signal if and only if it satisfies the *reproduction property*

$$f(b', a) = c_\psi^{-1} \iint_{\mathbb{R}^2_+} \langle \psi_{b',a'} | \psi_{b,a} \rangle\, f(b,a)\, \frac{db\, da}{a^2}\,. \tag{7.16}$$

This means that the CWT is a highly redundant representation!

The last statement implies that the full information must be contained in a small subset of the half-plane. This property may be exploited in two ways: either one takes a discrete subset, which leads to the theory of *frames*, or one considers only the lines of local maxima, called *ridges*.

8. Discrete wavelet transforms in 1-D

8.1. Discretization of the CWT

Of course, the CWT must be discretized for numerical implementation. This raises the question of choosing an adequate sampling grid. Typically, one says that a discrete lattice $\Gamma = \{b_{jk}, a_j, j, k \in \mathbb{Z}\} \subset \mathbb{R}^2_+$ yields a good discretization if the following *exact* relation holds:

$$s = \sum_{j,k \in \mathbb{Z}} \langle \psi_{jk} | s \rangle \widetilde{\psi}_{jk}, \tag{8.1}$$

with $\psi_{jk} := \psi_{b_{jk},a_j}$ and $\widetilde{\psi}_{jk}$ is a function explicitly constructible from ψ_{jk}. However, this approach in general leads to *frames*, not bases. Let us recall the basic definition (adapted to the present case). A frame in the Hilbert space $\mathcal{H}$ is a family $\{\psi_{jk}\}$ of vectors for which there exist two constants $\mathsf{m} > 0, \mathsf{M} < \infty$ such that

$$\mathsf{m}\, \|s\|^2 \leqslant \sum_{j,k \in \mathbb{Z}} |\langle \psi_{jk} | s \rangle|^2 \leqslant \mathsf{M}\, \|s\|^2, \ \forall\, s \in \mathcal{H}.$$

The upper bound means that the map $s \mapsto \{\langle\psi_{jk}|s\rangle\}$ is continuous from $\mathcal{H}$ to l^2, whereas the lower bound guarantees the numerical stability of the inverse map [17]. The constants m, M are called the frame bounds. If $\mathsf{m} = \mathsf{M} \neq 1$, the frame is said to be tight. Clearly, if $\mathsf{m} = \mathsf{M} = 1$ and $\|\psi_{jk}\| = 1$, for all $j, k \in \mathbb{Z}$, this reduces to an orthonormal basis.

The main question now is, given a wavelet ψ, to find a lattice Γ such that $\{\psi_{jk}\}$ is a good frame. A detailed analysis then shows that the expansion (8.1) converges essentially as a power series in the quantity $|\mathsf{M}/\mathsf{m} - 1|$, thus a good lattice must satisfy $|\mathsf{M}/\mathsf{m} - 1| \ll 1$. The key to the solution is to choose a lattice adapted to the geometry, for instance the *dyadic* grid $a_j = 2^{-j}$, $b_{jk} = k \cdot 2^{-j}$, which leads to

$$\psi_{jk}(x) = 2^{j/2}\psi(2^j\, x - k), \quad j, k \in \mathbb{Z}.$$

For example, both the Mexican hat and the Morlet wavelets give good, but non-tight frames.

8.2. The discrete WT (DWT)

For many applications, such as the description and analysis of signals, a tight frame is as good as an orthonormal basis. Yet, a tight frame is still redundant, which implies that the coefficients are not unique, nor statistically independent. Therefore, when it comes to managing large amounts of data and compressing them, an orthonormal basis becomes mandatory, and this requires a different approach (almost orthogonal to the previous one).

One of the successes of the WT was the discovery that it is possible to construct functions ψ for which $\{\psi_{jk}, j, k \in \mathbb{Z}\}$ is indeed an orthonormal basis of $L^2(\mathbb{R})$, while keeping the good properties of wavelets, including space *and* frequency localization. In addition, this approach yields fast algorithms, and this is the key to the usefulness of wavelets in many applications.

The construction is based on two facts. First, almost all examples of orthonormal bases of wavelets can be derived from a multiresolution analysis, and then the whole construction may be transcripted into the language of Quadrature Mirror Filters (QMF), familiar in the signal processing literature.

A *multiresolution analysis* of $L^2(\mathbb{R})$ is an increasing sequence of closed subspaces

$$\cdots \subset V_{-2} \subset V_{-1} \subset V_0 \subset V_1 \subset V_2 \subset \cdots$$

with $\bigcap_{j\in\mathbb{Z}} V_j = \{0\}$ and $\bigcup_{j\in\mathbb{Z}} V_j$ dense in $L^2(\mathbb{R})$, and such that

(1) $f(x) \in V_j \Leftrightarrow f(2x) \in V_{j+1}$;
(2) There exists a function $\phi \in V_0$, called a *scaling function*, such that the family $\{\phi(x-k), k \in \mathbb{Z}\}$ is an orthonormal basis of V_0.

Combining conditions (1) and (2), one sees that $\{\phi_{jk}(x) := 2^{j/2}\phi(2^j x - k),\, k \in \mathbb{Z}\}$ is an orthonormal basis of V_j. The space V_j can be interpreted as an *approximation* space at resolution 2^j. Defining now

$$V_j \oplus W_j = V_{j+1}, \tag{8.2}$$

we see that W_j contains the additional *details* needed to improve the resolution from 2^j to 2^{j+1}. Thus one gets the decomposition

$$L^2(\mathbb{R}) = \bigoplus_{j\in\mathbb{Z}} W_j = V_{j_0} \oplus \left(\bigoplus_{j=j_0}^{\infty} W_j \right), \tag{8.3}$$

if one chooses a lowest resolution level j_0. In practice, one starts from a sampled signal, taken in some V_J, and then the decomposition (8.3) is replaced by the finite representation

$$V_J = V_{j_o} \oplus \left(\bigoplus_{j=j_0}^{J-1} W_j \right). \tag{8.4}$$

The crucial theorem then asserts the existence of a function ψ, sometimes called the *mother wavelet*, explicitly computable from ϕ, such that $\{\psi_{jk}(x) := 2^{j/2}\psi(2^j x - k), k \in \mathbb{Z}\}$ is an orthonormal basis of W_j. Then $\{\psi_{jk}, j, k \in \mathbb{Z}\}$ constitutes an orthonormal basis of $L^2(\mathbb{R})$: these are the *orthonormal wavelets.* Well-known examples include the Haar wavelets, the B-splines, and the various Daubechies wavelets.

Now a natural question is to decide which version should be used in a concrete case, the (discretized) CWT or the DWT? As usual in a wavelet context, the answer depends on the application at hand:

- For feature detection, the CWT is preferable, since no *a priori* choice is made for a, b; the CWT is more flexible and also more robust to noise, but only frames will be available in general.
- For large amount of data or data compression, one should use the DWT; it yields orthonormal bases, it is faster, but it is also more rigid.

This last aspect explains why a number of generalizations have been introduced in order to alleviate these drawbacks of the DWT, such as biorthogonal wavelets, wavelet packets, continuous wavelet packets (integrated wavelets), redundant WT (on a rectangular lattice) or "Second generation" wavelets (the so-called lifting scheme). For all these, we refer to the textbooks [10, 17, 24].

9. Wavelet analysis of 2-D images

9.1. Basic formulas and interpretation

Now we turn to the two-dimensional CWT, which has become a major tool in image processing. In this context, an image is a two-dimensional signal of finite energy, represented by a complex-valued function $s \in L^2(\mathbb{R}^2, \mathrm{d}^2\vec{x})$. Given an image, all the geometric operations we want to apply to it are obtained by combining three elementary transformations, namely, rigid translations in the plane of the image, dilations or scaling (global zooming in and out) and rotations. Explicitly, the transformations act on $\vec{x} \in \mathbb{R}^2$ in the familiar way:

(i) translation by $\vec{b} \in \mathbb{R}^2 : \vec{x} \mapsto \vec{x}\,' = \vec{x} + \vec{b}$,
(ii) dilation by a factor $a > 0 : \vec{x} \mapsto \vec{x}\,' = a\vec{x}$,
(iii) rotation by an angle $\theta : \vec{x} \mapsto \vec{x}\,' = r_\theta(\vec{x})$,

where

$$r_\theta \equiv \begin{pmatrix} \cos\theta & -\sin\theta \\ \sin\theta & \cos\theta \end{pmatrix}, \; 0 \leqslant \theta < 2\pi,$$

is the familiar 2×2 rotation matrix.

Then dilations, translations, and rotations together constitute the *similitude group* of the plane $\mathrm{SIM}(2) = \mathbb{R}^2 \rtimes (\mathbb{R}_*^+ \times \mathrm{SO}(2))$, with action $\vec{y} = (\vec{b}, a, \theta)\vec{x} = ar_\theta\vec{x} + \vec{b}$. This transformation is realized by the following unitary map on finite energy signals:

$$\big[U(\vec{b}, a, \theta)s\big](\vec{x}) = s_{\vec{b},a,\theta}(\vec{x}) := a^{-1} s(a^{-1} r_{-\theta}(\vec{x} - \vec{b})) \tag{9.1}$$

and U is a unitary irreducible representation of SIM(2) in $L^2(\mathbb{R}^2, \mathrm{d}^2\vec{x})$, unique up to unitary equivalence. In addition, U is *square integrable* and a function ψ is admissible if and only if

$$\iiint_{\mathrm{SIM}(2)} \left|\langle U(\vec{b}, a, \theta)\psi|\psi\rangle\right|^2 \, \mathrm{d}^2\vec{b}\, \frac{\mathrm{d}a}{a^3}\, \mathrm{d}\theta \; < \infty, \tag{9.2}$$

where $a^{-3}\,\mathrm{d}^2\vec{b}\,\mathrm{d}a\,\mathrm{d}\theta$ is the left Haar measure on SIM(2).

In terms of this action, the basic formulas for the 2-D CWT read

$$S(\vec{b}, a, \theta) := \langle \psi_{\vec{b},a,\theta}|s\rangle = a^{-1} \int_{\mathbb{R}^2} \overline{\psi(a^{-1} r_{-\theta}(\vec{x} - \vec{b}))}\, s(\vec{x}) \; \mathrm{d}^2\vec{x} \tag{9.3}$$

$$= a \int_{\mathbb{R}^2} e^{i\vec{b}\cdot\vec{k}}\, \overline{\widehat{\psi}(ar_{-\theta}(\vec{k}))}\, \widehat{s}(\vec{k}) \; \mathrm{d}^2\vec{k}. \tag{9.4}$$

As in 1-D, besides the usual conditions of localization, we have to impose an *admissibility* condition on the wavelet ψ, namely,

$$c_\psi := (2\pi)^2 \int_{\mathbb{R}^2} \frac{|\widehat{\psi}(\vec{k})|^2}{|\vec{k}|^2} \; \mathrm{d}^2\vec{k} < \infty, \tag{9.5}$$

which again may be replaced in practice by the following necessary condition:

$$\widehat{\psi}(\vec{0}) = 0 \iff \int_{\mathbb{R}^2} \psi(\vec{x})\, \mathrm{d}^2\vec{x} = 0. \tag{9.6}$$

Besides the previous requirements, one can improve the efficiency of the wavelets by imposing additional properties, such as restrictions on the support of ψ and of $\widehat{\psi}$, or *vanishing moments*, up to order $N \geqslant 1$ ($N = 0$ yields the admissibility condition):

$$\int_{\mathbb{R}^2} x^\alpha\, y^\beta\, \psi(\vec{x}) \; \mathrm{d}^2\vec{x} \;= 0, \quad \vec{x} = (x, y), \quad 0 \leqslant \alpha + \beta \leqslant N.$$

This will improve the efficiency at detecting singularities in the signal, since the transform is then blind to the smoothest part of the signal, i.e., a polynomial of degree up to N, which contains little information, in general. For instance, if $N = 1$, the transform erases any linear *trend* in the signal, such as a linear gradient of luminosity.

Altogether, the mathematical properties of the 2-D CWT are essentially the same as in the 1-D case, so we list them without further comment.

(1) *Energy conservation*

$$c_{\psi}^{-1} \iiint_{\mathrm{SIM}(2)} |S(\vec{b}, a, \theta)|^2 \, \mathrm{d}^2\vec{b} \, \frac{\mathrm{d}a}{a^3} \, \mathrm{d}\theta = \int_{\mathbb{R}^2} |s(\vec{x})|^2 \, \mathrm{d}^2\vec{x}, \tag{9.7}$$

i.e., the 2-D CWT is an isometry from the space of signals $L^2(\mathbb{R}^2)$ onto a closed subspace of $L^2(\mathrm{SIM}(2))$, the space of wavelet transforms.

(2) *Reconstruction formula*

Inverting the CWT by the adjoint map, we get

$$s(\vec{x}) = c_{\psi}^{-1} \iiint_{\mathrm{SIM}(2)} \psi_{\vec{b},a,\theta}(\vec{x}) \, S(\vec{b}, a, \theta) \, \mathrm{d}^2\vec{b} \, \frac{\mathrm{d}a}{a^3} \, \mathrm{d}\theta, \tag{9.8}$$

i.e., we have a decomposition of the signal in terms of the analyzing wavelets $\psi_{\vec{b},a,\theta}$, with coefficients $S(\vec{b}, a, \theta)$.

(3) *Reproduction property (reproducing kernel)*

$$S(\vec{b}', a', \theta') = c_{\psi}^{-1} \iiint_{\mathrm{SIM}(2)} \langle \psi_{\vec{b}',a',\theta'} | \psi_{\vec{b},a,\theta} \rangle \, S(\vec{b}, a, \theta) \, \mathrm{d}^2\vec{b} \, \frac{\mathrm{d}a}{a^3} \, \mathrm{d}\theta. \tag{9.9}$$

(4) *Covariance*

The CWT is *covariant* under translations, dilations and rotations, that is, under SIM(2): the correspondence $W_{\psi} : s(\vec{x}) \mapsto S(\vec{b}, a, \theta)$ implies the following ones:

$$\begin{aligned} s(\vec{x} - \vec{b}_o) &\mapsto S(\vec{b} - \vec{b}_o, a, \theta), \\ a_o^{-1} s(a_o^{-1}\vec{x}) &\mapsto S(a_o^{-1}\vec{b}, a_o^{-1} a\theta), \\ s(r_{\theta_o}(\vec{x})) &\mapsto S(r_{-\theta_o}(\vec{b}), a, \theta - \theta_o). \end{aligned}$$

Note that translation covariance ("shift invariance") is lost in the standard formulation of the discrete WT, based on multiresolution, which creates problems in pattern recognition, for instance.

Since all the formulas are almost identical in 1-D and in 2-D, the interpretation of the CWT as a singularity analyzer still holds. In particular, the analysis is most efficient at high spatial frequencies or small scales, thus it provides a good detection of *discontinuities* in images, such as point singularities (contours, corners) or directional features (edges, segments, ...). Hence the 2-D CWT may be seen as a mathematical directional microscope with optics ψ, global magnification $1/a$, and orientation tuning parameter θ; it is a detector and analyzer of singularities.

As a matter of fact, the same analysis may be performed almost *verbatim* for n-dimensional signals and wavelets. The group is now the similitude group of $\mathbb{R}^n$, $\mathrm{SIM}(n) = \mathbb{R}^n \rtimes (\mathbb{R}_*^+ \times \mathrm{SO}(n))$, and the CWT is constructed in the same way as for $n = 2$, all formulas are basically the same. For instance, the admissibility condition (9.5) is essentially the same, the only difference is that the factor $|\vec{k}|^2$ in the denominator is replaced by $|\vec{k}|^n$. Therefore, the interpretation of this n-dimensional CWT will also remain the same. The case $n = 3$ may find applications in fluid dynamics, for example.

9.2. Choice of the analyzing wavelet

Even more than in 1-D, it is important to choose a wavelet that is well adapted to the problem at hand. One can distinguish two classes, isotropic wavelets and direction sensitive wavelets.

(i) *Isotropic wavelets*

If one decides to perform a pointwise analysis, or if directions are irrelevant, it is sufficient to use a rotation invariant wavelet. Two standard examples are:

1. *The 2-D Mexican hat wavelet*

   This wavelet (originally introduced by Marr in his pioneering work on vision [25]) is simply the Laplacian of a Gaussian:

$$\psi_{\mathrm{H}}(\vec{x}) = (2 - |\vec{x}|^2)\exp(-\tfrac{1}{2}|\vec{x}|^2), \quad \widehat{\psi_{\mathrm{H}}}(\vec{k}) = |\vec{k}|^2 \exp(-\tfrac{1}{2}|\vec{k}|^2). \tag{9.10}$$

2. *The Difference-of-Gaussians or DOG wavelet*

$$\psi_{\mathrm{D}}(\vec{x}) = \tfrac{1}{2\alpha^2}\exp(-\tfrac{1}{2\alpha^2}|\vec{x}|^2) - \exp(-\tfrac{1}{2}|\vec{x}|^2) \qquad (0 < \alpha < 1). \tag{9.11}$$

   This is a very good substitute for the Mexican hat, for $\alpha^{-1} = 1.6$, they are almost undistinguishable. It is widely used in psychophysics.

Here again, in both cases, the efficiency may be improved by adding a width parameter in the Gaussian.

(ii) *Directional wavelets*

On the other hand, if the goal is to detect directional features in an image, or to perform directional filtering, clearly one should resort to a direction sensitive wavelet. The most efficient solution is a *directional* wavelet, that is, a wavelet $\psi(\vec{x})$ such that the numerical support of its Fourier transform $\widehat{\psi}(\vec{k})$ is contained in a convex cone with apex at the origin. Two useful examples are as follows.

1. *The 2-D Morlet wavelet*

$$\widehat{\psi}_{\mathrm{M}}(\vec{k}) = \sqrt{\epsilon}\,\Big(\exp(-\tfrac{1}{2}|A^{-1}(\vec{k} - \vec{k}_0)|^2) - h(\vec{k})\Big), \tag{9.12}$$

   where $A = \mathrm{diag}[\epsilon^{-1/2}, 1], \epsilon \geqslant 1$, is a 2×2 anisotropy matrix and, as in 1-D, the correction term is required in order to satisfy the admissibility condition, and here too, it is negligible for $|\vec{k}_0| \geqslant 5.6$. The Morlet wavelet is directional, but has poor aperture selectivity. It is shown in Figure 3 for $\epsilon = 2, \vec{k}_0 = (0, 6)$.

2. *Conical wavelets*, with support in the convex cone

$$\mathcal{C}(-\alpha,\alpha) := \{\vec{k} \in \mathbb{R}^2 : -\alpha \leqslant \arg \vec{k} \leqslant \alpha,\ \alpha < \pi/2\}.$$

A very useful example is the Gaussian conical wavelet:

$$\widehat{\psi}^{\rm GC}_{lm} = \begin{cases} (\vec{k}\cdot\vec{e}_{-\tilde{\alpha}})^l (\vec{k}\cdot\vec{e}_{\tilde{\alpha}})^m \ \exp(-\frac{1}{2}k_x^2),\ \vec{k} \in \mathcal{C}(-\alpha,\alpha) \\ 0,\ \text{otherwise} \end{cases} \tag{9.13}$$

where $\vec{e}_\phi$ is the unit vector in the direction ϕ and $\tilde{\alpha} = -\alpha + \pi/2$, so that $\vec{e}_{-\alpha}\cdot\vec{e}_{\tilde{\alpha}} = \vec{e}_\alpha \cdot \vec{e}_{-\tilde{\alpha}} = 0$. The first two factors play the role of vanishing moments on the boundaries of the cone. The frequency space version of this so-called Gaussian conical wavelet is shown in Figure 3, in the case $l = m = 4$, $\alpha = 10°$. This wavelet will be used in Section 11 for motion estimation.

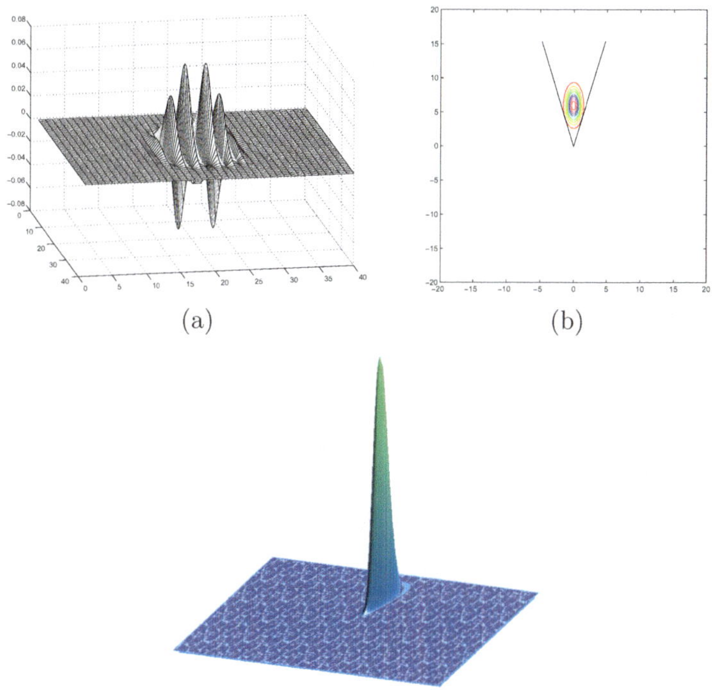

(a) (b) (c)

FIGURE 3. Two directional wavelets: The 2-D Morlet wavelet, (a) in position space and (b) in spatial frequency space; (c) The Gaussian conical wavelet, in spatial frequency space.

10. Extending the CWT to curved manifolds

10.1. Generalities

Many situations in physics yield data on non-flat manifolds, for instance:

- A *sphere:* geophysics, cosmology (CMB), statistics, ...;
- A *two-sheeted hyperboloid:* cosmology (an open expanding model of the universe), optics (catadioptric image processing, where a sensor overlooks a hyperbolic mirror);
- A *paraboloid:* optics (catadioptric image processing on a parabolic mirror);
- A *torus:* nuclear fusion (Tokamak), loop quantum gravity.

How can one find suitable analysis tools?

A possible solution is to extend the *continuous* wavelet transform. Translation of the wavelet may be achieved by an isometry of the manifold, i.e., an element of SO(3) or SO(1,2), for the two first cases above (although this does not work always, see below). As for dilations *on* the manifold, they have to be defined in each case. This being done, dilation controls locality of the CWT. However, in practice, the usual CWT works with discrete frames, hence one needs also discrete wavelet frames on the manifold. An alternative solution is to extend the *discrete* wavelet transform.

Let us look first at the general problem. Given a manifold $\mathcal{M}$, how to derive a CWT on it? First, one should identify the operations to be performed on the finite energy signals living on $\mathcal{M}$, i.e., functions in $L^2(\mathcal{M}, \mathrm{d}\mu)$, with μ a suitable measure on $\mathcal{M}$, namely,

(i) Motions are provided by the isometries of $\mathcal{M}$;
(ii) Dilations *on* $\mathcal{M}$ (zoom in/out) should form a one-parameter abelian group $A \sim \mathbb{R}_+$.

If $\mathcal{M}$ admits a global isometry group $G_{\rm iso}$, merge it with the dilation group A into a global group G. Next, find a unitary irreducible, square integrable representation U of G in $L^2(\mathcal{M}, \mathrm{d}\mu)$ and write (we assume that μ is G-invariant)

$$\psi_g(\zeta) := [U(g)\psi](\zeta) = \psi(g^{-1}\zeta),\ g \in G, \zeta \in \mathcal{M}.$$

Then a function $\psi \in L^2(\mathcal{M}, \mathrm{d}\mu)$ is an admissible wavelet if

$$c_\psi := \|\psi\|^{-2} \int_G |\langle U(g)\psi|\psi\rangle|^2 \, \mathrm{d}\mu_\ell(g) < \infty, \tag{10.1}$$

where μ_ℓ is the left Haar measure on G.

Then the CWT of $f \in L^2(\mathcal{M}, \mathrm{d}\mu)$ with respect to the admissible wavelet ψ is defined as

$$W_\psi f(g) := \langle \psi_g | f\rangle = \int_{\mathcal{M}} \overline{\psi(g^{-1}\zeta)}\, f(\zeta)\, \mathrm{d}\mu(\zeta),\ g \in G\,.$$

The corresponding *reconstruction formula* is then given by

$$f(\zeta) = c(\psi)^{-1/2} \int_G W_\psi f(g)\, \psi_g(\zeta)\, \mathrm{d}\mu_\ell(g)\,.$$

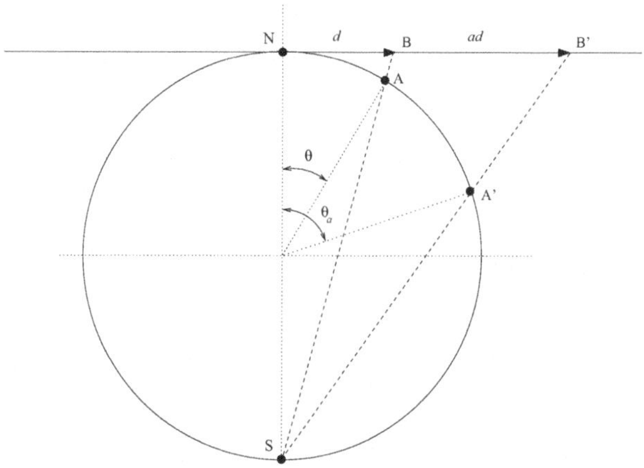

FIGURE 4. The stereographic dilation D_a : A $\mapsto$ A$'$ around the North Pole.

Note, however, that the method does not always work. Some manifolds (e.g., a paraboloid) do not have a global isometry group and, even if the latter exists, the resulting global group G need not have a square integrable UIR.

10.2. The WT on the two-sphere

This is the approach we have followed for the 1-D and the 2-D CWT. Now we illustrate the procedure by constructing a CWT on the two-sphere [7, 8, 9, 13], following the scheme developed in Section 3.3.

As usual, the origin of the spherical CWT consists in the affine transformations on $\mathbb{S}^2$, namely, motions, realized by rotations $\varrho \in \mathrm{SO}(3)$, and dilations. The problem is that they do *not* commute! Moreover, one cannot build a semidirect product of SO(3) and $\mathbb{R}_*^+$; the only extension of SO(3) by $\mathbb{R}_*^+$ is their *direct product.* A way out of this dilemma is to embed the two factors into the Lorentz group $\mathrm{SO}_o(3,1)$, by the Iwasawa decomposition

$$\mathrm{SO}_o(3,1) = \mathrm{SO}(3) \cdot A \cdot N,$$

where $A \sim \mathrm{SO}_o(1,1) \sim \mathbb{R} \sim \mathbb{R}_*^+$ consists of boosts in the z-direction and $N \sim \mathbb{C}$. The justification of this move is that the Lorentz group $\mathrm{SO}_o(3,1)$ is the *conformal* group both of the sphere $\mathbb{S}^2$ and of the tangent plane $\mathbb{R}^2$, say, at the North Pole.

The Lorentz group $\mathrm{SO}_o(3,1)$ acts transitively on $\mathbb{S}^2$. Then an explicit computation (with help of the Iwasawa decomposition) shows that a boost in z-direction corresponds to a pure dilation, namely, a *stereographic dilation* D_a, as shown in Figure 4. The dilation proceeds in three steps: (i) Project the initial point A stereographically onto the plane tangent at the North Pole N and get B; (ii) Dilate B radially around N by a factor a, to B$'$; (iii) Project B$'$ back to the sphere and get A$'$. Then D_a : A $\mapsto$ A$'$ is the required dilation on the sphere.

The Lorentz group $\mathrm{SO}_o(3,1)$ has a natural UIR in the Hilbert space $L^2(\mathbb{S}^2, \mathrm{d}\mu)$, where $\mathrm{d}\mu := \sin\theta\, \mathrm{d}\theta\, \mathrm{d}\varphi$ is the usual measure on the sphere, viz.

$$[U(g)f](\omega) = \lambda(g,\omega)^{1/2} f\left(g^{-1}\omega\right), \; g \in \mathrm{SO}_o(3,1), \; f \in L^2(\mathbb{S}^2, \mathrm{d}\mu),$$

where $\lambda(g,\omega)$ is a Radon–Nikodym derivative, required by the noninvariance of the measure μ under dilation.

Now the parameter space of the spherical CWT is $X := \mathrm{SO}_o(3,1)/N \simeq \mathrm{SO}(3) \cdot \mathbb{R}_*^+$. Thus we introduce a section $\sigma : X \to \mathrm{SO}_o(3,1)$ and consider the reduced representation $U(\sigma(\varrho, a))$. The natural (Iwasawa) section is $\sigma_I(\varrho, a) = \varrho\, a, \; \varrho \in \mathrm{SO}(3), \; a \in A$, and it yields

$$U(\sigma_I(\varrho, a)) = U(\varrho\, a) = U(\varrho)U(a) = R_\varrho\, D_a,$$

where R_ϱ is the unitary operator representing $\varrho \in \mathrm{SO}(3)$.

The UIR U is square integrable on X, that is, there exists nonzero (admissible) vectors $\psi \in L^2(\mathbb{S}^2, \mathrm{d}\mu)$ such that [$\mathrm{d}\varrho$ is the left Haar measure on SO(3)]

$$\int_X |\langle U(\sigma_I(\varrho, a))\psi|\phi\rangle|^2 \,\mathrm{d}\varrho\, \frac{\mathrm{d}a}{a^3} := \langle \phi | A_\psi \phi\rangle < \infty, \; \forall\, \phi \in L^2(\mathbb{S}^2, \mathrm{d}\mu)\,,$$

where the frame operator A_ψ is bounded, positive and invertible. Actually A_ψ is a Fourier multiplier, i.e., it is a multiplication operator in the Fourier representation. Then, given an admissible vector ψ, the corresponding CWT reads, with $U(\sigma_I(\varrho, a)) = R_\varrho\, D_a$:

$$\begin{aligned} F_\psi(\varrho, a) &:= \langle U(\sigma_I(\varrho, a))\psi | f\rangle \\ &= \int_{\mathbb{S}^2} \overline{[R_\varrho\, D_a \psi](\zeta)}\, f(\zeta)\, \mathrm{d}\mu(\zeta), \quad f \in L^2(\mathbb{S}^2, \mathrm{d}\mu). \end{aligned} \tag{10.2}$$

Moreover, given any admissible axisymmetric wavelet ψ, the family $\{\psi_{\varrho,a} := R_\varrho\, D_a \psi, \; (\varrho, a) \in X\}$ is a continuous frame, that is, there exist two constants $\mathsf{m} > 0$ and $\mathsf{M} < \infty$ such that

$$\mathsf{m}\, \|\phi\|^2 \leqslant \int_X |\langle \psi_{\varrho,a} | \phi\rangle|^2 \,\mathrm{d}\varrho\, \frac{\mathrm{d}a}{a^3} \leqslant \mathsf{M}\, \|\phi\|^2, \; \forall\, \phi \in L^2(\mathbb{S}^2, \mathrm{d}\mu).$$

The relation (10.2) yields a full spherical CWT, with correct Euclidean limit, i.e., the spherical CWT on a sphere of radius R tends to the usual 2-D CWT in the tangent plane when $R \to \infty$ (here one uses the technique of group contraction for evaluating the limit $X = \mathrm{SO}(3) \cdot \mathbb{R}_*^+ \to \mathrm{SIM}(2)$ and similarly for the corresponding representations). In addition, one may construct spherical frames, both semi-continuous (only the scale is discrete) and fully discrete, but one needs a generalized notion of frame (weighted, controlled frames). In addition, the method extends to the n-sphere.

As for the spherical DWT, there are many methods in the literature on approximation theory. An alternative consists in lifting the plane 2-D DWT from the tangent plane at the North Pole onto $\mathbb{S}^2$ by inverse stereographic projection. In that way, one obtains locally supported orthonormal wavelet bases on $\mathbb{S}^2$ [27].

10.3. The CWT on other manifolds

The techniques developed for the two-sphere may be generalized to other manifolds, in particular, the so-called *conic sections* (the sphere, the two-sheeted hyperboloid, the paraboloid). In all cases, a key ingredient for designing a dilation *on* the manifold is a suitable projection on a simpler one (a tangent cone, a tangent plane) [11, 12]. A detailed description may be found in [3, Chap.15].

1. *The two-sheeted hyperboloid*

   Since the two-sheeted hyperboloid $\mathbb{H}^2$ is the dual of the two-sphere (in the sense of differential geometry), it is not surprising that a group-theoretical construction similar to the one on $\mathbb{S}^2$ works well, using the isometry group SO(2,1) instead of SO(3). Thus one gets a fully developed CWT, but no results about frames are known [14].
2. *The paraboloid*
   The paraboloid $\mathbb{P}^2$ is a singular case, which can be obtained from both $\mathbb{S}^2$ and $\mathbb{H}^2$ by a limiting procedure. However, no (large) isometry group is available, so that the previous method does not work. An alternative construction is possible by lifting the CWT (or the DWT) vertically from the tangent plane onto the paraboloid.
3. *The two-torus*
   The two-torus $\mathbb{T}^2$ is isomorphic to the product of two circles, $\mathbb{T}^2 \simeq \mathbb{S}^1 \times \mathbb{S}^1 \simeq$ SO(2) $\times$ SO(2). This fact suggests to exploit the WT on circle, which is analogous to that on the two-sphere, by the same construction, replacing SO(3,1) by SO(2,1). Now, since SO(2,2) $\simeq \mathrm{SO}(2,1) \times \mathrm{SO}(2,1)/\mathbb{Z}_2$, one may go one step further and use SO(2,2). However, this will introduce *two* independent dilations instead of a single, global one. This difficulty may then be circumvented by combining SO(2,2) with the modular group SL$(2,\mathbb{Z})$. In that way one may obtain both a CWT and discrete wavelet frames on $\mathbb{T}^2$ [16].
4. *More general manifolds*
   In the case of a general (two-dimensional) manifold $\mathcal{M}$, a possible approach is to lift *locally* the plane CWT from the tangent plane along the local normal, thus defining a genuine notion of local dilation on $\mathcal{M}$. In this way, one obtains a local CWT on $\mathcal{M}$. It remains to glue the local patches by standard methods of differential geometry, but that is not always easy [12].
5. *A unified point of view*
   Let us come back for a moment to the group structure underlying the CWT. In 1-D, the group is the affine group of the line $G_{\mathrm{aff}}^{\mathbb{R}} = \mathbb{R} \rtimes \mathbb{R}_*$. In 2-D, one gets the similitude group of the plane SIM(2) $= \mathbb{R}^2 \rtimes (\mathbb{R}_*^+ \times \mathrm{SO}(2))$. Now, using polar coordinates, the right-hand factor $\mathbb{R}_*^+ \times \mathrm{SO}(2)$ may be identified with the pointed plane $\mathbb{R}_*^2$. Further identifying the real plane $\mathbb{R}^2$ with the complex plane $\mathbb{C}$, we finally obtain that the group underlying the 2-D CWT is $G_{\mathrm{aff}}^{\mathbb{C}} = \mathbb{C} \rtimes \mathbb{C}_*$.

This approach may be pursued one step further, following Ali [6]. The idea is to define *quaternionic wavelets.* Let $\mathbb{H}$ denote the field of all quaternions and $\mathbb{H}_*$ the group (under quaternionic multiplication) of all invertible quaternions. It turns out that the group $\mathbb{H}_*$ is isomorphic to the *affine* $SU(2)$ *group*, i.e., $\mathbb{R}_*^+ \times \mathrm{SU}(2)$, that is, the group SU(2) together with all (non-zero) dilations. As a set $\mathbb{H}_* \simeq \mathbb{R}_*^+ \times \mathbb{S}^3$, where $\mathbb{S}^3$ is the surface of the unit sphere in $\mathbb{R}^4$, or more simply, $\mathbb{H}_* \simeq \mathbb{R}^4 \setminus \{\mathbf{0}\}$. Exactly as in the two other cases, $\mathbb{H}_*$ acts transitively on $\mathbb{H}$ and from this one can construct quaternionic wavelets, with underlying group $G_{\mathrm{aff}}^{\mathbb{H}} = \mathbb{H} \rtimes \mathbb{H}_*$.

In this way, one obtains a useful and elegant approach to the three cases, with the group $G_{\mathrm{aff}}^{\mathbb{K}} = \mathbb{K} \rtimes \mathbb{K}_*$, where $\mathbb{K} = \mathbb{R}, \mathbb{C}$ or $\mathbb{H}$. We refer to [6] for the development of this type of wavelets.

11. Spatio-temporal wavelets and motion estimation

There exist many methods for motion estimation: Optical flow, Block matching, Phase difference, ... An alternative approach is provided by the motion-tuned continuous wavelet transform, developed by M. Duval-Destin (1991), R.Murenzi (1992) and F. Mujica (1999). The idea is to adapt to the (2D+T) space-time the general formalism of the continuous wavelet transform on a manifold, described in Section 10.1 above. The rest of this section is based on our paper [15].

Technically speaking, one builds a time-dependent wavelet, separable in frequency space:

$$\widehat{\psi}_{ST}(\vec{k}, \omega) = \underbrace{\widehat{\psi}_S(k_x, k_y)}_{\text{2-D wavelet}} \cdot \underbrace{\widehat{\psi}_T(\omega)}_{\text{1-D wavelet}} \quad . \tag{11.1}$$

Then one acts on it by the space-time (2D+T) group, containing space and time translations, space and time dilations, space rotations, namely, $G = \mathrm{SIM}(2) \times G_{\mathrm{aff}}^+$, via the usual unitary irreducible, square integrable, representations in the space of signals (image sequences) $L^2(\mathbb{R}^2 \times \mathbb{R}, \mathrm{d}\vec{x}\, \mathrm{d}t)$. Writing the Fourier transform as

$$\widehat{s}(\vec{k}, \omega) = (2\pi)^{-3/2} \iint_{\mathbb{R}^2 \times \mathbb{R}} e^{-i(\vec{k} \cdot \vec{x} + \omega t)} s(\vec{x}, t)\, \mathrm{d}\vec{x}\, \mathrm{d}t \,,$$

the Fourier space is simply $L^2(\mathbb{R}^2 \times \mathbb{R}, \mathrm{d}\vec{k}\, \mathrm{d}\omega)$.

Finally, one replaces the separate space and time dilations a_s, a_t by a *global* dilation a and a *speed tuning* parameter c.

The principle underlying motion estimation is that speed detection and quantization is done in the Fourier space, because the wavelet measures the inclination of the signal spectrum. Indeed the latter increases with speed, as shown in Fig. 5: a static object $\widehat{s}$ lives in the plane $\omega = 0$ of zero frequency, whereas an object $\widehat{s}$ moving with constant speed $\vec{v}$ lives in the plane $\vec{k} \cdot \vec{v} + \omega = 0$.

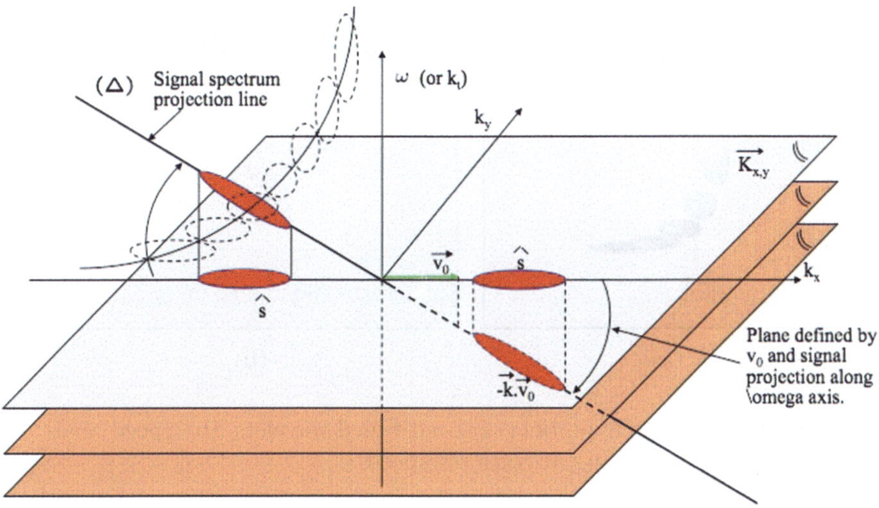

FIGURE 5. Inclination of a signal spectrum as a function of speed.

The next step is to identify the unitary motion operators acting in the Fourier space $L^2(\mathbb{R}^2 \times \mathbb{R}, \mathrm{d}\vec{k}\,\mathrm{d}\omega)$, namely,

$$\begin{aligned}
\text{Dilation :} \qquad & [\widehat{D}^a\, \widehat{\psi}](\vec{k},\omega) &=& \; a^{3/2}\widehat{\psi}(a\vec{k}, a\,\omega) \\
\text{Translation :} \qquad & [\widehat{T}^{\vec{b},\tau}\, \widehat{\psi}](\vec{k},\omega) &=& \; e^{-i(\vec{k}\cdot\vec{b}+\omega\tau)}\widehat{\psi}(\vec{k},\omega) \\
\text{Rotation :} \qquad & [\widehat{R}^{\theta}\, \widehat{\psi}](\vec{k},\omega) &=& \; \widehat{\psi}(r_{-\theta}\vec{k},\omega) \\
\text{Speed tuning :} \qquad & [\widehat{\Lambda}^{c}\, \widehat{\psi}](\vec{k},\omega) &=& \; \widehat{\psi}(c^{q}\vec{k}, c^{-p}\omega).
\end{aligned}$$

In the definition of speed tuning, we have the additional constraints that $\widehat{\Lambda}^c$ must map the $\vec{v}_o$-plane, $\vec{k}\cdot\vec{v}_o+\omega=0$, into the $c\vec{v}_o$-plane and must be unitary, which implies that $p=2/3$ and $q=1/3$. This corresponds to the psycho-visual effect. If an object moves fast, only *large* details can be detected, but, if it moves slowly, then *small* details can be detected. Hence, a high speed object must be large to be "captured" and a low speed object can be small. Therefore wavelets must be speed-tuned (distorted and elongated) to capture a moving object, which means that wavelets move on a hyperbola-like curve with increasing speed, as seen on Figure 6 (a). We show on Figure 6 (b) the speed analysis of an object moving at constant speed. Capture is achieved when the signal spectrum (red) intersects the family of speed-tuned wavelets (blue).

The next step is to choose adequate wavelets for space and time components. The standard 2D+T wavelet is the Duval–Destin–Murenzi (DDM) wavelet

$$\widehat{\psi}_{\rm DDM}(\vec{k},\omega) = \underbrace{\widehat{\psi}_{\rm M}(k_x,k_y)}_{\text{2-D Morlet wavelet}} \cdot \underbrace{\widehat{\psi}_{\rm M}(\omega)}_{\text{1-D Morlet}}$$

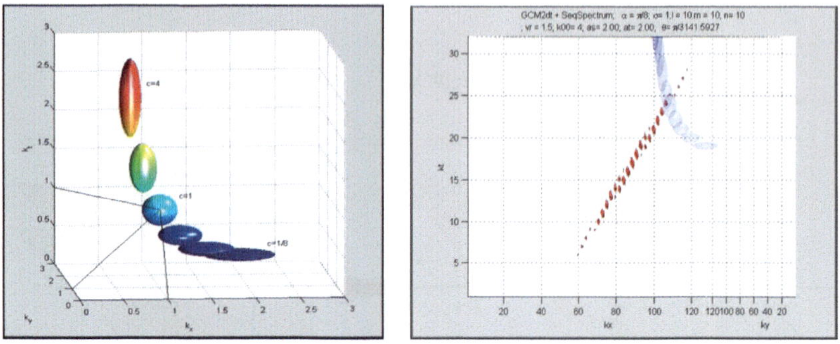

(a) (b)

FIGURE 6. (a) The hyperbola of speed-tuned wavelets. (b) Speed analysis of an object moving at constant speed.

where the 1-D Morlet wavelet is

$$\widehat{\psi}_{\mathrm{M}}(\omega) = \exp(-\tfrac{1}{2}(\omega - \omega_0)^2) - \widehat{h}(\omega),$$

with the correction term $\widehat{h}(\omega)$ negligible in practice (for $\omega_0 \gtrsim 5.5$).

Since the 2-D Morlet wavelet has poor selectivity properties, we replace it by a *Gaussian-conical wavelet*, as given in (9.13), and get a GCM 2D+T wavelet

$$\widehat{\psi}_{lm}^{\mathrm{GCM}}(\vec{k},\omega) = \begin{cases} \underbrace{\widehat{\psi}_{lm}^{\mathrm{GC}}(k_x,k_y)}_{\text{2-D Gaussian-Conical}} \cdot \underbrace{\widehat{\psi}_{\mathrm{M}}(\omega)}_{\text{1-D Morlet}}, & \vec{k} \in \mathcal{C}(-\alpha,\alpha), \\ 0, \ \text{otherwise.} \end{cases}$$

Explicitly

$$\widehat{\psi}_{lm}^{\mathrm{GCM}}(\vec{k},\omega) = \begin{cases} (\vec{k}\cdot\vec{e}_{-\widetilde{\alpha}})^l \, (\vec{k}\cdot\vec{e}_{\widetilde{\alpha}})^m e^{-\frac{\sigma}{2}(k_x - \chi(\sigma))^2} \, e^{-\frac{1}{2}(\omega-\omega_0)^2}, & \vec{k} \in \mathcal{C}(-\alpha,\alpha), \\ 0, \ \text{otherwise} \end{cases}$$

and the so-called center correction term $\chi(\sigma) = \sqrt{l + m\frac{\sigma-1}{\sigma}}$ controls the radial support of the wavelet $\widehat{\psi}_{lm}^{\mathrm{GCM}}$.

Several experiments [15] prove that the GCM 2D+T wavelet has a better angular resolving power than the DDM 2D+T wavelet, it shows an extreme efficiency in directional speed selectivity down to very small angle apertures, it has a good capacity of radial stability and adjustment with respect to variation of the aperture and, last but not least, it has a good stability with respect to noise.

In conclusion, we may say that the GCM 2D+T wavelet is a highly directionally selective speed-tuned wavelet, which provides a much more powerful tool than the DDM 2D+T wavelet for the recognition and tracking of spectral signatures.

References

[1] S.T. Ali and M. Engliš, *Quantization methods: a guide for physicists and analysts*, Rev. Math. Phys. **17** (2005), 391–490.

[2] S.T. Ali, J.-P. Antoine, J.-P. Gazeau and U.A. Mueller, *Coherent states and their generalizations: A mathematical overview*, Reviews Math. Phys. **7** (1995), 1013–1104.

[3] S.T. Ali, J.-P. Antoine and J.-P. Gazeau, *Coherent States, Wavelets and Their Generalizations*, 2nd ed., Springer-Verlag, New York et al., 2014.

[4] S.T. Ali, T. Bhattacharyya and S.S. Roy, *Coherent states on Hilbert modules*, J. Phys. A: Math. Theor. **44** (2011), 275202.

[5] S.T. Ali, F. Bagarello and J.-P. Gazeau, *Quantizations from reproducing kernel spaces*, Ann. Phys. **332** (2013), 127–142.

[6] S.T. Ali and K. Thirulogasanthar, *The quaternionic affine group and related continuous wavelet transforms on complex and quaternionic Hilbert spaces*, J. Math. Phys. **55** (2014), 063501.

[7] J.-P. Antoine and P. Vandergheynst, *Wavelets on the n-sphere and related manifolds*, J. Math. Phys. **39** (1998), 3987–4008.

[8] J.-P. Antoine and P. Vandergheynst, *Wavelets on the 2-sphere: A group-theoretical approach*, Appl. Comput. Harmon. Anal. **7** (1999), 262–291.

[9] J.-P. Antoine, L. Demanet, L. Jacques, and P. Vandergheynst, *Wavelets on the sphere: Implementation and approximations*, Appl. Comput. Harmon. Anal. **13** (2002), 177–200.

[10] J.-P. Antoine, R. Murenzi, P. Vandergheynst and S.T. Ali, *Two-dimensional Wavelets and Their Relatives*, Cambridge University Press, Cambridge (UK), 2004.

[11] J.-P. Antoine, I. Bogdanova, and P. Vandergheynst, *The continuous wavelet transform on conic sections*, Int. J. Wavelets, Multires. and Inform. Proc., **6** (2007) 137–156.

[12] J.-P. Antoine, D. Roşca, and P. Vandergheynst, *Wavelet transform on manifolds: Old and new approaches*, Appl. Comput. Harmon. Anal. **28** (2010), 189–202.

[13] I. Bogdanova, P. Vandergheynst, J.-P. Antoine, L. Jacques, and M. Morvidone, *Stereographic wavelet frames on the sphere*, Appl. Comput. Harmon. Anal. **19** (2005), 223–252.

[14] I. Bogdanova, P. Vandergheynst, and J.-P. Gazeau, *Continuous wavelet transform on the hyperboloid*, Appl. Comput. Harmon. Anal. **23** (2007), 286–306.

[15] P. Brault and J.-P. Antoine, *A spatio-temporal Gaussian-Conical wavelet with high aperture selectivity for motion and speed analysis*, Appl. Comput. Harmon. Anal. **34** (2013), 148–161.

[16] M. Calixto, J. Guerrero, and D. Roşca, *Wavelet transform on the torus: a group-theoretical approach*, Appl. Comput. Harmon. Anal. **38** (2015), 32–49.

[17] I. Daubechies, *Ten Lectures on Wavelets*, SIAM, Philadelphia, 1992.

[18] D. Gabor, *Theory of communication*, J. Inst. Electr. Engrg. (London) **93** (1946), 429–457.

[19] J.-P. Gazeau, *Coherent States in Quantum Physics*, Wiley-VCH, Berlin, 2009.

[20] R. Gilmore, *Geometry of symmetrized states*, Ann. Phys. (NY) **74** (1972), 391–463; *On properties of coherent states*, Rev. Mex. Fis. **23** (1974), 143–187.

[21] A. Grossmann, J. Morlet and T. Paul, *Integral transforms associated to square integrable representations. I. General results*, J. Math. Phys. **26** (1985), 2473–2479; *II. Examples,* Ann. Inst. H. Poincaré **45** (1986), 293–309.

[22] C. Heil and D. Walnut (eds.), *Fundamental Papers in Wavelet Theory*, Princeton University Press, Princeton, NJ, 2006.

[23] E.C. Lance, *Hilbert C*-modules*, London Math. Soc. Lect. Notes Series 210, Cambridge UP, Cambridge, 1995.

[24] S.G. Mallat, *A Wavelet Tour of Signal Processing, 3rd ed.: The Sparse Way*, Academic Press, San Diego, 2008.

[25] D. Marr, *Vision*, Freeman, San Francisco, 1982.

[26] A.M. Perelomov, *Coherent states for arbitrary Lie group*, Commun. Math. Phys. **26** (1972), 222–236.

[27] D. Roşca and J.-P. Antoine, *Locally supported orthogonal wavelet bases on the sphere via stereographic projection*, Math. Probl. Engineering, Vol. **2009** (2009), 124904.

[28] K. Thirulogasanthar and S.T. Ali, *A class of vector coherent states defined over matrix domains*, J. Math. Phys. **44** (2003), 5070–5083.

[29] K. Thirulogasanthar, G. Honnouvo and A. Krzyżak, *Coherent states and Hermite polynomials on quaternionic Hilbert spaces*, J. Phys. A: Math. Theor. **43** (2010), 385205.

[30] J. Ville, *Théorie et applications de la notion de signal analytique*, Câbles et Transm. **2ème A** (1948), 61–74.

[31] E.P. Wigner, *On the quantum correction for thermodynamic equilibrium*, Phys. Rev. **40** (1932), 749–759.

Jean-Pierre Antoine
Institut de Recherche en Mathématique et Physique
Université catholique de Louvain
B-1348 Louvain-la-Neuve, Belgium
e-mail: `jean-pierre.antoine@uclouvain.be`

Operator Theory:
Advances and Applications, Vol. 247, 157–171
© Springer International Publishing Switzerland 2015

On Some Families of Complex Hermite Polynomials and their Applications to Physics

F. Balogh, Nurisya M. Shah and S. Twareque Ali

Abstract. In this paper we study certain families of complex Hermite polynomials and construct deformed versions of them, using a $GL(2,\mathbb{C})$ transformation. This construction leads to the emergence of biorthogonal families of deformed complex Hermite polynomials, which we then study in the context of a two-dimensional model of noncommutative quantum mechanics.

Mathematics Subject Classification (2010). 42C05, 33C45, 33C50, 81S99.

Keywords. Complex orthogonal polynomials, deformed Hermite polynomials, noncommutative quantum mechanics.

1. Introduction

Complex orthogonal polynomials and their quaternionic extensions have recently received considerable attention in various branches of physics and mathematics. A small sampling of the relevant literature may be found in [1, 6, 7, 9, 12, 13, 14, 15, 16, 23, 24, 25]. In this paper we look at certain classes of complex Hermite polynomials which, in a somewhat different form, have earlier found applications to quantum optics [23, 24], and for which we now show a relationship to the recently developed and very popular theory of non-commutative quantum mechanics.

To put things in perspective, we recall a few basic facts about non-commutative quantum mechanics, or more properly, a two-dimensional version of it. The motivating factor here is the belief that a modification of standard quantum mechanics is needed to model physical space-time at very short distances. One way to introduce such a modification is to alter the canonical commutation relations of quantum mechanics. One can then study, for example, the effect of such a modification on well-known Hamiltonians, such as the harmonic oscillator or the Landau problem and their energy spectra. For some recent work in this direction see, e.g., [4, 5, 8, 17, 22, 20, 21]. In this paper we work with a similar version of noncommutative quantum mechanics, i.e., one describing a system with two degrees of freedom and look at some associated classes of complex biorthogonal polynomials, arising as a consequence of the altered commutation relations. These biorthogonal polynomials, appear in much the same way as the *complex Hermite polynomials* [9, 13, 14, 18, 19] arise in the standard quantum mechanics of a system with two degrees of freedom.

We start with the usual quantum mechanical commutation relations

$$[Q_i, P_j] = i\hbar\delta_{ij}I, \quad i, j = 1, 2. \tag{1}$$

Here the Q_i, P_j are the quantum mechanical position and momentum observables, respectively. In non-commutative quantum mechanics one imposes the additional commutation relation

$$[Q_1, Q_2] = i\vartheta I, \tag{2}$$

where ϑ is a small, positive parameter which measures the additionally introduced noncommutativity between the observables of the two spatial coordinates. The limit $\vartheta = 0$ then corresponds to standard (two-dimensional) quantum mechanics. One could also impose a second non-commutativity between the two momentum operators:

$$[P_1, P_2] = i\gamma I, \tag{3}$$

where γ is yet another positive parameter. Physically, such a commutator would mean that there is a magnetic field in the system.

The Q_i and P_i, $i = 1, 2$, satisfying the modified commutation relations (2) and (3) can be written in terms of the standard quantum mechanical position and momentum operators $\hat{q}_i, \hat{p}_i$, $i = 1, 2$, with

$$[\hat{q}_i, \hat{p}_j] = i\delta_{ij}, \quad [\hat{q}_i, \hat{q}_j] = [\hat{p}_i, \hat{p}_j] = 0.$$

One possible representation is

$$\begin{aligned} Q_1 &= \hat{q}_1 - \frac{\vartheta}{2}\hat{p}_2 \quad P_1 = c\hat{p}_1 + d\hat{q}_2 \quad c = \frac{1}{2}(1 \pm \sqrt{\kappa}), \quad d = \frac{1}{\vartheta}(1 \mp \sqrt{\kappa}) \\ Q_2 &= \hat{q}_2 + \frac{\vartheta}{2}\hat{p}_1 \quad P_2 = c\hat{p}_2 - d\hat{q}_1 \quad \kappa = 1 - \gamma\vartheta, \qquad \gamma \neq \frac{1}{\vartheta}. \end{aligned} \tag{4}$$

In this paper, we shall assume such a non-commutative system, however, with the additional restriction

$$\vartheta = \gamma.$$

Then, introducing the annihilation and creation operators,

$$A_i = \frac{1}{\sqrt{2}}(Q_i + iP_i), \quad A_i^\dagger = \frac{1}{\sqrt{2}}(Q_i - iP_i), \quad i = 1, 2, \tag{5}$$

we have the modified commutation relations,

$$[A_i, A_i^\dagger] = 1, \quad [A_i, A_j] = 0, \quad [A_1, A_2^\dagger] = i\vartheta, \quad i, j = 1, 2. \tag{6}$$

Note, in particular that

$$[A_1, A_2] = 0 \quad \Longrightarrow \quad [A_1^\dagger, A_2^\dagger] = 0,$$

which means that one still has two independent bosons and the operators A_1 and A_2 still have a common ground state, which we may conveniently denote by $|0,0\rangle$.

To connect the above discussion with Hermite polynomials, we note first that the usual Hermite polynomials in a real variable are naturally associated to the commutation relations $[a, a^\dagger] = 1$ for a single bosonic degree of freedom. Writing them as $H_n(x)$, $n = 0, 1, 2, \ldots$, $x \in \mathbb{R}$, they satisfy the orthogonality relations:

$$\int_{\mathbb{R}} H_m(x) H_n(x)\, e^{-x^2}\, dx = \sqrt{\pi} 2^n n!\, \delta_{mn}, \tag{7}$$

and are obtainable using the formula:

$$H_n(x) = (-1)^n e^{x^2} \left(\frac{d}{dx}\right)^n e^{-x^2}, \tag{8}$$

or using the generating function,

$$e^{2xz - z^2} = \sum_{n=0}^{\infty} \frac{z^n}{n!} H_n(x). \tag{9}$$

On the Hilbert space $L^2(\mathbb{R}, e^{-x^2} dx)$, the operators of creation and annihilation are,

$$a^\dagger = \frac{1}{\sqrt{2}}\left(2x - \frac{d}{dx}\right), \qquad a = \frac{1}{\sqrt{2}} \frac{d}{dx}, \qquad [a, a^\dagger] = 1. \tag{10}$$

On this space, the normalized Hermite polynomials h_n can be obtained as:

$$h_n(x) := \frac{1}{[\sqrt{\pi}\, 2^n\, n!]^{\frac{1}{2}}} H_n(x) \quad \text{and} \quad h_n = \frac{(a^\dagger)^n}{\sqrt{n!}} h_0, \qquad \langle h_m \mid h_n \rangle_{\mathfrak{H}_{rhp}} = \delta_{mn}, \tag{11}$$

where h_0, the ground state, is the constant function,

$$h_0(x) = \frac{1}{\pi^{\frac{1}{4}}}, \qquad x \in \mathbb{R}.$$

The vectors h_n form an orthonormal basis of $L^2(\mathbb{R}, e^{-x^2} dx)$. All this, of course, is standard and well known.

A second representation of the commutation relation $[a, a^\dagger] = 1$, and the one that will be more pertinent to the present work, is on the Hilbert space (Fock–Bargmann space) $L^2_{\text{analyt}}(\mathbb{C}, d\nu(z, \overline{z}))$ of all analytic functions of a complex variable $z = x + iy$, which are square integrable with respect to the measure

$$d\nu(z, \overline{z}) = e^{-|z|^2}\, \frac{dz \wedge d\overline{z}}{i2\pi} = e^{-[x^2 + y^2]}\, \frac{dx\, dy}{\pi}.$$

On this space the creation and annihilation operators take the form $a^\dagger = z$ (operator of multiplication by z) and $a = \partial_z$, respectively. The normalized ground state is the constant function $h_0(z) = 1$, $z \in \mathbb{C}$. The orthonormal basis, built again as in (11), is now given by the monomials

$$h_n(z) = \frac{(a^\dagger)^n}{\sqrt{n!}} h_0 = \frac{z^n}{\sqrt{n!}}. \tag{12}$$

For two independent bosons, the two sets of creation and annihilation operators $a_i^\dagger, a_i,\ \ i=1,2$, satisfy the commutation relations $[a_i, a_j^\dagger] = \delta_{ij}$, and this set can be irreducibly represented on the full Hilbert space $\mathfrak{H}(\mathbb{C}) = L^2(\mathbb{C}, d\nu(z,\overline{z}))$, via the operators (see, for example, [9])

$$a_1 = \partial_z, \ \ a_1^\dagger = z - \partial_{\overline{z}}, \qquad a_2 = \partial_{\overline{z}}, \ \ a_2^\dagger = \overline{z} - \partial_z. \tag{13}$$

Note that $L^2_{\text{anal}}(\mathbb{C}, d\nu(z,\overline{z}))$ is a proper subspace of $\mathfrak{H}(\mathbb{C})$. Using again the ground state $h_{0,0}(z,\overline{z}) \equiv 1$, one can build an orthonormal basis for $\mathfrak{H}(\mathbb{C})$:

$$h_{m,n}(z,\overline{z}) = \frac{(a_1^\dagger)^m\ (a_2^\dagger)^n}{\sqrt{m!\ n!}}\ h_{0,0} \ = \ \frac{(z-\partial_{\overline{z}})^m\ (\overline{z}-\partial_z)^n}{\sqrt{m!\ n!}}\ 1, \tag{14}$$

$m,n = 0,1,2,\ldots,\infty$. It is clear that $h_{m,0}(z,\overline{z}) = h_m(z,\overline{z})$. The functions $h_{m,n}(z,\overline{z})$ have the explicit forms:

$$h_{m,n}(z,\overline{z}) = \sqrt{m!\ n!}\sum_{j=0}^{m\curlyvee n}\frac{(-1)^j}{j!}\frac{(z)^{m-j}}{(m-j)!}\frac{\overline{z}^{n-j}}{(n-j)!}, \tag{15}$$

where $m \curlyvee n$ denotes the smaller of the two numbers m and n. Moreover, it is easy to verify that the functions $H_{m,n}((z,\overline{z})) = \sqrt{m!n!}\ h_{m,n}((z,\overline{z}))$ are also obtainable as

$$H_{m,n}(\overline{z}, z) = (-1)^{m+n}\ e^{|z|^2}\partial_z^m\partial_{\overline{z}}^n\ e^{-|z|^2}, \tag{16}$$

a relation which should be compared to (8). By analogy, the functions $H_{m,n}$ (of which the $h_{m,n}$ are just normalized versions) are called *complex Hermite polynomials* [9, 13].

A large number of interesting facts are known about these polynomials [3, 12, 24]. Below we list a few.

First, there is the following direct relationship between the real and complex Hermite polynomials:

$$H_{m,n}(z,\overline{z}) = \frac{(-1)^n}{\sqrt{2^{m+n}}}\int_{\mathbb{R}} e^{-(u+\frac{\overline{z}}{\sqrt{2}})^2} H_m(u) H_n(u + \frac{1}{\sqrt{2}}(z+\overline{z}))\ \frac{du}{\sqrt{\pi}}. \tag{17}$$

Similarly, defining the $(L+1)\times(L+1)$ matrix $M(L)$, with elements

$$M(L)_{rm} = \frac{1}{2^L}\sum_{k=\max\{0,r+m-L\}}^{\min\{r,m\}}(-1)^{L-m-r+k}(i)^{L-r}\binom{m}{k}\binom{L-m}{r-k},$$

we have a second relationship between the real and complex Hermite polynomials,

$$H_{m,L-m}(z,\overline{z}) = \sum_{r=0}^{L} M(L)_{rm} H_r(x) H_{L-r}(y)\ , \quad z = x+iy. \tag{18}$$

Another useful and interesting result is:

$$H_{m,n}(z,\overline{z}) = e^{-\partial_z\partial_{\overline{z}}}(z^m\ \overline{z}^n). \tag{19}$$

Additionally, one has the rather amazing generalization of the Kibble–Slepian formula [12]:

Theorem 1. *Let $Z = (z_1, z_2, \ldots, z_N)$, and $H = (h_{j,k})$ an $N \times N$ Hermitian matrix with $\|H\| < 1$. Let I_N denote the $N \times N$ identity matrix. Then*

$$\begin{aligned}\mathrm{Det}[(I_N - H)^{-1}]\, e^{\left(ZH(I_N-H)^{-1}Z^*\right)} \\ = \sum_K \prod_{1\le j,k\le N} \frac{(h_{j,k})^{n_{j,k}}}{n_{j,k}!} H_{r_1,c_1}(z_1,\overline{z_1}) \cdots H_{r_N,c_N}(z_N,\overline{z_N}),\end{aligned} \tag{20}$$

where $K = (n_{j,k} : 1 \le j,k \le N)$ is a general matrix with nonnegative integer entries, c_k is the sum of the elements of K in column k and r_j is the sum of the elements of K in row j, that is

$$c_k = \sum_{j=1}^{N} n_{j,k}, \qquad r_j = \sum_{k=1}^{N} n_{j,k}. \tag{21}$$

In the sequel we shall basically "deform" the relation (14) to obtain families of *generalized biorthogonal Hermite polynomials*, in which the operators $a_i\ a_i^\dagger,\ \ i = 1,2$ will be replaced by operators similar to those in (5) of noncommutative quantum mechanics.

The rest of this paper is organized as follows. In Section 2 we lay down some abstract preliminaries connected with Hermite polynomials and construct generating functions, using an operator technique. In Section 3 we introduce the deformed complex Hermite polynomials, obtain some of their immediate properties and work out the representation of $GL(2,\mathbb{C})$ which gives rise to the deformed polynomials. In Section 4 we introduce the families of biorthogonal deformed complex Hermite polynomials. Section 5 is devoted to a study of the pertinence of the above results to a two-dimensional model of noncommutative quantum mechanics. Finally, in Section 6 we look at some second-order generators built out of the deformed creation and annihilation operators introduced earlier and identify the Lie algebras generated by them.

2. Some abstract preliminaries and generating functions

We go back to the algebra associated to two independent bosons, generated by the usual lowering and raising operators a_1, a_2 and $a_1^\dagger, a_2^\dagger$ respectively, satisfying the commutation relations

$$[a_i, a_j] = 0,\ [a_i^\dagger, a_j^\dagger] = 0,\ [a_i, a_j^\dagger] = \delta_{ij}, \quad i,j = 1,2. \tag{22}$$

Assuming an irreducible representation of this system in an abstract Hilbert space $\mathfrak{H}$, the lowering operators annihilate the vacuum state $|0,0\rangle$,

$$a_i\,|0,0\rangle = 0, \qquad i = 1,2 \tag{23}$$

the Hilbert space $\mathfrak{H}$ is then spanned by the orthonormal basis set,

$$|k,l\rangle = \frac{1}{\sqrt{k!l!}}(a_1^\dagger)^k (a_2^\dagger)^l\,|0,0\rangle\,, \quad k,l = 0,1,2,\ldots, \tag{24}$$

with

$$a_1^\dagger |k,l\rangle = \sqrt{k+1}\,|k+1,l\rangle\,,\ \ a_2^\dagger |k,l\rangle = \sqrt{l+1}\,|k,l+1\rangle\,.$$

The vector-valued function

$$F(u,\overline{u}) = \sum_{k,l=0}^{\infty} \frac{u^k \overline{u}^l}{\sqrt{k!\,l!}}\,|k,l\rangle \tag{25}$$

will serve as a useful book-keeping device in the subsequent calculations based on the identities

$$\frac{\partial F}{\partial u}(u,\overline{u}) = a_1^\dagger F(u,\overline{u}) \text{ and } \frac{\partial F}{\partial \overline{u}}(u,\overline{u}) = a_2^\dagger F(u,\overline{u}). \tag{26}$$

Obviously,

$$F(u,\overline{u}) = e^{ua_1^\dagger + \overline{u}a_2^\dagger}\,|0,0\rangle\,. \tag{27}$$

On the Hilbert space $\mathfrak{H}(\mathbb{C}) = L^2(\mathbb{C}, d\nu(z,\overline{z}))$, introduced above,

$$|0,0\rangle \mapsto h_{0,0}(z,\overline{z}) = 1, \qquad \text{for all} \quad z,\overline{z}, \tag{28}$$

and for any vector $f \in \mathfrak{H}(\mathbb{C})$

$$[e^{ua_1^\dagger} f](z,\overline{z}) = e^{uz} f(z,\overline{z}-u)$$
$$[e^{\overline{u}a_2^\dagger} f](z,\overline{z}) = e^{\overline{uz}} f(z-\overline{u},\overline{z})$$

and therefore

$$F(u,\overline{u}) = e^{ua_1^\dagger + \overline{u}a_2^\dagger}\,|0,0\rangle = e^{uz+\overline{uz}-u\overline{u}}. \tag{29}$$

Expanding this with respect to u and $\overline{u}$ and comparing with (14) gives the *generating function* for the complex Hermite polynomials in (15)–(16),

$$e^{uz+\overline{uz}-u\overline{u}} = \sum_{k,l=0}^{\infty} h_{k,l}(z,\overline{z}) \frac{u^k\overline{u}^l}{\sqrt{k!\,l!}} = \sum_{k,l=0}^{\infty} H_{k,l}(z,\overline{z}) \frac{u^k\overline{u}^l}{k!\,l!}. \tag{30}$$

Consider another well-known representation of the real Hermite polynomials [11] on the Hilbert space $\mathfrak{H} = L^2(\mathbb{R}^2, dx_1dx_2)$, with

$$|0,0\rangle \mapsto h_{0,0}(x_1,x_2) = e^{-\frac{1}{2}(x_1^2+x_2^2)} \tag{31}$$

and

$$[a_i^\dagger f](x_1,x_2) = \frac{1}{\sqrt{2}}\left(x_i - \frac{\partial}{\partial x_i}\right) f(x_1,x_2) \quad i=1,2,$$
$$[a_i f](x_1,x_2) = \frac{1}{\sqrt{2}}\left(x_i + \frac{\partial}{\partial x_i}\right) f(x_1,x_2) \quad i=1,2.$$

Then, by the Baker–Campbell–Hausdorff formula,

$$[e^{ua_1^\dagger} f](x_1, x_2) = e^{ux_1 - \frac{1}{2}u^2} f(x_1 - u, x_2)$$
$$[e^{\overline{u}a_2^\dagger} f](x_1, x_2) = e^{\overline{u}x_2 - \frac{1}{2}\overline{u}^2} f(x_1, x_2 - \overline{u})$$

and therefore

$$\begin{aligned} F(u, \overline{u}) &= e^{ua_1^\dagger + \overline{u}a_2^\dagger} |0, 0\rangle \\ &= e^{ux_1 - \frac{1}{2}u^2 + \overline{u}x_2 - \frac{1}{2}\overline{u}^2 - \frac{1}{2}(x_1 - u)^2 - \frac{1}{2}(x_2 - \overline{u})^2} \\ &= e^{2ux_1 - u^2 + 2\overline{u}x_2 - \overline{u}^2 - \frac{1}{2}x_1^2 - \frac{1}{2}x_2^2}. \end{aligned}$$

Expanding this with respect to u and $\overline{u}$ and comparing with (9) gives the generating function for products of real *Hermite polynomials* in two variables

$$F(u, \overline{u}) = e^{-\frac{1}{2}[x_1^2 + x_2^2]} \sum_{k,l=0}^{\infty} \frac{u^k \overline{u}^l}{k!\, l!} H_k(x_1)\, H_l(x_2). \tag{32}$$

3. Deformed generalized Hermite polynomials

Going back to (14), we now *deform* the complex Hermite polynomials $h_{m,n}$ essentially by replacing the operators $a_1^\dagger$ and $a_2^\dagger$ by linear combinations of these. Specifically, we define the operators

$$\begin{aligned} a_1^{g\dagger} &= g_{11}a_1^\dagger + g_{21}a_2^\dagger, \qquad & a_2^{g\dagger} &= g_{12}a_1^\dagger + g_{22}a_2^\dagger, \\ a_1^g &= \overline{g_{11}}a_1 + \overline{g_{21}}a_2^\dagger, \qquad & a_2^g &= \overline{g_{12}}a_1 + \overline{g_{22}}a_2, \end{aligned} \tag{33}$$

parametrized by a 2×2 invertible complex matrix

$$g = \begin{bmatrix} g_{11} & g_{12} \\ g_{21} & g_{22} \end{bmatrix} \in GL(2, \mathbb{C}). \tag{34}$$

The *g-deformed basis elements* are then defined to be

$$|k, l\rangle_g = \frac{1}{\sqrt{k!l!}} (a_1^{g\dagger})^k (a_2^{g\dagger})^l\, |0, 0\rangle \qquad k, l = 0, 1, \ldots. \tag{35}$$

The generating function of the g-deformed basis is given by

$$F_g(u, \overline{u}) = \sum_{k,l=0}^{\infty} \frac{u^k \overline{u}^l}{k!\, l!} (a_1^{g\dagger})^k (a_2^{g\dagger})^l\, |0, 0\rangle = e^{ua_1^{g\dagger} + \overline{u}a_2^{g\dagger}}\, |0, 0\rangle \tag{36}$$

which can be written as

$$\begin{aligned} F_g(u, \overline{u}) &= e^{ua_1^{g\dagger} + \overline{u}a_2^{g\dagger}}\, |0, 0\rangle \\ &= e^{(g_{11}u + g_{12}\overline{u})a_1^\dagger + (g_{21}u + g_{22}\overline{u})a_2^\dagger}\, |0, 0\rangle \\ &= F(g_{11}u + g_{12}\overline{u}, g_{21}u + g_{22}\overline{u}). \end{aligned}$$

In particular, we get the analogue of (30):

$$\begin{aligned} F_g(u,\overline{u}) &= \exp\left((g_{11}u+g_{12}\overline{u})z+(g_{21}u+g_{22}\overline{u})\overline{z}-(g_{11}u+g_{12}\overline{u})(g_{21}u+g_{22}\overline{u})\right) \\ &= \sum_{k,l=0}^{\infty} h^g_{k,l}(z,\overline{z})\frac{u^k\overline{u}^l}{\sqrt{k!\,l!}} = \sum_{k,l=0}^{\infty} H^g_{k,l}(z,\overline{z})\frac{u^k\overline{u}^l}{k!\,l!}, \end{aligned} \tag{37}$$

where the polynomials

$$h^g_{k,l}(z,\overline{z}) := \frac{(a_1^{g\dagger})^k\,(a_2^{g\dagger})^l}{\sqrt{k!\,l!}}\,h_{0,0}(z,\overline{z}) = \frac{H^g_{k,l}(z,\overline{z})}{\sqrt{k!\,l!}}, \qquad k,l=0,1,\ldots \tag{38}$$

are the *g-deformed complex Hermite polynomials.*

Our aim is to describe the operator T_g defined by

$$T_g|k,l\rangle = |k,l\rangle_g \qquad k,l=0,1,\ldots \tag{39}$$

in terms of the group element $g\in GL(2,\mathbb{C})$. Consider the map

$$P\colon \mathfrak{H}\to\mathbb{C}[s,t] \qquad |k,l\rangle \mapsto \frac{1}{\sqrt{k!l!}}s^kt^l, \tag{40}$$

where $\mathbb{C}[s,t]$ denotes the set of all complex polynomials in the two variables s and t. Then

$$Pa_1^\dagger = M_sP, \qquad Pa_2^\dagger = M_tP, \tag{41}$$

where M_s and M_t stand for the operators of multiplication by s and t respectively. Therefore

$$T_g|k,l\rangle_g = \frac{1}{\sqrt{k!l!}}(g_{11}s+g_{21}t)^k(g_{12}s+g_{22}t)^l. \tag{42}$$

Let

$$R_g\ :\ \mathbb{C}[s,t]\to\mathbb{C}[s,t] \qquad [R_gf](s,t) = f(g_{11}s+g_{21}t, g_{12}s+g_{22}t). \tag{43}$$

Then the following intertwining relation holds:

$$PT_g = R_gP. \tag{44}$$

To summarize we have a commutative diagram:

$$\begin{array}{ccc} H & \xrightarrow{P} & \mathbb{C}[s,t] \\ \Big\downarrow{\scriptstyle T_g} & & \Big\downarrow{\scriptstyle R_g} \\ H & \xrightarrow{P} & \mathbb{C}[s,t]. \end{array} \tag{45}$$

The operators R_g realize a representation of $GL(2,\mathbb{C})$ on the space $\mathbb{C}[s,t]$ and it splits into an infinite direct sum of irreducible representations

$$\mathbb{C}[s,t] = \bigoplus_{L=0}^{\infty}\mathbb{C}_L[s,t], \tag{46}$$

where $\mathbb{C}_L[s,t]$ stands for the subspace of homogeneous polynomials of degree L:

$$\mathbb{C}_L[s,t] = \operatorname{span}\{s^kt^{L-k}\colon k=0,\ldots,L\}. \tag{47}$$

If we take

$$V = \mathbb{C}_1[s,t] = \mathrm{span}\{s,t\}, \tag{48}$$

then we see that $R_g|_V$ is the standard representation of $GL(2,\mathbb{C})$ and that

$$\mathbb{C}_L[s,t] \simeq \mathrm{Sym}^L V \tag{49}$$

(L fold symmetric tensor product of V). This representation is irreducible. A straightforward calculation gives

$$\begin{aligned}
&(g_{11}s+g_{21}t)^k(g_{12}s+g_{22}t)^l \\
&= \sum_{i=0}^{k}\sum_{j=0}^{l}\binom{k}{i}\binom{l}{j} g_{11}^i g_{21}^{k-i} g_{12}^j g_{22}^{l-j} s^{i+j} t^{k+l-i-j} \\
&= \sum_{r=0}^{k+l}\left[\sum_{q=\max\{0,r-l\}}^{\min\{r,k\}}\binom{k}{q}\binom{l}{r-q} g_{11}^q g_{21}^{k-q} g_{12}^{r-q} g_{22}^{l+q-r}\right] s^r t^{k+l-r}
\end{aligned}$$

where we used the substitution $r = i + j$ and $q = i$. If we choose the basis in $\mathbb{C}_L[s,t]$ as

$$f_k(s,t) = s^k t^{L-k} \qquad k = 0,1,\ldots,L, \tag{50}$$

then

$$R_g f_k = \sum_{r=0}^{L}\left[\sum_{q=\max\{0,r+k-L\}}^{\min\{r,k\}}\binom{k}{q}\binom{L-k}{r-q} g_{11}^q g_{21}^{k-q} g_{12}^{r-q} g_{22}^{L-k+q-r}\right] f_r. \tag{51}$$

So the matrix $M(g,L)$ of $R_g|_{C_L[s,t]}$ in the basis $\{f_k\}_{k=0}^L$ is given by the matrix elements

$$M(g,L)_{rk} = \sum_{q=\max\{0,r+k-L\}}^{\min\{r,k\}}\binom{k}{q}\binom{L-k}{r-q} g_{11}^q g_{21}^{k-q} g_{12}^{r-q} g_{22}^{L-k+q-r} \quad 0 \le r,k \le L. \tag{52}$$

A useful relation that we read off from the above is that

$$M(g,L)^* = M(g^*,L), \tag{53}$$

the star denoting the adjoint matrix in each case.

If furthermore, λ_1, λ_2 are non-zero eigenvalues of $M(g,2)$, corresponding to non-zero eigenvectors f_1, f_2, it is possible to show that the eigenvalues of $M(g,L)$ are

$$\Lambda_L = \{\lambda_1^k \lambda_2^{L-k}\,;\, k = 0,1,\ldots,L\}. \tag{54}$$

This is another useful result since (54) provides complete information regarding any particular choice of roots, using which their characteristic polynomials can easily be obtained.

From the above discussion it is clear that when $\mathfrak{H} = \mathfrak{H}(\mathbb{C})$, $T_g h_{m,n} = h_{m,n}^g$ and T_g leaves the $(L+1)$-dimensional subspace of $\mathfrak{H}(\mathbb{C})$ spanned by the vectors

$$\mathfrak{S}(L) = \{h_{L,0},\ h_{L-1,1},\ h_{L-2,2},\ \ldots,\ h_{0,L}\} \tag{55}$$

invariant. Let $T(g,L)$ denote the restriction of T_g to this subspace. Then the matrix elements of $T(g,L)$ in the $\mathfrak{S}(L)$-basis are just the $M(g,L)_{rk}$ in (52).

There is an interesting intertwining relation between $M(g,L)$ and $T(g,L)$ [12] that is worth mentioning here. Note first, that using (19) one immediately finds

$$h_{m,n}(z,\overline{z}) = e^{-\partial_z \partial_{\overline{z}}} p_{m,n}(z,\overline{z}), \tag{56}$$

where

$$p_{m,n}(z,\overline{z}) = \frac{z^m \overline{z}^n}{\sqrt{m!n!}}.$$

From this and the preceding discussion it is straightforward to verify that

$$e^{-\partial_z \partial_{\overline{z}}} M(g,L) = T(g,L) e^{-\partial_z \partial_{\overline{z}}}. \tag{57}$$

4. Biorthogonal families of polynomials

From the way they were constructed in (14), it follows that the normalized complex Hermite polynomials $h_{m.n}(z,\overline{z})$ form an orthonormal basis of $\mathfrak{H}(\mathbb{C})$,

$$\int_{\mathbb{C}} \overline{h_{m,n}(z,\overline{z})} h_{k,l}(z,\overline{z})\, d\nu(z,\overline{z}) = \delta_{mk}\, \delta_{nl}, \tag{58}$$

where the bar denotes complex conjugation. Moreover, two subspaces generated by bases $\mathfrak{S}(L)$ and $\mathfrak{S}(M)$, with $L \neq M$, are mutually orthogonal. On the other hand the g-deformed polynomials $h^g_{m,n}$ cannot be expected to form an orthogonal set, except in very special cases. However, as we demonstrate below, it is possible to construct a dual family of deformed polynomials $\widetilde{h}^g_{m,n}$, $m+n=L$, which are *biorthogonal* with respect to the $h^g_{m,n}$, $m+n=L$, Indeed, we have the result:

Theorem 4.1. *The basis dual to* $h_{m,n}(z,\overline{z})$, $m+n=L$, $L=0,1,2,\ldots$, *in* $\mathfrak{H}(\mathbb{C})$, *consists of the deformed polynomials*

$$\widetilde{h}^g_{m,n} = [T(g,L)^*]^{-1} h_{m,n} = h^{(g^*)^{-1}}_{m,n}, \qquad m+n=L, \tag{59}$$

which are biorthogonal with respect to the $h^g_{m,n}$, $m+n=L$, *i.e.*,

$$\int_{\mathbb{C}} \overline{\widetilde{h}^g_{L-n,n}(z,\overline{z})} h^g_{M-k,k}(z,\overline{z})\, d\nu(z,\overline{z}) = \delta_{LM}\, \delta_{nk}, \tag{60}$$

where, $n=0,1,2,\ldots,L$, $k=0,1,2,\ldots,M$.

Proof. Since the matrix $T(g,L)$, with matrix elements $M(g,L)_{rk}$, constitutes a representation of $GL(2,\mathbb{C})$ on the subspace $\mathfrak{H}_L(\mathbb{C})$ of $\mathfrak{H}(\mathbb{C})$, generated by the basis $\mathfrak{S}(L)$, we know that $T(g,L)^{-1} = T(g^{-1},L)$. From (53) it also follows that $T(g,L)^* = T(g^*,L)$. Thus the second equality in (60) follows. An easy computation establishes the biorthogonality relation (60). ☐

In fact, as shown in [2], a somewhat more general orthogonality relation can be obtained. To summarize, the Hilbert space $\mathfrak{H}(\mathbb{C})$ decomposes into the orthogonal direct sum

$$\mathfrak{H}(\mathbb{C}) = \bigoplus_{L=0}^{\infty} \mathfrak{H}_L(\mathbb{C}),$$

of $(L+1)$-dimensional subspaces $\mathfrak{H}_L(\mathbb{C})$, spanned by the orthonormal basis vectors $\mathfrak{S}(L)$, consisting of the complex Hermite polynomials $h_{L-k.k}$, $k = 0, 1, 2, \ldots, L$. On each such subspace the operators $T(g, L)$, $g \in GL(2, \mathbb{C})$ define an $(L+1) \times (L+1)$-matrix representation of $GL(2, \mathbb{C})$. For each $g \in GL(2, \mathbb{C})$ one obtains a set of g-deformed complex Hermite polynomials $h^g_{L-k.k} = T(g, L) h_{L-k,k}$, $k = 0, 1, 2, \ldots, L$, in $\mathfrak{H}_L(\mathbb{C})$ and a biorthogonal set $\widetilde{h}^g_{L-k.k}$, $k = 0, 1, 2, \ldots, L$, which constitute a family of g'-deformed complex Hermite polynomials, with $g' = (g^{-1})^*$. In particular, when g is the identity matrix, the two sets coincide with the (undeformed) complex Hermite polynomials $h_{m,n}$.

5. Back to noncommutative quantum mechanics

Let us specialize to Hermitian group elements $g \in GL(2, \mathbb{C})$ of the type

$$g = \begin{pmatrix} \alpha & \beta \\ \overline{\beta} & \alpha \end{pmatrix}, \quad \alpha \in \mathbb{R}, \ \ 0 < |\alpha| < 1, \quad \beta = i\sqrt{1 - \alpha^2}. \tag{61}$$

For such a matrix we denote the deformed operators a^g_i and $a^{g\dagger}_i$ by a^α_i and $a^{\alpha\dagger}_i$, respectively. They are seen to obey the commutation relations

$$[a^\alpha_i, a^{\alpha\dagger}_i] = 1, \quad [a^\alpha_i, a^\alpha_j] = 0, \quad [a^\alpha_1, a^{\alpha\dagger}_2] = 2i\alpha\sqrt{1 - \alpha^2}, \quad i, j = 1, 2. \tag{62}$$

These would be the same commutation relations as obeyed by the operators A_i, $A^\dagger_i$ in (6) if we were to set $A_i = a^\alpha_i$, $A^\dagger_i = a^{\alpha\dagger}_i$ and $\vartheta = 2\alpha\sqrt{1 - \alpha^2}$. In other words a matrix of the type (61) is characteristic of a noncommutative model of quantum mechanics, obeying commutation relations of the type given in (1)–(3) with $\gamma = \vartheta$. Denoting the associated deformed polynomials $h^g_{m,n}$ by $h^\alpha_{m,n}$ we say that the biorthogonal system of deformed complex Hermite polynomials $\{h^\alpha_{m,n}, \widetilde{h}^\alpha_{m,n}, m, n = 0, 1, 2, \ldots\}$, is naturally associated to this model of noncommutative quantum mechanics, in the same way as the orthonormal set of polynomials $h_{m,n}$ is associated to the standard quantum mechanics of two degrees of freedom. Note, however, that the matrix associated to $\widetilde{h}^\alpha_{m,n}$, which is the inverse of g in (61) is not of the same type as g. However, had we allowed a somewhat more general biorthogonality relation, e.g., of the type

$$\int_{\mathbb{C}} \overline{\widetilde{h}^\alpha_{L-n,n}(z, \overline{z})} h^\alpha_{M-k,k}(z, \overline{z}) \, d\nu(z, \overline{z}) = \lambda_{L,n} \delta_{LM} \, \delta_{nk},$$

where the $\lambda_{L,n}$ are positive constants, it would have been possible to find dual matrices under the action of which the commutation relations of noncommutative quantum mechanics would be preserved.

As an example, consider the Hermitian matrix

$$g' = \begin{pmatrix} \alpha & -\beta \\ -\overline{\beta} & \alpha. \end{pmatrix}$$

Comparing with (61), we see that $g'g = \Delta\mathbb{I}_2$, where Δ is the determinant of g (or g') and $\mathbb{I}_2$ the 2×2 identity matrix. Defining the polynomials $\widetilde{h}^{\alpha}_{L-n,n}$ using this matrix, it is not hard to see from (52) that $T(g', L)T(g, L) = \Delta^L\mathbb{I}_{L+1}$, so that $\lambda_L, n = \Delta^L$.

To end this section let us note that we discussed here the model of non-linear quantum mechanics in which we took $\vartheta = \gamma$, which means that the noncommutativity in the two position and the two momentum operators in (2) and (3) are of the same amount. This meant that we had a system of two independent bosons as described by the commutation relations ((6), in particular the relation $[A_1, A_2] = 0$. On the other hand this condition was necessary to ensure that the ground state $|0,0\rangle$ remained the same after the transformation (33).

6. A note on some associated algebras of bilinear generators

We compile in this section some interesting results on algebras built out of bilinear combinations of the creation and annihilation operators $a_i^{\alpha\dagger}$, a_i^{α}, $i = 1, 2$, for a fixed matrix of the type (61). Note first that the operators

$$J_1 = \frac{1}{2}\left(a_1^{\dagger}a_2 + a_2^{\dagger}a_1\right), \quad J_2 = \frac{1}{2i}\left(a_1^{\dagger}a_2 - a_2^{\dagger}a_1\right), \quad J_3 = \frac{1}{2}\left(a_1^{\dagger}a_1 - a_2^{\dagger}a_2\right), \tag{63}$$

obey the commutation relations of the generators of $SU(2)$, i.e., $[J_1, J_2] = iJ_3$ (cyclic). If we add to this set $J_4 = \frac{1}{2}\left(a_1^{\dagger}a_1 + a_2^{\dagger}a_2\right)$, we see that it commutes with the other three. We would like to study the group(s) generated by the *deformed* operators

$$J_1^{\alpha} = \frac{1}{2}\left(a_1^{\alpha\dagger}a_2^{\alpha} + a_2^{\alpha\dagger}a_1^{\alpha}\right), \ J_2^{\alpha} = \frac{1}{2i}\left(a_1^{\alpha\dagger}a_2^{\alpha} - a_2^{\alpha\dagger}a_1^{\alpha}\right), \ J_3^{\alpha} = \frac{1}{2}\left(a_1^{\alpha\dagger}a_1^{\alpha} - a_2^{\alpha\dagger}a_2^{\alpha}\right), \tag{64}$$

built by replacing the $a_i^{\dagger}$, a_i by the $a_i^{\alpha\dagger}$, a_i^{α}, which obey (see also [22]) the commutation relations

$$[a_j^{\alpha}, a_k^{\alpha\dagger}] = \delta_{jk} + \varepsilon_{jk}2i\alpha\sqrt{1-\alpha^2}, \tag{65}$$

where ε_{jk} is the usual anti-symmetric two-tensor. Using (65), we get

$$\begin{aligned} &[J_1^{\alpha}, J_2^{\alpha}] = iJ_3^{\alpha}, && [J_2^{\alpha}, J_3^{\alpha}] = iJ_1^{\alpha}, \\ &[J_3^{\alpha}, J_4^{\alpha}] = i\vartheta J_1^{\alpha}, && [J_4^{\alpha}, J_1^{\alpha}] = i\vartheta J_3^{\alpha}, \\ &[J_3^{\alpha}, J_1^{\alpha}] = iJ_2^{\alpha} + i\vartheta J_4^{\alpha}, && [J_2^{\alpha}, J_4^{\alpha}] = 0, \end{aligned} \tag{66}$$

where again $\vartheta = 2\alpha\sqrt{1-\alpha^2}$ and J_4^{α} is defined as

$$J_4^{\alpha} = \frac{1}{2}\left(a_1^{\alpha\dagger}a_1^{\alpha} + a_2^{\alpha\dagger}a_2^{\alpha}\right). \tag{67}$$

The above commutation relations are taken to hold for $0 < \vartheta < 1$ and $0 < \alpha < 1$.

In order to analyze the Lie algebra generated by the commutation relations (66), which we denote by $\mathfrak{g}$, it is convenient to make a basis change. We identify two mutually commuting subalgebras $\mathfrak{g}_1$ and $\mathfrak{g}_2$:

$$\{X_1^\vartheta, X_2^\vartheta, X_3^\vartheta\} \equiv \{iJ_1^\alpha, iJ_3^\alpha, i(J_2^\alpha + \vartheta J_4^\alpha)\} \in \mathfrak{g}_1,$$
$$\{Y^\vartheta\} \equiv \{\vartheta J_2^\alpha + J_4^\alpha\} \in \mathfrak{g}_2.$$

We then have the commutation relations,

$$[X_1^\vartheta, X_2^\vartheta] = X_3^\vartheta, \qquad [X_2^\vartheta, X_3^\vartheta] = 1 - \vartheta^2)X_1^\vartheta, \qquad [X_3^\vartheta, X_1^\vartheta] = 1 - \vartheta^2)X_2^\vartheta,$$
$$[Y^\vartheta, X_i^\vartheta] = 0, \tag{68}$$

so that $\mathfrak{g} = \mathfrak{g}_1 \oplus \mathfrak{g}_2$.

In the limit $\vartheta \to 0$, (68) leads to

$$[X_i^0, X_j^0] = \varepsilon_{ijk} X_k^0, \quad [Y^0, X_i^0] = 0, \quad i, j = 1, 2, 3. \tag{69}$$

In other words, $\mathfrak{g}_1 = \mathfrak{su}(2)$ as expected and hence,

$$\mathfrak{g} = \mathfrak{su}(2) \oplus \mathfrak{u}(1), \qquad \vartheta = 0. \tag{70}$$

In the other limit, i.e., $\vartheta \to 1$ (the maximum value) we get

$$[X_1^1, X_3^1] = [X_2^1, X_3^1] = 0, \quad [X_1^1, X_2^1] = X_3^1, \quad [Y^1, X_i^1] = 0, \tag{71}$$

$i = 1, 2, 3$, which is a nilradical basis, isomorphic to the Heisenberg algebra $\mathfrak{h}$ [10]. Thus,

$$\mathfrak{g} = \mathfrak{h} \oplus \mathfrak{u}(1), \qquad \vartheta = 1. \tag{72}$$

Finally for $0 < \vartheta < 1$, with the re-scaled generators,

$$Z_1^\vartheta = \sqrt{1 - \vartheta^2} X_1^\vartheta, \quad Z_2^\vartheta = X_2^\vartheta, \quad Z_3^\vartheta = \sqrt{1 - \vartheta^2} X_3^\vartheta, \tag{73}$$

we again get

$$[Z_i^\vartheta, Z_j^\vartheta] = \varepsilon_{ijk} Z_k^\vartheta, \quad [Y^\vartheta, Z_i^\vartheta] = 0, \quad i, j = 1, 2, 3, \tag{74}$$

as in (69). Once again,

$$\mathfrak{g} = \mathfrak{su}(2) \oplus \mathfrak{u}(1), \qquad 0 < \vartheta < 1. \tag{75}$$

It is interesting note that, except in the case where $\vartheta = 1$, the algebra generated by the deformed generators is that of $\mathfrak{su}(2) \oplus \mathfrak{u}(1)$, exactly as in the undeformed case (63). In the limit of , $\vartheta = 1$, $\alpha^2 = \frac{1}{2}$ and the commutation relation $[a_1^\alpha, a_2^{\alpha\dagger}] = 2i\alpha\sqrt{1 - \alpha^2}$ in (62) becomes $[a_1^\alpha, a_2^{\alpha\dagger}] = i$.

Acknowledgement

We would like to thank P. Winternitz, CRM, Université de Montréal, for useful discussions. While working on this paper NMS would like to recognize the Ministry of Higher Education (MOHE) of Malaysia for the support she received under the Skim Latihan Akademik Institut Pengajian Tinggi (SLAI) Universiti Putra Malaysia (UPM) scholarship. One of us (STA) would like to acknowledge a grant from the Natural Science and Engineering Research Council (NSERC) of Canada.

References

[1] S.T. Ali, K. Górska, A. Horzela, and F.H. Szafraniec, *Squeezed states and Hermite polynomials in a complex variable*, J. Math. Phys. **55**: 012107 (2014).

[2] S.T. Ali, M.E.H. Ismail and N.M. Shah, *Deformed complex Hermite polynomials*, 12 pp., Concordia University preprint (2014).

[3] F. Balogh, N.M. Shah and S.T. Ali, *Some biorthogonal families of polynomials ariwsing in noncommutative quantum mechanics*, arXiv:1309.4163v1 [math-ph] 17 Sep. 2013.

[4] J.B. BenGeloun, J. Govaerts and M.N. Hounkonnou, *A (p,q)-deformed Landau problem in a spherical harmonic well: spectrum and non-commutative coordinates*, EPL, vol. 80 (2007).

[5] J. BenGeloun, S. Gangopadhyay and F.G. Scholtz, *Harmonic oscillator in a background magnetic field in noncommutative quantum phase space*, EPL, vol. 86 (2009).

[6] N. Cotfas, J.-P. Gazeau, K. Górska, *Complex and real Hermite polynomials and related quantizations*, J. Phys.A: Math. Theor. **43**: 305304 (2010).

[7] S.J.L. van Eindhoven and J.L.H. Meyers, *New orthogonality relations for the Hermite polynomials and related Hilbert spaces*, J. Math. Anal. Appl. **146**, 89 (1990).

[8] J. Gamboa, M. Loewe and J.C. Rojas, *Non-Commutative Quantum Mechanics*, Phys. Rev. D, vol. 64, issue 6 (2001).

[9] A. Ghanmi, *A class of generalized complex Hermite polynomials*, J. Math. Anal. and App. **340** 1395–1406 (2008).

[10] B.C. Hall, *Lie Groups, Lie Algebras, and Representations: An Elementary Introduction*. Springer, Graduate Texts in Mathematics, no. 222, (2003).

[11] K. Hannabuss, An Introduction to Quantum Theory, Oxford University Press, Oxford (2004).

[12] M.E.H. Ismail, *Analytic properties of complex Hermite polynomials*, Trans. Am. Math. Soc., to appear.

[13] A. Intissar and A. Intissar, *Spectral properties of the Cauchy transform on* $L_2(\mathbb{C}, e^{-|z|^2}\lambda(z))$, J. Math. Anal. and Applic. **313**, 400–418 (2006).

[14] K. Itô, *Complex multiple Wiener integral*, Jap. J. Math. **22**, 63–86 (1952).

[15] K. Thirulogasanthar and S. Twareque Ali, *Regular subspaces of a quaternionic Hilbert space from quaternionic Hermite polynomials and associated coherent states*, J. Math. Phys. **54**: 013506 (2013).

[16] A. Torre, *Generalized Zernike or disc polynomials: An application in quantum optics*, J. Comput. Appl. Math. **222**, 622–644 (2008).

[17] S. Jing and B. Lin, *A new kind of representations on noncommutative phase space*, (2009) math-ph/0902.3782v1.

[18] H. Matsumoto, *Quadratic Hamiltonians and associated orthogonal polynomials*, J. Funct. Anal. **136**, 214–225 (1996).

[19] W.F. McGee, *Complex Gaussian noise moments*, IEEE Trans. Inform. Theory **IT-17**, 149–157 (1971).

[20] B. Muthukumar and B. Mitra, *Non-commutative oscillators and the commutative limit*, Phys. Rev. D 66, 027701(2002).

[21] F.G. Scholtz, L. Gouba, A. Hafverand, and C.M. Rohwer, *Formulation, interpretation and application of non-commutative quantum mechanics*, J. Phys. A: Math. Theor. 42 (2009).

[22] K. Li, J. Wang and C. Chen, *Representation of Noncommutative Phase Space*, Modern Physics Letters A, **20**, 2165–2174 (2005).

[23] A. Wünsche, *Generalized Zernike or disc polynomials*, J. Comput. Appl. Math. **174**, 135–163 (2005).

[24] A. Wünsche, *General Hermite and Laguerre two-dimensional polynomials*, J. Phys. AL Math. Gen. **33**, 1603–1629 (2000).

[25] Y. Xu, *Complex versus real orthogonal polynomials of two variables*, arXiv:1307.7819v1 [math.CA] 30 Jul. 2013.

F. Balogh
Scuola Internazionale Superiore di Studi Avanzati
via Bonomea 265
I-34136 Trieste, Italy
e-mail: fbalogh@sissa.it

Nurisya M. Shah
Department of Physics
Faculty of Science
University Putra Malaysia
43000 UPM Serdang, Malaysia
e-mail: risya@upm.edu.my

S. Twareque Ali
Department of Mathematics and Statistics
Concordia University
1455 de Maisonneuve Blvd. West
Montreal, Quebec, Canada H3G 1M8

Operator Theory:
Advances and Applications, Vol. 247, 173–184
© Springer International Publishing Switzerland 2015

An Eigenvalue Inequality for Schrödinger Operators with δ- and δ'-interactions Supported on Hypersurfaces

Vladimir Lotoreichik and Jonathan Rohleder

Abstract. We consider self-adjoint Schrödinger operators in $L^2(\mathbb{R}^d)$ with a δ-interaction of strength α and a δ'-interaction of strength β, respectively, supported on a hypersurface, where α and β^{-1} are bounded, real-valued functions. It is known that the inequality $0 < \beta \leq 4/\alpha$ implies inequality of the eigenvalues of these two operators below the bottoms of the essential spectra. We show that this eigenvalue inequality is strict whenever $\beta < 4/\alpha$ on a nonempty, open subset of the hypersurface. Moreover, we point out special geometries of the interaction support, such as broken lines or infinite cones, for which strict inequality of the eigenvalues even holds in the borderline case $\beta = 4/\alpha$.

Mathematics Subject Classification (2010). 35P15, 35J20.

Keywords. δ- and δ'-interactions on a hypersurface, discrete spectrum, eigenvalue inequality.

1. Introduction

Schrödinger operators with δ- and δ'-interactions supported on hypersurfaces have attracted considerable attention in recent years, see the monograph [EK], the review paper [E08] and, e.g., [BEL14a, EI01, EJ13, EP14], as well as [BEL14b, BEW09, CDR08, DR13, EN03, L13] for interactions supported on hypersurfaces with special geometries. In this note we focus on the self-adjoint Schrödinger operators $-\Delta_{\delta,\alpha}$ and $-\Delta_{\delta',\beta}$ in $L^2(\mathbb{R}^d)$, $d \geq 2$, which are formally given by

$$-\Delta_{\delta,\alpha} = -\Delta - \alpha\langle\cdot,\delta_\Sigma\rangle\delta_\Sigma \quad \text{and} \quad -\Delta_{\delta',\beta} = -\Delta - \beta\langle\cdot,\delta'_\Sigma\rangle\delta'_\Sigma,$$

where Δ is the Laplacian and the support Σ of the interactions is a Lipschitz hypersurface; we emphasize that Σ is not required to be compact or connected,

The first author gratefully acknowledges financial support by the Austrian Science Fund (FWF), project P 25162-N26.

see Section 2.1 for the details. These operators can be defined rigorously, e.g., via quadratic forms, as is indicated in Section 2.2 below. We assume that the strengths α and β of the interactions are real-valued functions on Σ with $\alpha, \beta^{-1} \in L^\infty(\Sigma)$.

Let us denote by $\sigma_{\rm ess}(-\Delta_{\delta,\alpha})$ and $\sigma_{\rm ess}(-\Delta_{\delta',\beta})$ the essential spectra of $-\Delta_{\delta,\alpha}$ and $-\Delta_{\delta',\beta}$, respectively. Moreover, let

$$\lambda_1(-\Delta_{\delta,\alpha}) \le \lambda_2(-\Delta_{\delta,\alpha}) \le \cdots < \inf \sigma_{\rm ess}(-\Delta_{\delta,\alpha})$$

and

$$\lambda_1(-\Delta_{\delta',\beta}) \le \lambda_2(-\Delta_{\delta',\beta}) \le \cdots < \inf \sigma_{\rm ess}(-\Delta_{\delta',\beta})$$

be the eigenvalues of $-\Delta_{\delta,\alpha}$ and $-\Delta_{\delta',\beta}$, respectively, below the bottom of the essential spectrum, counted with multiplicities; for many choices of Σ the existence of such eigenvalues has been proved, see, e.g., [BEL14a, BEL14b, BEW09, EI01, EK03].

In [BEL14a, Theorem 3.6] for $0 < \beta \le \frac{4}{\alpha}$ the operator inequality

$$U^{-1}\big(-\Delta_{\delta',\beta}\big)U \le -\Delta_{\delta,\alpha}$$

was established, where U is a unitary transformation in $L^2(\mathbb{R}^d)$; cf. (3.1) below. This implies $\inf \sigma_{\rm ess}(-\Delta_{\delta',\beta}) \le \inf \sigma_{\rm ess}(-\Delta_{\delta,\alpha})$ as well as

$$\lambda_n(-\Delta_{\delta',\beta}) \le \lambda_n(-\Delta_{\delta,\alpha})$$

for all $n \in \mathbb{N}$ such that $\lambda_n(-\Delta_{\delta,\alpha}) < \inf \sigma_{\rm ess}(-\Delta_{\delta',\beta})$. The aim of this note is to sharpen the latter inequality as follows.

Theorem A. *Let $0 < \beta \le 4/\alpha$ and assume that $\beta|_\sigma < 4/\alpha|_\sigma$ on a nonempty, open set $\sigma \subset \Sigma$. Then*

$$\lambda_n(-\Delta_{\delta',\beta}) < \lambda_n(-\Delta_{\delta,\alpha})$$

holds for all $n \in \mathbb{N}$ such that $\lambda_n(-\Delta_{\delta,\alpha}) < \inf \sigma_{\rm ess}(-\Delta_{\delta',\beta})$.

If the hypersurface Σ is compact, it is known that $\sigma_{\rm ess}(-\Delta_{\delta',\beta}) = \sigma_{\rm ess}(-\Delta_{\delta,\alpha}) = [0,\infty)$; cf. [BEL14a, Theorem 4.2]. Therefore in this case Theorem A implies strict inequality between all negative eigenvalues of $-\Delta_{\delta,\alpha}$ and $-\Delta_{\delta',\beta}$; note that if Σ is compact and sufficiently regular, these operators have only finitely many negative eigenvalues, see [BEKS94, Theorem 4.2] and [BLL13, Theorem 3.14].

Corollary. *Let the assumptions of Theorem A be satisfied and let, additionally, Σ be compact and C^∞-smooth. Then*

$$\lambda_n(-\Delta_{\delta',\beta}) < \lambda_n(-\Delta_{\delta,\alpha}), \quad n = 1, \dots, N(-\Delta_{\delta,\alpha}),$$

holds, where $N(-\Delta_{\delta,\alpha})$ denotes the number of negative eigenvalues of $-\Delta_{\delta,\alpha}$.

Our proof of Theorem A is based on an idea which was suggested by Filonov in [F05] and which was used and modified later on in various spectral problems, see [FL10, GM09, K10, R14]. We remark that the result of Theorem A can be proved analogously for the more general case of Σ being a Lipschitz partition of $\mathbb{R}^d$ as considered in [BEL14a]. However, in order to avoid technicalities we restrict ourselves to the case of a hypersurface.

Besides the general result of Theorem A, which is proved in Section 3, in Section 4 we discuss several examples of special geometries of Σ for which the strict inequality of Theorem A holds even in the borderline case $\beta = 4/\alpha$, for constant strengths α, β. Among these examples there are the cases of a broken line in $\mathbb{R}^2$ and an infinite cone in $\mathbb{R}^3$.

2. Preliminaries

2.1. Lipschitz hypersurfaces and weak normal derivatives

Let us first recall some basic facts and notions. For an arbitrary open set $\Omega \subset \mathbb{R}^d$, $d \geq 2$, we write $(\cdot,\cdot)_\Omega$ for both the inner products in the spaces $L^2(\Omega)$ and $L^2(\Omega, \mathbb{C}^d)$ of scalar and vector-valued square-integrable functions, respectively, without any danger of confusion; the associated norms are denoted by $\|\cdot\|_\Omega$. As usual, $H^1(\Omega)$ is the Sobolev space of order one and $H^1_0(\Omega)$ denotes the closure of the space of smooth functions with compact supports in $H^1(\Omega)$.

In the following we understand Lipschitz domains in the general sense of, e.g., [St, §VI.3]; in particular, we allow noncompact boundaries. We write Σ for the boundary of a Lipschitz domain Ω and denote the inner product in $L^2(\Sigma)$ by $(\cdot,\cdot)_\Sigma$ and the corresponding norm by $\|\cdot\|_\Sigma$. For $u \in H^1(\Omega)$ we denote by $u|_\Sigma$ the trace of u on Σ, which extends the restriction map of smooth functions to Σ as a bounded linear operator from $H^1(\Omega)$ to $L^2(\Sigma)$.

For our purposes it is convenient to deal with the Laplacian as well as the normal derivatives of appropriate Sobolev functions in the following weak sense; such definitions can be found, e.g., in the textbook [McL].

Definition 2.1. Let $\Omega \subset \mathbb{R}^d$ be a Lipschitz domain.

(i) Let $u \in H^1(\Omega)$. If there exists $f \in L^2(\Omega)$ with

$$(\nabla u, \nabla v)_\Omega = (f, v)_\Omega \quad \text{for all } v \in H^1_0(\Omega),$$

we say $\Delta u \in L^2(\Omega)$ and set $-\Delta u := f$.

(ii) Let $u \in H^1(\Omega)$ with $\Delta u \in L^2(\Omega)$. If there exists $b \in L^2(\Sigma)$ with

$$(\nabla u, \nabla v)_\Omega - (-\Delta u, v)_\Omega = (b, v|_\Sigma)_\Sigma \quad \text{for all } v \in H^1(\Omega),$$

we say $\partial_\nu u|_\Sigma \in L^2(\Sigma)$ and set $\partial_\nu u|_\Sigma := b$.

We remark that $\partial_\nu u|_\Sigma$ is unique if it exists. For each sufficiently smooth $u \in L^2(\Omega)$ the function $\partial_\nu u|_\Sigma$ on Σ is the usual derivative in the direction of the outer unit normal, which follows immediately from the first Green identity.

We call $\Sigma \subset \mathbb{R}^d$ a *Lipschitz hypersurface* if Σ coincides with the boundary of a Lipschitz domain $\Omega_1 \subset \mathbb{R}^d$. In this case also $\Omega_2 := \mathbb{R}^d \setminus \overline{\Omega_1}$ is a Lipschitz domain with the same boundary Σ, and Σ separates $\mathbb{R}^d$ into Ω_1 and Ω_2. Note that we do not require Ω_1, Ω_2, or Σ to be connected; see, e.g., Figure 3 in Example 4.3 below.

For a Lipschitz hypersurface Σ and the corresponding Lipschitz domains Ω_1 and Ω_2 as above we occasionally write a function $u \in L^2(\mathbb{R}^d)$ as $u = u_1 \oplus u_2$, where $u_j = u|_{\Omega_j}$, $j = 1, 2$, referring to the orthogonal decomposition $L^2(\mathbb{R}^d) =$

$L^2(\Omega_1) \oplus L^2(\Omega_2)$. Moreover, we write $\partial_{\nu_j} u_j|_\Sigma$, $j = 1, 2$, for the normal derivative of u_j in Definition 2.1 (ii).

For the following definition cf. [BEL14a, Section 2.3].

Definition 2.2. Let Σ be a Lipschitz hypersurface which separates $\mathbb{R}^d$ into two Lipschitz domains Ω_1 and Ω_2. Let $u = u_1 \oplus u_2 \in H^1(\mathbb{R}^d)$ with $\Delta u_j \in L^2(\Omega_j)$, $j = 1, 2$. If there exists $\widetilde{b} \in L^2(\Sigma)$ such that

$$\big(\nabla u, \nabla v\big)_{\mathbb{R}^d} - \big((-\Delta u_1) \oplus (-\Delta u_2), v\big)_{\mathbb{R}^d} = (\widetilde{b}, v|_\Sigma)_\Sigma \quad \text{for all } v \in H^1(\mathbb{R}^d),$$

we say $[\partial_\nu u]_\Sigma \in L^2(\Sigma)$ and set $[\partial_\nu u]_\Sigma := \widetilde{b}$.

Note that $[\partial_\nu u]_\Sigma$ is unique if it exists; cf. [BEL14a, Section 2.3]. The interpretation of $[\partial_\nu u]_\Sigma$ is provided in the following lemma.

Lemma 2.3. *Let Σ be a Lipschitz hypersurface which separates $\mathbb{R}^d$ into two Lipschitz domains Ω_1 and Ω_2. Let $u = u_1 \oplus u_2 \in H^1(\mathbb{R}^d)$ with $\Delta u_j \in L^2(\Omega_j)$ and $\partial_{\nu_j} u_j|_\Sigma \in L^2(\Sigma)$, $j = 1, 2$. Then $[\partial_\nu u]_\Sigma \in L^2(\Sigma)$ and*

$$[\partial_\nu u]_\Sigma = \partial_{\nu_1} u_1|_\Sigma + \partial_{\nu_2} u_2|_\Sigma.$$

Proof. Let us fix an arbitrary $v \in H^1(\mathbb{R}^d)$. Clearly $v_j \in H^1(\Omega_j)$ holds for $j = 1, 2$. Thus employing Definition 2.1 (ii) we get

$$\begin{aligned}
&\big(\nabla u, \nabla v\big)_{\mathbb{R}^d} - \big((-\Delta u_1) \oplus (-\Delta u_2), v\big)_{\mathbb{R}^d} \\
&\quad = \Big[\big(\nabla u_1, \nabla v_1\big)_{\Omega_1} - \big(-\Delta u_1, v_1\big)_{\Omega_1}\Big] + \Big[\big(\nabla u_2, \nabla v_2\big)_{\Omega_2} - \big(-\Delta u_2, v_2\big)_{\Omega_2}\Big] \\
&\quad = (\partial_{\nu_1} u_1|_\Sigma + \partial_{\nu_2} u_2|_\Sigma, v|_\Sigma)_\Sigma
\end{aligned}$$

and the claim follows from Definition 2.2. □

2.2. Schrödinger operators with δ- and δ'-interactions

In this paragraph we recall the mathematically rigorous definitions of the self-adjoint Schrödinger operators with δ- and δ'-interactions supported on a Lipschitz hypersurface Σ. For the required material on semibounded, closed sesquilinear forms and corresponding self-adjoint operators we refer the reader to [K, Chapter VI].

Definition 2.4. Let Σ be a Lipschitz hypersurface which separates $\mathbb{R}^d$ into two Lipschitz domains Ω_1 and Ω_2.

(i) The Schrödinger operator $-\Delta_{\delta,\alpha}$ in $L^2(\mathbb{R}^d)$ with a δ-interaction supported on Σ of strength $\alpha : \Sigma \to \mathbb{R}$ with $\alpha \in L^\infty(\Sigma)$ is the unique self-adjoint operator in $L^2(\mathbb{R}^d)$ which corresponds to the densely defined, symmetric, lower semibounded, closed sesquilinear form

$$\mathfrak{a}_{\delta,\alpha}[u, v] = (\nabla u, \nabla v)_{\mathbb{R}^d} - (\alpha u|_\Sigma, v|_\Sigma)_\Sigma, \quad \operatorname{dom} \mathfrak{a}_{\delta,\alpha} = H^1(\mathbb{R}^d), \tag{2.1}$$

(cf. [BEKS94, Section 2] for C^1-smooth Σ and [BEL14a, Proposition 3.1] for the Lipschitz case).

(ii) The Schrödinger operator $-\Delta_{\delta',\beta}$ in $L^2(\mathbb{R}^d)$ with a δ'-interaction supported on Σ of strength $\beta : \Sigma \to \mathbb{R}$ with $\beta^{-1} \in L^\infty(\Sigma)$ is the self-adjoint operator in $L^2(\mathbb{R}^d)$ which corresponds to the densely defined, symmetric, lower semibounded and closed sesquilinear form

$$\mathfrak{a}_{\delta',\beta}[u,v] = (\nabla u_1, \nabla v_1)_{\Omega_1} + (\nabla u_2, \nabla v_2)_{\Omega_2} - \Big(\frac{1}{\beta}(u_1|_\Sigma - u_2|_\Sigma), (v_1|_\Sigma - v_2|_\Sigma)\Big)_\Sigma,$$

$$\operatorname{dom} \mathfrak{a}_{\delta',\beta} = H^1(\mathbb{R}^d \setminus \Sigma), \tag{2.2}$$

(cf. [BEL14a, Proposition 3.1]).

The actions and domains of the operators $-\Delta_{\delta,\alpha}$ and $-\Delta_{\delta',\beta}$ can be characterized in the following way, using the weak Laplacians and normal derivatives from Definition 2.1 and Definition 2.2.

Proposition 2.5. [BEL14a, Theorem 3.3] *Let Σ be a Lipschitz hypersurface which separates $\mathbb{R}^d$ into two Lipschitz domains Ω_1 and Ω_2. Moreover, let $\alpha, \beta : \Sigma \to \mathbb{R}$ be functions such that $\alpha, \beta^{-1} \in L^\infty(\Sigma)$. Then the self-adjoint operators $-\Delta_{\delta,\alpha}$ and $-\Delta_{\delta',\beta}$ in Definition 2.4 have the following representations.*

(i) $-\Delta_{\delta,\alpha} u = (-\Delta u_1) \oplus (-\Delta u_2)$, *and* $u = u_1 \oplus u_2 \in \operatorname{dom}(-\Delta_{\delta,\alpha})$ *if and only if*

(a) $u \in H^1(\mathbb{R}^d)$,

(b) $\Delta u_j \in L^2(\Omega_j)$, $j = 1, 2$, *and*

(c) $[\partial_\nu u]_\Sigma \in L^2(\Sigma)$ *exists in the sense of Definition 2.2 and*

$$[\partial_\nu u]_\Sigma = \alpha u|_\Sigma.$$

(ii) $-\Delta_{\delta',\beta} u = (-\Delta u_1) \oplus (-\Delta u_2)$, *and* $u = u_1 \oplus u_2 \in \operatorname{dom}(-\Delta_{\delta',\beta})$ *if and only if*

(a′) $u_j \in H^1(\Omega_j)$, $j = 1, 2$,

(b′) $\Delta u_j \in L^2(\Omega_j)$, $j = 1, 2$, *and*

(c′) $\partial_{\nu_j} u_j|_\Sigma \in L^2(\Sigma)$ *exist in the sense of Definition 2.1 (ii), $j = 1, 2$, and*

$$u_1|_\Sigma - u_2|_\Sigma = \beta \partial_{\nu_1} u_1|_\Sigma = -\beta \partial_{\nu_2} u_2|_\Sigma.$$

3. Proof of Theorem A

In this section we provide a proof of Theorem A. As a first step we show the following proposition. In its formulation the unitary operator

$$U: L^2(\mathbb{R}^d) \to L^2(\mathbb{R}^d), \quad U(u_1 \oplus u_2) := u_1 \oplus (-u_2), \quad u_1 \in L^2(\Omega_1), u_2 \in L^2(\Omega_2), \tag{3.1}$$

appears, which was already mentioned in the introduction.

Proposition 3.1. *Let $0 < \beta \le 4/\alpha$. If*

$$W_\mu := U\big(\operatorname{dom}(-\Delta_{\delta,\alpha})\big) \cap \ker(-\Delta_{\delta',\beta} - \mu) = \{0\} \tag{3.2}$$

holds for each $\mu < \inf \sigma_{\rm ess}(-\Delta_{\delta',\beta})$ *then*

$$\lambda_n(-\Delta_{\delta',\beta}) < \lambda_n(-\Delta_{\delta,\alpha})$$

holds for all $n \in \mathbb{N}$ *such that* $\lambda_n(-\Delta_{\delta,\alpha}) < \inf \sigma_{\rm ess}(-\Delta_{\delta',\beta})$.

Proof. Let $N_{\delta,\alpha}(\cdot)$ and $N_{\delta',\beta}(\cdot)$ be the counting functions for the eigenvalues below the bottom of the essential spectrum of the operators $-\Delta_{\delta,\alpha}$ and $-\Delta_{\delta',\beta}$, respectively, that is,

$$N_{\delta,\alpha}(\mu) := \#\{k \in \mathbb{N} : \lambda_k(-\Delta_{\delta,\alpha}) \le \mu\}, \quad \mu < \inf \sigma_{\rm ess}(-\Delta_{\delta,\alpha}),$$

and

$$N_{\delta',\beta}(\mu) := \#\{k \in \mathbb{N} : \lambda_k(-\Delta_{\delta',\beta}) \le \mu\}, \quad \mu < \inf \sigma_{\rm ess}(-\Delta_{\delta',\beta}).$$

It follows from the min-max principle, see [BS, Chapter 10] or [S, Chapter 12], that these functions can be expressed as

$$N_{\delta,\alpha}(\mu) = \max\left\{\dim L \colon L \text{ subspace of } H^1(\mathbb{R}^d),\ \mathfrak{a}_{\delta,\alpha}[u] \le \mu\|u\|^2_{\mathbb{R}^d},\ u \in L\right\}$$

and

$$N_{\delta',\beta}(\mu) = \max\left\{\dim L \colon L \text{ subspace of } H^1(\mathbb{R}^d \setminus \Sigma),\ \mathfrak{a}_{\delta',\beta}[u] \le \mu\|u\|^2_{\mathbb{R}^d},\ u \in L\right\},$$

where $\mathfrak{a}_{\delta,\alpha}$ and $\mathfrak{a}_{\delta',\beta}$ are the sesquilinear forms in (2.1) and (2.2), respectively. Let $\mu < \inf \sigma_{\rm ess}(-\Delta_{\delta',\beta}) \le \inf \sigma_{\rm ess}(-\Delta_{\delta,\alpha})$ and define

$$F := U(\mathrm{span}\{\ker(-\Delta_{\delta,\alpha} - \lambda) \colon \lambda \le \mu\})$$

with U as in (3.1). Then $\dim F = N_{\delta,\alpha}(\mu)$ and

$$\mathfrak{a}_{\delta,\alpha}[U^{-1}u] \le \mu\|u\|^2_{\mathbb{R}^d}, \qquad u \in F, \tag{3.3}$$

where we have used the abbreviation $\mathfrak{a}_{\delta,\alpha}[w] := \mathfrak{a}_{\delta,\alpha}[w,w]$ for $w \in \mathrm{dom}\,\mathfrak{a}_{\delta,\alpha}$. For $u \in F$ and $v \in \ker(-\Delta_{\delta',\beta} - \mu)$ we have $u_1|_\Sigma = -u_2|_\Sigma$ and it follows from (2.2) that $u, v \in \mathrm{dom}\,\mathfrak{a}_{\delta',\beta}$ and

$$\begin{aligned}\mathfrak{a}_{\delta',\beta}[u+v] &= \|\nabla(u_1+v_1)\|^2_{\Omega_1} + \|\nabla(u_2+v_2)\|^2_{\Omega_2} \\ &\quad - \Big(\frac{1}{\beta}(2u_1|_\Sigma + (v_1|_\Sigma - v_2|_\Sigma)), 2u_1|_\Sigma + (v_1|_\Sigma - v_2|_\Sigma)\Big)_\Sigma \\ &= I + J + K,\end{aligned} \tag{3.4}$$

where

$$I := \|\nabla v_1\|^2_{\Omega_1} + \|\nabla v_2\|^2_{\Omega_2} - \Big(\frac{1}{\beta}(v_1|_\Sigma - v_2|_\Sigma), v_1|_\Sigma - v_2|_\Sigma\Big)_\Sigma,$$

$$J := \|\nabla u_1\|^2_{\Omega_1} + \|\nabla u_2\|^2_{\Omega_2} - \Big(\frac{4}{\beta}u_1|_\Sigma, u_1|_\Sigma\Big)_\Sigma,$$

and

$$K := 2\mathrm{Re}\Big[(\nabla u_1, \nabla v_1)_{\Omega_1} + (\nabla u_2, \nabla v_2)_{\Omega_2} - \Big(\frac{2}{\beta}u_1|_\Sigma, (v_1|_\Sigma - v_2|_\Sigma)\Big)_\Sigma\Big].$$

According to the choices of u an v and due to (3.3) we get

$$I = \mathfrak{a}_{\delta',\beta}[v] = \mu\|v\|^2_{\mathbb{R}^d} \quad \text{and} \quad J = \mathfrak{a}_{\delta,4/\beta}[U^{-1}u] \le \mathfrak{a}_{\delta,\alpha}[U^{-1}u] \le \mu\|u\|^2_{\mathbb{R}^d}, \tag{3.5}$$

since $\alpha \leq 4/\beta$. Moreover, Definition 2.1 (ii) and Proposition 2.5 (ii) give us

$$K = 2\mathrm{Re}\Big[\mu(u,v)_{\mathbb{R}^d} + \big(u_1|_\Sigma, \partial_{\nu_1} v_1|_\Sigma\big)_\Sigma + \big(u_2|_\Sigma, \partial_{\nu_2} v_2|_\Sigma\big)_\Sigma - \Big(\frac{2}{\beta} u_1|_\Sigma, (v_1|_\Sigma - v_2|_\Sigma)\Big)_\Sigma\Big],$$

where we have used that $-\Delta v_j = \mu v_j$, $j = 1, 2$. By Proposition 2.5 we have

$$u_1|_\Sigma = -u_2|_\Sigma \quad \text{and} \quad \partial_{\nu_1} v_1|_\Sigma = -\partial_{\nu_2} v_2|_\Sigma = \frac{1}{\beta}(v_1|_\Sigma - v_2|_\Sigma),$$

and hence we obtain

$$\begin{aligned} K &= 2\mathrm{Re}\Big[\mu(u,v)_{\mathbb{R}^d} + \Big(\frac{2}{\beta} u_1|_\Sigma, (v_1|_\Sigma - v_2|_\Sigma)\Big)_\Sigma - \Big(\frac{2}{\beta} u_1|_\Sigma, (v_1|_\Sigma - v_2|_\Sigma)\Big)_\Sigma\Big] \\ &= 2\mu \mathrm{Re}(u,v)_{\mathbb{R}^d}. \end{aligned}$$

Combining the above expression for K with (3.4) and (3.5) we arrive at

$$\mathfrak{a}_{\delta',\beta}[u+v] \leq \mu\|u\|^2_{\mathbb{R}^d} + 2\mu \mathrm{Re}(u,v)_{\mathbb{R}^d} + \mu\|v\|^2_{\mathbb{R}^d} = \mu\|u+v\|^2_{\mathbb{R}^d} \tag{3.6}$$

for all $u \in F$ and all $v \in \ker(-\Delta_{\delta',\beta} - \mu)$. From the assumption (3.2) we conclude

$$\dim\big(F + \ker(-\Delta_{\delta',\beta} - \mu)\big) = N_{\delta,\alpha}(\mu) + \dim\ker(-\Delta_{\delta',\beta} - \mu)$$

and thus (3.6) implies

$$N_{\delta',\beta}(\mu) \geq N_{\delta,\alpha}(\mu) + \dim\ker(-\Delta_{\delta',\beta} - \mu).$$

Hence,

$$\#\big\{k \in \mathbb{N} \colon \lambda_k(-\Delta_{\delta',\beta}) < \mu\big\} = N_{\delta',\beta}(\mu) - \dim\ker(-\Delta_{\delta',\beta} - \mu) \geq N_{\delta,\alpha}(\mu).$$

Choosing $\mu = \lambda_n(-\Delta_{\delta,\alpha})$ for an arbitrary $n \in \mathbb{N}$ such that $\mu < \inf \sigma_{\mathrm{ess}}(-\Delta_{\delta',\beta})$, it follows that

$$\#\big\{k \in \mathbb{N} \colon \lambda_k(-\Delta_{\delta',\beta}) < \lambda_n(-\Delta_{\delta,\alpha})\big\} \geq n.$$

Thus $\lambda_n(-\Delta_{\delta',\beta}) < \lambda_n(-\Delta_{\delta,\alpha})$ for all n with $\lambda_n(-\Delta_{\delta,\alpha}) < \inf \sigma_{\mathrm{ess}}(-\Delta_{\delta',\beta})$. This completes the proof of the proposition. □

We will now apply Proposition 3.1 in order to prove Theorem A.

Proof of Theorem A. Let $\sigma \subset \Sigma$ be a nonempty open set such that

$$\beta|_\sigma < (4/\alpha)|_\sigma. \tag{3.7}$$

By Proposition 3.1, in order to prove Theorem A it suffices to verify (3.2) for each $\mu < \inf \sigma_{\mathrm{ess}}(-\Delta_{\delta',\beta})$. Let us fix such a μ and let $u \in W_\mu$. Proposition 2.5 (ii) yields

$$-\Delta u_j = \mu u_j, \quad j = 1, 2, \tag{3.8}$$

and

$$u_1|_\Sigma - u_2|_\Sigma = \beta \partial_{\nu_1} u_1|_\Sigma = -\beta \partial_{\nu_2} u_2|_\Sigma. \tag{3.9}$$

On the other hand, from Proposition 2.5 (i) and Lemma 2.3 we obtain

$$u_1|_\Sigma + u_2|_\Sigma = 0 \quad \text{and} \quad \partial_{\nu_1} u_1|_\Sigma - \partial_{\nu_2} u_2|_\Sigma = \alpha u_1|_\Sigma. \tag{3.10}$$

The conditions (3.9) and (3.10) yield

$$\partial_{\nu_1} u_1|_\Sigma = \alpha u_1|_\Sigma + \partial_{\nu_2} u_2|_\Sigma = \frac{\alpha\beta}{2}\partial_{\nu_1} u_1|_\Sigma + \partial_{\nu_2} u_2|_\Sigma = \Big(\frac{\alpha\beta}{2} - 1\Big)\partial_{\nu_1} u_1|_\Sigma. \quad (3.11)$$

By (3.7) we have $\frac{\alpha\beta}{2}|_\sigma - 1 < 1$ on σ, hence (3.11) implies $\partial_{\nu_1} u_1|_\sigma = 0$. With the help of (3.9) and (3.10) it follows that $u_1|_\sigma = \frac{\beta}{2}\partial_{\nu_1} u_1|_\sigma = 0$. Let now Ω be a connected component of Ω_1 such that $\partial\Omega \cap \sigma \neq \varnothing$. As in the proof of [BR12, Proposition 2.5] let us choose a connected Lipschitz domain $\widetilde{\Omega}$ such that $\Omega \subset \widetilde{\Omega}$, $\partial\Omega \setminus \sigma \subset \partial\widetilde{\Omega}$, and $\widetilde{\Omega} \setminus \Omega$ has a nonempty interior. Then the function $\widetilde{u}$ with $\widetilde{u} = u_1$ on Ω and $\widetilde{u} = 0$ on $\widetilde{\Omega} \setminus \Omega$ belongs to $L^2(\widetilde{\Omega})$ and satisfies $-\Delta\widetilde{u} = \mu\widetilde{u}$ on $\widetilde{\Omega}$. Indeed, $\widetilde{u} \in H^1(\widetilde{\Omega})$ since $u_1|_\sigma = 0$. Moreover, for each $\widetilde{v} \in H^1_0(\widetilde{\Omega})$ we have

$$(\nabla\widetilde{u}, \nabla\widetilde{v})_{\widetilde{\Omega}} = (\nabla u_1, \nabla v)_\Omega = (-\Delta u_1, v)_\Omega + (\partial_{\nu_1} u_1|_{\partial\Omega}, v|_{\partial\Omega})_{\partial\Omega},$$

where v denotes the restriction of $\widetilde{v}$ to Ω. Since $v|_{\partial\Omega\setminus\sigma} = 0$ and $\partial_{\nu_1} u_1|_\sigma = 0$ it follows with the help of (3.8)

$$(\nabla\widetilde{u}, \nabla\widetilde{v})_{\widetilde{\Omega}} = (\mu u_1, v)_\Omega = (\mu\widetilde{u}, \widetilde{v})_{\widetilde{\Omega}},$$

thus $-\Delta\widetilde{u} = \mu\widetilde{u}$ by Definition 2.1 (i). As $\widetilde{u}$ vanishes on the nonempty interior of $\widetilde{\Omega}\setminus\Omega$, a unique continuation argument implies $\widetilde{u} = 0$, see, e.g., [RS-IV, Theorem XIII.63]. Hence u_1 is identically equal to zero on the connected component Ω of Ω_1.

It remains to conclude from this that $u = 0$ identically on $\mathbb{R}^d$. Indeed, since Σ separates $\mathbb{R}^d$ into the Lipschitz domains Ω_1 and Ω_2, there exists a connected component Λ of Ω_2 such that $\tau := \partial\Omega \cap \partial\Lambda \neq \varnothing$. Since $u_1|_\tau = \partial_{\nu_1} u_1|_\tau = 0$ it follows with the help of (3.9) that $u_2|_\tau = \partial_{\nu_2} u_2|_\tau = 0$; another application of unique continuation implies $u_2|_\Lambda = 0$. Repeating the same argument successively for the respective neighboring connected components finally it follows that $u = 0$ on all of $\mathbb{R}^d$, which completes the proof of the theorem. □

4. The borderline case $\beta = 4/\alpha$

In this section we present various examples with explicit geometries of the interaction support Σ, where $\beta = 4/\alpha$ and the strict eigenvalue inequality in Theorem A remains valid. In all the following examples the strengths of interactions α and β are constants.

Example 4.1. In this example we consider the broken line

$$\Sigma := \big\{(x, \cot(\theta)|x|) \in \mathbb{R}^2 \colon x \in \mathbb{R}\big\}, \qquad \theta \in (0, \pi/2),$$

which splits $\mathbb{R}^2$ into the two domains

$$\Omega_1 = \big\{(x,y) \in \mathbb{R}^2 \colon x \in \mathbb{R}, y > \cot(\theta)|x|\big\}$$

and

$$\Omega_2 = \big\{(x,y) \in \mathbb{R}^2 \colon x \in \mathbb{R}, y < \cot(\theta)|x|\big\};$$

cf. Figure 1. Moreover, we assume that $\beta = 4/\alpha > 0$ is constant. Then

$$\sigma_{\rm ess}(-\Delta_{\delta,\alpha}) = \big[-\alpha^2/4, +\infty\big) = \big[-4/\beta^2, +\infty\big) = \sigma_{\rm ess}(-\Delta_{\delta',\beta}),$$

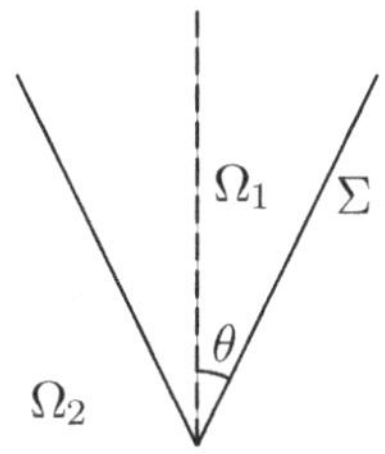

FIGURE 1. A broken line Σ with angle $\theta \in (0, \pi/2)$, which splits $\mathbb{R}^2$ into two wedge-type domains Ω_1 and Ω_2.

see [EN03, Proposition 5.4] and [BEL14a, Corollary 4.11], and the discrete spectra of both operators are nonempty, see [EI01, Theorem 5.2] and [BEL14a, Corollary 4.12].

We are going to apply Proposition 3.1. Let $\mu < -\alpha^2/4$ and $u \in W_\mu$, see (3.2). By Proposition 2.5 (ii) we have

$$u_1|_\Sigma - u_2|_\Sigma = (4/\alpha)\partial_{\nu_1} u_1|_\Sigma = -(4/\alpha)\partial_{\nu_2} u_2|_\Sigma, \tag{4.1}$$

and from Proposition 2.5 (i) and Lemma 2.3 we obtain

$$u_1|_\Sigma + u_2|_\Sigma = 0 \quad \text{and} \quad \partial_{\nu_1} u_1|_\Sigma - \partial_{\nu_2} u_2|_\Sigma = \alpha u_1|_\Sigma. \tag{4.2}$$

Combining (4.1) and (4.2) yields

$$\partial_{\nu_j} u_j|_\Sigma = (\alpha/2) u_j|_\Sigma, \qquad j = 1, 2. \tag{4.3}$$

It was shown in [LP08, Lemma 2.8] that the bottom of the spectrum of the self-adjoint Laplacian in $L^2(\Omega_2)$, subject to the Robin boundary condition (4.3), equals $-\alpha^2/4$. Since $-\Delta u_2 = \mu u_2$ on Ω_2 and $\mu < -\alpha^2/4$, it follows that $u_2 = 0$ identically. Plugging this into (4.2) implies $u_1|_\Sigma = 0$. Recall that $\mu < -\alpha^2/4$ and that the function u_1 satisfies $-\Delta u_1 = \mu u_1$ in Ω_1. Since the self-adjoint Dirichlet Laplacian on Ω_1 is non-negative, we get $u_1 = 0$ identically as well, hence $u = 0$. Thus it follows from Proposition 3.1 that

$$\lambda_n(-\Delta_{\delta',\beta}) < \lambda_n(-\Delta_{\delta,\alpha})$$

holds for all $n \in \mathbb{N}$ such that $\lambda_n(-\Delta_{\delta,\alpha}) < -\alpha^2/4$.

Example 4.2. Another example of a similar flavour is given by the infinite cone

$$\Sigma := \left\{ (x, y, \cot(\theta)\sqrt{x^2 + y^2}) \in \mathbb{R}^3 \colon (x, y) \in \mathbb{R}^2 \right\}, \qquad \theta \in (0, \pi/2); \tag{4.4}$$

cf. Figure 2. For constant $\alpha > 0$ it was shown in [BEL14b, Theorem 2.1] that $\sigma_{\rm ess}(-\Delta_{\delta,\alpha}) = [-\alpha^2/4, +\infty)$, and the discrete spectrum of $-\Delta_{\delta,\alpha}$ was proved in [BEL14b, Theorem 3.2] to be nonempty and even infinite. Following the lines of Example 4.1 and referring to [LP08, Example 2.9] instead of [LP08, Lemma 2.8]

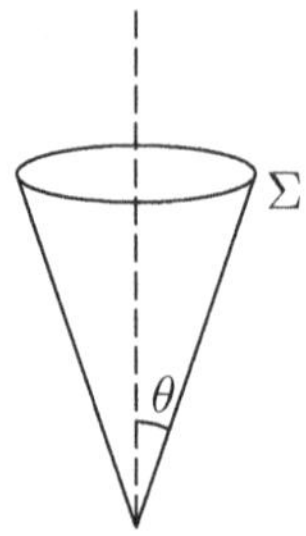

FIGURE 2. An infinite cone Σ with angle $\theta \in (0, \pi/2)$.

it follows for constant $\beta = 4/\alpha > 0$ that

$$\lambda_n(-\Delta_{\delta',\beta}) < \lambda_n(-\Delta_{\delta,\alpha}) \tag{4.5}$$

for all $n \in \mathbb{N}$ such that $\lambda_n(-\Delta_{\delta,\alpha}) < \inf \sigma_{\text{ess}}(-\Delta_{\delta',\beta})$.[1]

Example 4.3. In this example we consider an unconnected hypersurface Σ. Let $\mathbb{R}^d_\pm := \{(x', x_d) : x' \in \mathbb{R}^{d-1}, x_d \in \mathbb{R}_\pm\}$, let $\Omega' \subset \mathbb{R}^d_+$ be a bounded Lipschitz domain with positive distance to $\mathbb{R}^d_-$ and let

$$\Sigma := \{(x', 0) : x' \in \mathbb{R}^{d-1}\} \cup \partial\Omega'. \tag{4.6}$$

The surface Σ splits $\mathbb{R}^d$ into the two Lipschitz domains

$$\Omega_1 = \Omega' \cup \mathbb{R}^d_- \quad \text{and} \quad \Omega_2 = \mathbb{R}^d_+ \setminus \overline{\Omega'};$$

cf. Figure 3. As in the previous examples we consider constant interaction strengths α, β with $\beta = 4/\alpha > 0$. According to [BEL14a, Corollary 4.9] for constants $\alpha, \beta > 0$

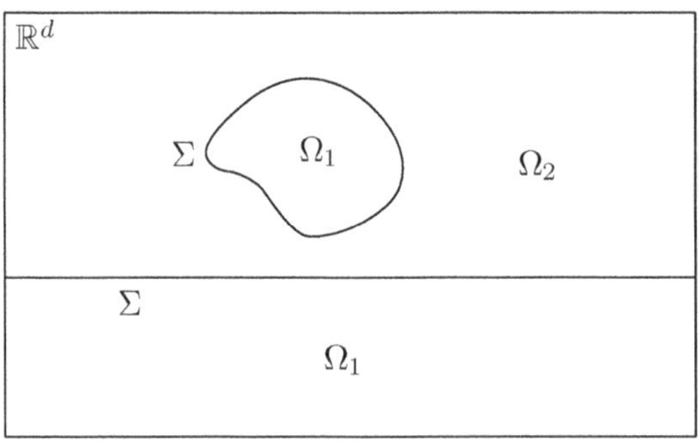

FIGURE 3. The unconnected hypersurface Σ splits $\mathbb{R}^d$ into two domains Ω_1 and Ω_2, and Ω_1 consists of two connected components.

[1]In fact we expect that one can prove $\inf \sigma_{\text{ess}}(-\Delta_{\delta',\beta}) = -4/\beta^2$ using the arguments in the proof of [BEL14b, Theorem 2.1]. This would imply that (4.5) holds for all $n \in \mathbb{N}$.

we have

$$\sigma_{\rm ess}(-\Delta_{\delta,\alpha}) = \big[-\alpha^2/4, +\infty\big) = \big[-4/\beta^2, +\infty\big) = \sigma_{\rm ess}(-\Delta_{\delta',\beta}).$$

We are going to conclude from Proposition 3.1 that

$$\lambda_n(-\Delta_{\delta',\beta}) < \lambda_n(-\Delta_{\delta,\alpha}) \tag{4.7}$$

holds for all $n \in \mathbb{N}$ such that $\lambda_n(-\Delta_{\delta,\alpha}) < -\alpha^2/4$. In order to do so, let $\mu < -\alpha^2/4$ and $u \in W_\mu$ with W_μ as in (3.2). As in Example 4.1 we find that u satisfies the conditions (4.1), (4.2), and (4.3). Since the spectrum of the self-adjoint Laplacian on $\mathbb{R}^d_-$ satisfying the Robin boundary condition (4.3) equals $[-\alpha^2/4, +\infty)$, we conclude from (4.3) and $-\Delta u_1 = \mu u_1$ that $u_1|_{\mathbb{R}^d_-} = 0$ identically. Together with (4.2) and a unique continuation argument it follows as in the proof of Theorem A that $u_2 = 0$ identically on Ω_2. Finally, after another application of (4.3) and of the unique continuation principle we arrive at $u_1|_{\Omega'} = 0$, hence $u = 0$. Therefore Proposition 3.1 yields the eigenvalue inequality (4.7).

References

[BEL14a] J. Behrndt, P. Exner, and V. Lotoreichik, Schrödinger operators with δ- and δ'-interactions on Lipschitz surfaces and chromatic numbers of associated partitions, *Rev. Math. Phys.* **26** (2014), 1450015.

[BEL14b] J. Behrndt, P. Exner, and V. Lotoreichik, Schrödinger operators with δ-interactions on conical surfaces, *J. Phys. A: Math. Theor.* **47** (2014), 355202.

[BLL13] J. Behrndt, M. Langer, and V. Lotoreichik, Schrödinger operators with δ and δ'-potentials supported on hypersurfaces, *Ann. Henri Poincaré* **14** (2013), 385–423.

[BR12] J. Behrndt and J. Rohleder, An inverse problem of Calderón type with partial data, *Comm. Partial Differential Equations* **37** (2012), 1141–1159.

[BS] M.Sh. Birman and M.Z. Solomjak, *Spectral theory of selfadjoint operators in Hilbert spaces*, Dordrecht, Holland, 1987.

[BEKS94] J.F. Brasche, P. Exner, Yu.A. Kuperin, and P. Šeba, Schrödinger operators with singular interactions, *J. Math. Anal. Appl.* **184** (1994), 112–139.

[BEW09] B.M. Brown, M.S.P. Eastham, and I.G. Wood, Estimates for the lowest eigenvalue of a star graph, *J. Math. Anal. Appl.* **354** (2009), 24–30.

[CDR08] H. Cornean, P. Duclos, and B. Ricaud, On the skeleton method and an application to a quantum scissor, in *Analysis on graphs and its applications, Proc. Sympos. Pure Math.* Amer. Math. Soc. Providence, 2008.

[DR13] V. Duchene and N. Raymond, Spectral asymptotics of a broken δ-interaction, *J. Phys. A* **47** (2014), 155203, 19 pp.

[E08] P. Exner, Leaky quantum graphs: a review, in: *Analysis on graphs and its applications.* Selected papers based on the Isaac Newton Institute for Mathematical Sciences programme, Cambridge, UK, 2007. *Proc. Symp. Pure Math.* **77** (2008), 523–564.

[EI01] P. Exner and T. Ichinose, Geometrically induced spectrum in curved leaky wires, *J. Phys. A* **34** (2001), 1439–1450.

[EJ13] P. Exner and M. Jex, Spectral asymptotics of a strong δ'-interaction on a planar loop, *J. Phys. A* **46** (2013), 345201, 12 pp.

[EK03] P. Exner and S. Kondej, Bound states due to a strong δ-interaction supported by a curved surface, *J. Phys. A* **36** (2003), 443–457.

[EK] P. Exner and H. Kovařík, *Quantum Waveguides*, Springer International Publishing, 2015.

[EN03] P. Exner and K. Němcová, Leaky quantum graphs: approximations by point-interaction Hamiltonians, *J. Phys. A* **36** (2003), 10173–10193.

[EP14] P. Exner and K. Pankrashkin, Strong coupling asymptotics for a singular Schrödinger operator with an interaction supported by an open arc, *Comm. Partial Differential Equations* **39** (2014), 193–212.

[F05] N. Filonov, On an inequality between Dirichlet and Neumann eigenvalues for the Laplace operator, *St. Petersburg Math. Journal* **16** (2005), 413–416.

[FL10] R.L. Frank and A. Laptev, Inequalities between Dirichlet and Neumann eigenvalues on the Heisenberg group, *Int. Math. Res. Not. IMRN* (2010), 2889–2902.

[GM09] F. Gesztesy and M. Mitrea, Nonlocal Robin Laplacians and some remarks on a paper by Filonov on eigenvalue inequalities, *J. Differential Equations* **247** (2009), 2871–2896.

[K] T. Kato, *Perturbation theory for linear operators*, Springer-Verlag, Berlin, 1995.

[K10] J. Kelliher, Eigenvalues of the Stokes operator versus the Dirichlet Laplacian in the plane, *Pacific J. Math.* **244** (2010), 99–132.

[LP08] M. Levitin and L. Parnovski, On the principal eigenvalue of a Robin problem with a large parameter, *Math. Nachr.* **281** (2008), 272–281.

[L13] V. Lotoreichik, Note on 2D Schrödinger operators with δ-interactions on angles and crossing lines, *Nanosystems: Phys. Chem. Math.* **4** (2013), 1–7.

[McL] W. McLean, *Strongly Elliptic Systems and Boundary Integral Equations*, Cambridge University Press, Cambridge, 2000.

[RS-IV] M. Reed and B. Simon: Methods of modern mathematical physics: IV Analysis of Operators, *Academic press inc.*, 1978.

[R14] J. Rohleder, Strict inequality of Robin eigenvalues for elliptic differential operators on Lipschitz domains, *J. Math. Anal. Appl.* **418** (2014), 978–984.

[S] K. Schmüdgen, *Unbounded Self-adjoint Operators on Hilbert Space*, Springer, Dordrecht, 2012.

[St] E. Stein, *Singular integrals and differentiability properties of functions*, Princeton University Press, Princeton, 1970.

Vladimir Lotoreichik, Nuclear Physics Institute CAS
CZ-25068 Řež, Czech Republic
e-mail: `lotoreichik@ujf.cas.cz`

Jonathan Rohleder, Institut für Numerische Mathematik
Technische Universität Graz, Steyrergasse 30, A-8010 Graz
e-mail: `rohleder@tugraz.at`

Operator Theory:
Advances and Applications, Vol. 247, 185–194
© Springer International Publishing Switzerland 2015

A Note on Submaximal Operator Space Structures

Vinod Kumar P. and M.S. Balasubramani

Abstract. In this note, we consider the smallest submaximal space structure $\mu(X)$ on a Banach space X. We derive a characterization of $\mu(X)$ up to complete isometric isomorphism in terms of a universal property. Also, we show that an injective Banach space has a unique submaximal space structure and we explore some duality relations of μ-spaces.

Mathematics Subject Classification (2010). 46L07, 47L25.

Keywords. Operator spaces, maximal operator spaces, submaximal spaces, μ-spaces.

1. Introduction and preliminaries

An *operator space* consists of a Banach space X and an isometric embedding $J : X \to B(\mathcal{H})$, for some Hilbert space $\mathcal{H}$. In contrast to the Banach space case, an operator space carries not just a complete norm on X, but also a sequence of complete norms on $M_n(X)$, the space of $n \times n$ matrices on X, for every $n \in \mathbb{N}$. These matrix norms are obtained via the natural identification of $M_n(X)$ as a subspace of $M_n(B(\mathcal{H})) \approx B(\mathcal{H}^n)$, where $\mathcal{H}^n$ is the Hilbert space direct sum of n copies of $\mathcal{H}$. An operator space can be described in an abstract way [14], as a pair $(X, \{\|.\|_n\}_{n\in\mathbb{N}})$ consisting of a linear space X and a complete norm $\|.\|_n$ on $M_n(X)$ for every $n \in \mathbb{N}$, such that $(R1)$ $\|\alpha x\beta\|_n \leq \|\alpha\| \, \|x\|_n \, \|\beta\|$ for all $\alpha, \beta \in M_n$ and for all $x \in M_n(X)$, and $(R2)$ $\|x \oplus y\|_{m+n} = \max\{\|x\|_m, \|y\|_n\}$ for all $x \in M_m(X)$, and for all $y \in M_n(X)$, where $x \oplus y$ denotes the matrix $\begin{bmatrix} x & 0 \\ 0 & y \end{bmatrix}$ in $M_{m+n}(X)$ and where 0 stands for zero matrices of appropriate orders. Here, the conditions $(R1)$ and $(R2)$ are called *Ruan's axioms* and the sequence of matrix norms $\{\|.\|_n\}_{n\in\mathbb{N}}$ is called an *operator space structure* on the linear space X. An operator space structure on a Banach space $(X, \|.\|)$ will usually mean a sequence of matrix norms

Vinod Kumar. P. is the corresponding author.

$\{\|.\|_n\}_{n\in\mathbb{N}}$ as above, but with $\|.\|_1 = \|.\|$ and in that case, we say $\{\|.\|_n\}_{n\in\mathbb{N}}$ is an *admissible* operator space structure on X.

If X is a Banach space, the closed unit ball $\{x \in X; \|x\| \leq 1\}$ is denoted by Ball(X). If X and Y are operator spaces and $\varphi : X \to Y$ is a linear map, $\varphi^{(n)} : M_n(X) \to M_n(Y)$, given by $[x_{ij}] \to [\varphi(x_{ij})]$, with $[x_{ij}] \in M_n(X)$ and $n \in \mathbb{N}$, determines a linear map from $M_n(X)$ to $M_n(Y)$. The *complete bound norm* (in short *cb-norm*) of φ is defined as $\|\varphi\|_{cb} = \sup\{\|\varphi^{(n)}\|; n \in \mathbb{N}\}$. The map φ is *completely bounded* if $\|\varphi\|_{cb} < \infty$ and φ is a *complete isometry* if each map $\varphi^{(n)} : M_n(X) \to M_n(Y)$ is an isometry. If φ is a complete isometry, then $\|\varphi\|_{cb} = 1$. If $\|\varphi\|_{cb} \leq 1$, φ is said to be a complete contraction. If $\varphi : X \to Y$ is a completely bounded linear bijection and if its inverse is also completely bounded, then φ is said to be a *complete isomorphism*. Two operator spaces are considered to be the same if there is a complete isometric isomorphism from X onto Y. In that case, we write $X \approx Y$ *completely isometrically* .

A Banach space Z is *injective* if for any Banach spaces X and Y where Y contains X as a closed subspace, and for any bounded linear map $\varphi : X \to Z$, there exists a bounded linear extension $\tilde{\varphi} : Y \to Z$ such that $\tilde{\varphi}|_X = \varphi$ and $\|\tilde{\varphi}\| = \|\varphi\|$. In a similar manner, an operator space Z is *injective* [3] if for any operator spaces X and Y where Y contains X as a closed subspace, and for any completely bounded linear map $\varphi : X \to Z$, there exists a completely bounded extension $\tilde{\varphi} : Y \to Z$ such that $\tilde{\varphi}|_X = \varphi$ and $\|\tilde{\varphi}\|_{cb} = \|\varphi\|_{cb}$. An operator space X is *homogeneous* [9] if each bounded linear operator φ on X is completely bounded with $\|\varphi\|_{cb} = \|\varphi\|$. More information about operator spaces and completely bounded mappings may be found in the books [4], [8] and [10].

2. Minimal and maximal operator space structures

A given Banach space has in general many realizations as an operator space and a very basic question in operator space theory is to exhibit some particular operator space structures on a given Banach space X. In the most general situation, Blecher and Paulsen [2] achieved this by noting that the set of all operator space structures admissible on a given Banach space X admits a minimal and maximal element. These structures were further investigated in [6] and [7]. By the Hahn–Banach theorem, it follows that any subspace of a minimal operator space is again minimal. Quotients of minimal operator spaces are called Q-spaces [12], and they need not be minimal. Also, the category of Q-spaces is stable under taking quotients and subspaces. An operator space X is said to be *submaximal* [10] if it embeds completely isometrically into a maximal operator space Y. Generally, a submaximal space need not be maximal, but maximality passes to quotients[10]. The maximal operator spaces with the property that all submaximal spaces turn out to be maximal are called *hereditarily maximal* spaces [13]. Any two Banach isomorphic subspaces of a hereditarily maximal space will be completely isomorphic as operator spaces. The subspace structure of various maximal operator spaces was further studied in [5].

If X is a Banach space, then there is a *minimal* operator space structure on X, denoted by $\mathrm{Min}(X)$, and this quantization is characterized by the property that for any arbitrary operator space Y and for any bounded linear map $\varphi : Y \to \mathrm{Min}(X)$ is completely bounded and satisfies $\|\varphi : Y \to \mathrm{Min}(X)\|_{cb} = \|\varphi : Y \to X\|$. An operator space X is minimal if $\mathrm{Min}(X) = X$. Also, an operator space is minimal if and only if it is completely isometric to a subspace of a commutative C*-algebra [4]. If X is a Banach space, there is a maximal way to consider it as an operator space. The matrix norms given by $\|[x_{ij}]\|_n = \sup\{\|[\varphi(x_{ij})]\| \,;\, \varphi \in \mathrm{Ball}(B(X,Y))\}$ where the supremum is taken over all operator spaces Y and all linear maps $\varphi \in \mathrm{Ball}(B(X,Y))$, makes X an operator space. This operator space is denoted by $\mathrm{Max}(X)$ and is called the *maximal* operator space structure on X. For $[x_{ij}] \in M_n(X)$, we write $\|[x_{ij}]\|_{\mathrm{Max}(X)}$ to denote its norm as an element of $M_n(\mathrm{Max}(X))$. An operator space X is maximal if $\mathrm{Max}(X) = X$. By Ruan's theorem [14], we also have $\|[x_{ij}]\|_{\mathrm{Max}(X)} = \sup\{\|[\varphi(x_{ij})]\| \,;\, \varphi \in \mathrm{Ball}(B(X, B(\mathcal{H})))\}$ where the supremum is taken over all Hilbert spaces $\mathcal{H}$ and all linear maps $\varphi \in \mathrm{Ball}(B(X, B(\mathcal{H})))$. By the definition of $\mathrm{Max}(X)$, any operator space structure on X is smaller than $\mathrm{Max}(X)$. This maximal quantization of a normed space is characterized by the property that for any arbitrary operator space Y, any bounded linear map $\varphi : \mathrm{Max}(X) \to Y$ is completely bounded and satisfies $\|\varphi : \mathrm{Max}(X) \to Y\|_{cb} = \|\varphi : X \to Y\|$. If X is any operator space, then the identity map on X defines completely contractive maps $\mathrm{Max}(X) \to X \to \mathrm{Min}(X)$. For any Banach space X, we have the following duality relations [1]: $\mathrm{Min}(X)^* \approx \mathrm{Max}(X^*)$ and $\mathrm{Max}(X)^* \approx \mathrm{Min}(X^*)$ completely isometrically.

Just like every operator space embeds completely isometrically into $B(\mathcal{H})$ for some Hilbert space $\mathcal{H}$, every submaximal space embeds completely isometrically into $\mathrm{Max}(B(\mathcal{H}))$ for some Hilbert space $\mathcal{H}$. To see this, let $X \subset Y$, where Y is a maximal operator space. Also, let $\iota : X \to B(\mathcal{H})$ be a complete isometric inclusion. Since $B(\mathcal{H})$ is injective, the inclusion $\iota : X \to B(\mathcal{H})$ extends to a complete contraction $\varphi : Y \to B(\mathcal{H})$. Since Y is maximal,

$$\|\varphi : Y \to \mathrm{Max}(B(\mathcal{H}))\|_{cb} = \|\varphi : Y \to \mathrm{Max}(B(\mathcal{H}))\| \leq \|\varphi : Y \to B(\mathcal{H})\|_{cb} \leq 1.$$

Let $\widetilde{\iota} = \varphi|_X$, then $\|\widetilde{\iota} : X \to \mathrm{Max}(B(\mathcal{H}))\|_{cb} \leq 1$. If $\widetilde{\iota}(X) = \widetilde{X}$, by the definition of maximal operator spaces, we have $\left\|\widetilde{\iota}^{-1} : \widetilde{X} \to X\right\|_{cb} \leq 1$. Thus $\widetilde{\iota} : X \to \widetilde{X}$ is a completely isometric isomorphism.

In the following, we consider the smallest submaximal space structure on a Banach space X, namely the μ-space structure which is denoted by $\mu(X)$. We prove that $\mu(X)$ will be homogeneous. We also derive a universal property of μ-spaces which distinguishes it among other submaximal spaces. By making use of this property, we show that the class of μ-spaces is stable under taking subspaces. We also show that an injective Banach space X has a unique submaximal space structure and describes some equivalent conditions for the uniqueness of the submaximal space structure on a Banach space X. Finally, we explore the duality relations of μ-spaces.

3. Main results

Just like minimal and maximal operator space structures, we have a minimal and a maximal way to view a Banach space X as a submaximal space, which we denote as $\mathrm{Min}_S(X)$ and $\mathrm{Max}_S(X)$ respectively. From the definition of a submaximal space it follows that $\mathrm{Max}_S(X) = \mathrm{Max}(X)$. T. Oikhberg [5] introduced the μ-space structure on a Banach space X and proved that $\mathrm{Min}_S(X) = \mu(X)$. Suppose X is a Banach space. Note that $\mathrm{Min}(X)$ is the operator space structure on X inherited by regarding $X \subset C(K)$, where $K = \mathrm{Ball}(X^*)$, the closed unit ball of the dual space of X with its weak* topology. Also, from [1], $\mathrm{Max}(X)^* = \mathrm{Min}(X^*)$, so that $\mathrm{Max}(X)^{**} = (\mathrm{Min}(X^*))^* = \mathrm{Max}(X^{**})$ completely isometrically. Since $X \hookrightarrow X^{**}$, we have $\mathrm{Max}(X) \hookrightarrow \mathrm{Max}(X^{**})$ completely isometrically.

Definition 3.1. An operator space X is a μ-space if it embeds completely isometrically into $\mathrm{Max}(C(K)^{**})$, where $K = \mathrm{Ball}(X^*)$ the unit closed ball of the dual space of X with its weak* topology.

A Banach space X, with the above-defined μ-space structure is denoted by $\mu(X)$ and the corresponding sequence of matrix norms by $\{\|.\|_n^\mu\}_{n\in\mathbb{N}}$. Note that the μ-space structure on a Banach space X is an admissible operator space structure on X.

Remark 3.2. Suppose that X and Y are injective Banach spaces and E and F are isomorphic (isometric) closed subspaces of X and Y respectively. Let $\varphi : E \to F$ be an isomorphism. Since Y is injective, there exists a map $\tilde{\varphi} : \mathrm{Max}(X) \to \mathrm{Max}(Y)$ such that $\tilde{\varphi}|_E = \varphi$ and $\|\tilde{\varphi}\| = \|\varphi\|$. Since X has maximal operator space structure, we have $\|\tilde{\varphi}\|_{cb} = \|\varphi\|$. Thus, $\|\varphi\|_{cb} \le \|\tilde{\varphi}\|_{cb} = \|\tilde{\varphi}\| = \|\varphi\|$. This shows that φ is completely bounded. Similarly φ^{-1} is also completely bounded, so that φ is a complete isomorphism. Thus, E and F are completely (isometrically) isomorphic as operator subspaces of $\mathrm{Max}(X)$ and $\mathrm{Max}(Y)$ respectively. From this fact, it follows that μ-spaces can also be described as a submaximal subspace of an injective commutative C^*-algebra, because the operator space structure is independent of the particular embedding.

Now we give a direct proof, different from [5] of the fact that $\mu(X)$ is the *smallest* submaximal operator space structure on a given Banach space X.

Theorem 3.3. *Let X be a Banach space. Then $\mu(X)$ is the smallest submaximal space structure on X.*

Proof. Let j be a complete isometric embedding of X in $\mathrm{Max}(C(K)^{**})$ described in the definition of μ-spaces. Consider a complete isometric embedding $\varphi : X \to$

Max(Y). Denote the sequence of matrix norms on X obtained via this embedding by $\{\|.\|_n^Y\}_{n\in\mathbb{N}}$. Then for any $[x_{ij}] \in M_n(X)$,

$$\|[x_{ij}]\|_n^Y = \|[\varphi(x_{ij})]\|_{\mathrm{Max}(Y)} = \sup\{\|[u(\varphi(x_{ij}))]\|\,; u \in \mathrm{Ball}(B(Y, B(\mathcal{K})))\}$$

where the supremum is taken over all possible maps $u : Y \to B(\mathcal{K})$ and over all Hilbert spaces $\mathcal{K}$. Also, we have

$$\|[x_{ij}]\|_n^\mu = \|[x_{ij}]\|_{\mathrm{Max}(C(K)^{**})} = \sup\{\|[v(x_{ij})]\|\,; v \in \mathrm{Ball}(B(C(K)^{**}, B(\mathcal{H})))\}$$

where the supremum is taken over all possible maps $v : C(K)^{**} \to B(\mathcal{H})$ and over all Hilbert spaces $\mathcal{H}$. Consider the following diagram.

$$\begin{array}{ccccc} X & \xrightarrow{\varphi} & \mathrm{Max}(Y) & \xrightarrow{u} & B(\mathcal{K}) \\ \big\downarrow{\scriptstyle id} & & & & \\ X & \xrightarrow{j} & \mathrm{Max}(C(K)^{**}) & \xrightarrow{v} & B(\mathcal{H}) \end{array}$$

Since $\varphi^{-1} : \varphi(X) \subset Y \to X$ is bounded, $w = j \circ id \circ \varphi^{-1} : \varphi(X) \subset Y \to C(K)^{**}$ is bounded and $\|w\| = 1$. Since $C(K)^{**}$ is injective as a Banach space, w has a bounded extension $\widetilde{w} : Y \to C(K)^{**}$ with $\|\widetilde{w}\| = 1$. Therefore, for any map $v : C(K)^{**} \to B(\mathcal{H})$ with $\|v\| \leq 1$, $\tilde{v} = v \circ \widetilde{w} : Y \to B(\mathcal{H})$ is a bounded map and is a completely bounded map with $\|\tilde{v}\|_{cb} \leq 1$ when regarded as a map from Max(Y) to $B(\mathcal{H})$. Thus $\|[x_{ij}]\|_n^\mu \leq \|[x_{ij}]\|_n^Y$ for any $[x_{ij}] \in M_n(X)$. This shows that the μ-space structure on X is the smallest submaximal space structure on X. □

Maximal and minimal operator spaces are homogeneous, but in general, submaximal spaces need not be homogeneous. Now we show that μ-spaces are homogeneous.

Proposition 3.4. *Every μ-space is homogeneous.*

Proof. Let $\varphi : \mu(X) \to \mu(X)$ be a bounded linear map. Then φ extends to a bounded linear map $\tilde{\varphi}$ on $C(K)^{**}$, and it is then completely bounded on Max($C(K)^{**}$) and $\|\tilde{\varphi}\|_{cb} = \|\tilde{\varphi}\| = \|\varphi\|$. But $\|\varphi\|_{cb} \leq \|\tilde{\varphi}\|_{cb} = \|\varphi\|$. Hence $\mu(X)$ is homogeneous. □

Completely bounded Banach–Mazur distance between two operator spaces X and Y is defined as $d_{cb}(X, Y) = \inf\{\|\varphi\|_{cb} \|\varphi^{-1}\|_{cb} : \varphi : X \to Y$ is a complete isomorphism $\}$. If X is a Banach space, then the μ-space structure on X lies between the operator space structures Min(X) and Max(X). Now we shall show that the cb distance between these spaces can be realized as the cb-norm of the identity mapping between them.

Theorem 3.5. *For a Banach space X, we have:*

$$d_{cb}(\mathrm{Min}(X), \mu(X)) = \|id : \mathrm{Min}(X) \to \mu(X)\|_{cb}$$

and

$$d_{cb}(\mu(X), \mathrm{Max}(X)) = \|id : \mu(X) \to \mathrm{Max}(X)\|_{cb}.$$

Proof. Let $T : \mu(X) \to \mathrm{Min}(X)$ be a complete isomorphism. Let $\tilde{T}$ denote the same map regarded as a mapping from $\mu(X)$ to $\mu(X)$. Consider the following diagram:

$$\begin{array}{ccc} \mu(X) & \xrightarrow{\quad T \quad} & \mathrm{Min}(X) \\ \uparrow \tilde{T} & \nwarrow^{id} & \| \\ \mu(X) & \xleftarrow[T^{-1}]{\quad\quad} & \mathrm{Min}(X). \end{array}$$

Here id denotes the formal identity mapping regarded as a mapping from $\mathrm{Min}(X)$ to $\mu(X)$. From the diagram, we get

$$\|id : \mathrm{Min}(X) \to \mu(X)\|_{cb} = \left\|\tilde{T} \circ T^{-1}\right\|_{cb} \leq \left\|\tilde{T}\right\|_{cb} \|T^{-1}\|_{cb}.$$

Since $\mu(X)$ is homogeneous (by the above Proposition 3.4),

$$\left\|\tilde{T}\right\|_{cb} = \left\|\tilde{T}\right\| = \|T\| = \|T\|_{cb},$$

where the last equality is determined by the minimal operator space structure of the range space of T. Thus we have $\|id : \mathrm{Min}(X) \to \mu(X)\|_{cb} \leq \|T\|_{cb} \left\|T^{-1}\right\|_{cb}$. This shows that $d_{cb}(\mathrm{Min}(X), \mu(X)) = \|id : \mathrm{Min}(X) \to \mu(X)\|_{cb}$. Similarly, the other case follows. □

We show that among submaximal spaces, the μ-spaces are characterized by the following universal property.

Theorem 3.6. *A submaximal space X is a μ-space up to complete isometric isomorphism if and only if for any submaximal space Y, any bounded linear map $\varphi : Y \to X$ is completely bounded with $\|\varphi\|_{cb} = \|\varphi\|$.*

Proof. Assume that $X = \mu(X)$. By definition of μ-spaces,

$$X = \mu(X) \subset \mathrm{Max}(C(K)^{**}), \quad \text{where } K = \mathrm{Ball}(X^*).$$

Since Y is submaximal, we have $Y \subset \mathrm{Max}(Z)$ for some operator space Z. Now, $\varphi : Y \to \mu(X)$ can be regarded as a map Y to $\mathrm{Max}(C(K)^{**})$. Since the bidual of $C(K)$ is injective, there exists $\tilde{\varphi} : Z \to \mathrm{Max}(C(K)^{**})$ with $\|\tilde{\varphi}\| = \|\varphi\|$ and $\tilde{\varphi}|_Y = \varphi$. Considering $\tilde{\varphi} : \mathrm{Max}(Z) \to \mathrm{Max}(C(K)^{**})$, we see that $\tilde{\varphi}$ is completely bounded and $\|\tilde{\varphi}\|_{cb} = \|\tilde{\varphi}\|$. But $\|\varphi\|_{cb} \leq \|\tilde{\varphi}\|_{cb} = \|\tilde{\varphi}\| = \|\varphi\|$. This shows that $\|\varphi\|_{cb} = \|\varphi\|$.

Conversely, take $Y = \mu(X)$ and $\varphi = id : \mu(X) \to X$, the formal identity mapping. Then by assumption, $\|id\|_{cb} = \|id\| = 1$. Also, from the above part, $\left\|id^{-1}\right\|_{cb} = \|id : X \to \mu(X)\|_{cb} = \|id\| = 1$. Thus $id : X \to \mu(X)$ is a complete isometric isomorphism. □

Remark 3.7. This universal property shows that $\mu(X)$ is indeed the smallest submaximal operator space structure on a Banach space X. For, let $\{|||.|||_n\}_{n\in\mathbb{N}}$ be a submaximal operator space structure on X with $|||.|||_1 = \|.\|_1^{\mu}$, and let $\widetilde{X}$ denote X with this operator space structure. Then, $id : \widetilde{X} \to \mu(X)$ is a linear isometry and so by universal property of μ-spaces, $\|id\|_{cb} = \|id\| = 1$. But this says precisely that $\||.\||_n$ dominates the norm in $\mu(X)$.

We know that if X has minimal operator space structure, then every bounded linear map defined on another operator space with values in X is completely bounded. Also, we have shown that if X has the μ-space structure, then any bounded linear map from a submaximal space to X is completely bounded. Now, let X be endowed with any operator space structure $\{\|.\|_n\}_{n\in\mathbb{N}}$ such that $\|[x_{ij}]\|_n \leq \|[x_{ij}]\|_n^{\mu}$ for every $[x_{ij}] \in M_n(X)$, and for all $n \in \mathbb{N}$. In this case also, any bounded linear map φ from a submaximal space Y to X is completely bounded with $\|\varphi\|_{cb} = \|\varphi\|$. Here, the operator space structure on X is not submaximal, since the μ-space structure is the smallest submaximal structure on any normed space. To see this, consider the following diagram.

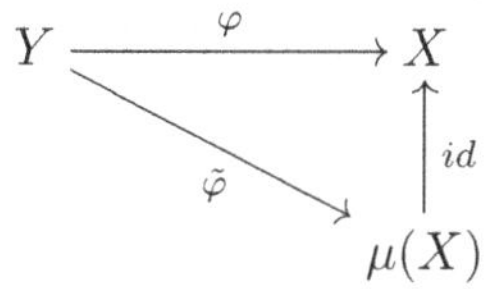

Since the identity mapping $id : \mu(X) \to X$ is a complete contraction, using Theorem 3.6, we have: $\|\varphi\|_{cb} \leq \|id\|_{cb} \|\tilde{\varphi}\|_{cb} \leq \|\tilde{\varphi}\| = \|\varphi\|$. Thus, $\|\varphi\|_{cb} = \|\varphi\|$.

Now we make use of the universal property of μ-spaces to show that the class of μ-spaces is stable under taking subspaces.

Corollary 3.8. *Let $Y \subset \mu(X)$. Then Y is also a μ-space.*

Proof. Let Z be any submaximal space. Then any bounded linear map $\varphi : Z \to Y$ can be regarded as a map from Z to $\mu(X)$. By the universal property of μ-spaces, we see that $\|\varphi\|_{cb} = \|\varphi\|$. This shows that Y is a μ-space. □

The following theorem gives a more general characterization of μ-spaces up to complete isomorphism.

Theorem 3.9. *A submaximal space X is completely isomorphic to a μ-space if and only if for any submaximal space Y, any completely bounded linear bijection $\varphi : X \to Y$ is a complete isomorphism.*

Proof. Let $\psi : X \to \mu(Z)$ be a complete isomorphism. Then for any completely bounded linear bijection $\varphi : X \to Y$, by Theorem 3.6, $\psi \circ \varphi^{-1} : Y \to \mu(Z)$ is completely bounded and $\|\psi \circ \varphi^{-1}\|_{cb} = \|\psi \circ \varphi^{-1}\|$. Therefore,

$$\|\varphi^{-1}\|_{cb} = \|\psi^{-1} \circ \psi \circ \varphi^{-1}\|_{cb} \leq \|\psi^{-1}\|_{cb} \|\psi \circ \varphi^{-1}\|_{cb} < \infty,$$

showing that φ is a complete isomorphism. For the converse, take Y as $\mu(X)$ and φ as the formal identity map $id : X \to \mu(X)$. □

Now we look at the case when the domain is endowed with the μ-space structure.

Theorem 3.10. *Let X be an operator space. Then the formal identity map $id : \mu(X) \to X$ is completely bounded if and only if for every submaximal space Y, every bounded linear map $\varphi : Y \to X$ is completely bounded. Moreover, we have: $\|id : \mu(X) \to X\|_{cb} = \sup\{\frac{\|u\|_{cb}}{\|u\|}\}$, where the supremum is taken over all bounded non zero linear maps $u : Y \to X$ and all submaximal spaces Y.*

Proof. Assume that $id : \mu(X) \to X$ is completely bounded with $\|id\|_{cb} = C$. Let Y be a submaximal space and $u : Y \to X$ be a bounded linear map. Let $\tilde{u}$ denote the same map u regarded as a map from Y to $\mu(X)$. Then by the universal property of μ-spaces, $\tilde{u}$ is completely bounded and $\|\tilde{u}\|_{cb} = \|\tilde{u}\| = \|u\|$. Since $u = id \circ \tilde{u}$, we have $\|u\|_{cb} \leq \|id\|_{cb}\|u\|_{cb} = C\|u\| < \infty$. Thus u is completely bounded. For the converse, take Y as $\mu(X)$, and u as the identity map. Also, from the above inequality, it follows that $\|id : \mu(X) \to X\|_{cb} = \sup\{\frac{\|u\|_{cb}}{\|u\|}\}$, where the supremum is taken over all bounded non zero linear maps $u : Y \to X$ and all submaximal spaces Y. □

The following theorem shows that an injective Banach space X has a unique submaximal space structure, or in other words $\mathrm{Min}_S(X) = \mathrm{Max}_S(X)$, if X is an injective Banach space.

Theorem 3.11. *If X is an injective Banach space, then $\mu(X)$ is completely isometrically isomorphic to* $\mathrm{Max}(X)$.

Proof. Consider the formal identity map $id : \mu(X) \to \mathrm{Max}(X)$. By definition of μ-spaces, $\mu(X) \subset \mathrm{Max}(C(K)^{**})$ and since X is injective as a Banach space, id extends to a bounded linear map $\widetilde{id} : \mathrm{Max}(C(K)^{**}) \to \mathrm{Max}(X)$ with $\|\widetilde{id}\| = \|id\| = 1$. Since its domain has maximal operator space structure, we have $\|\widetilde{id}\|_{cb} = 1$ and hence $\|id\|_{cb} = 1$. Also, $\|id^{-1}\|_{cb} = 1$, showing that $id : \mu(X) \to \mathrm{Max}(X)$ is a complete isometric isomorphism. □

We know that the converse of the above theorem is not true. For example, the space ℓ_1^2 has a unique operator space structure, but it is not an injective Banach space. The following theorem describes some equivalent conditions for the uniqueness of the submaximal space structures.

Theorem 3.12. *For a Banach space X, the following are equivalent.*

(1) *X has a unique submaximal space structure.*
(2) *$\mu(X) = \mathrm{Max}(X)$ completely isometrically.*
(3) *Any bounded linear map $\varphi : X \to B(\mathcal{H})$ admits a bounded extension $\widetilde{\varphi} : C(Ball(X^*)) \to B(\mathcal{H})$ with $\|\widetilde{\varphi}\| = \|\varphi\|$.*

Proof. It is clear from the definition that (1) ⇔(2). Now to prove (2) ⇒(3), regard φ as a map from $\mu(X) \to B(\mathcal{H})$, we see that φ is completely bounded and $\|\varphi\|_{cb} = \|\varphi\|$. Since $\mu(X) \hookrightarrow \mathrm{Max}(C(\mathrm{Ball}(X^*)))$ completely isometrically and since $B(\mathcal{H})$ is injective, there exists an extension $\widetilde{\varphi} : \mathrm{Max}(C(\mathrm{Ball}(X^*))) \to B(\mathcal{H})$ with $\|\widetilde{\varphi}\|_{cb} = \|\varphi\|_{cb}$. Thus $\widetilde{\varphi} : C(\mathrm{Ball}(X^*)) \to B(\mathcal{H})$ satisfies $\|\widetilde{\varphi}\| = \|\widetilde{\varphi}\|_{cb} = \|\varphi\|_{cb} = \|\varphi\|$.

Assume that any bounded linear map $\varphi : X \to B(\mathcal{H})$ admits a bounded extension $\widetilde{\varphi} : C(\mathrm{Ball}(X^*)) \to B(\mathcal{H})$ with $\|\widetilde{\varphi}\| = \|\varphi\|$. Clearly $\|[x_{ij}]\|_n^\mu \le \|[x_{ij}]\|_n^{\max}$ for any $[x_{ij}] \in M_n(X)$. By definition of maximal spaces,

$$\|[x_{ij}]\|_n^{\max} = \sup\{\|[\varphi(x_{ij})]\|\,;\varphi \in \mathrm{Ball}(B(X, B(\mathcal{H})))\}$$

where the supremum is taken over all possible bounded linear maps $\varphi : X \to B(\mathcal{H})$ and over all Hilbert spaces $\mathcal{H}$. Also,

$$\begin{aligned}\|[x_{ij}]\|_n^\mu &= \|[x_{ij}]\|_{\mathrm{Max}(C(\mathrm{Ball}(X^*)))}\\ &= \sup\{\|[v(x_{ij})]\|\,; v \in \mathrm{Ball}(B(C(\mathrm{Ball}(X^*)), B(\mathcal{H})))\}\end{aligned}$$

where the supremum is taken over all possible bounded linear maps

$$v : C(\mathrm{Ball}(X^*)) \to B(\mathcal{H})$$

and over all Hilbert spaces $\mathcal{H}$. By the assumed extension property of X, corresponding to any $u \in \mathrm{Ball}(B(X, B(\mathcal{H})))$, we have an extended function $\tilde{u} \in \mathrm{Ball}(B(C(\mathrm{Ball}(X^*)), B(\mathcal{H})))$, so that $\|[x_{ij}]\|_n^\mu \ge \|[x_{ij}]\|_n^{\max}$ for any $[x_{ij}] \in M_n(X)$. Thus $\mu(X) = \mathrm{Max}(X)$, showing that (3) $\Rightarrow$ (2). □

Remark 3.13. Since every injective Banach space has a unique submaximal space structure, every injective Banach space X has the above-described extension property.

A Q-space is an (operator) quotient of a minimal space [12]. Note that if X is a Q-space, then X^* is a submaximal space. Conversely, the dual of a submaximal space is a Q-space. Eric Ricard [11] introduced the maximal Q-space structure on a Banach space X denoted by $\mathrm{Max}_Q(X)$, where the matrix norms are defined as:

$$\|[x_{ij}]\| = \sup\{\|[u(x_{ij})]\|_{M_n(E)}\,; u : X \to E,\ E \text{ a } Q\text{-space and } \|u\| \le 1\ \}.$$

We now prove the duality relations between the $\mu(X)$ and $\mathrm{Max}_Q(X)$.

Theorem 3.14. *For any Banach space X, we have:* $(\mathrm{Max}_Q(X))^* = \mu(X^*)$ *and* $(\mu(X))^* = \mathrm{Max}_Q(X^*)$.

Proof. Note that $(\mathrm{Max}_Q(X))^*$ is a submaximal space structure on X^*, so that by Theorem 3.3, the formal identity map $id : (\mathrm{Max}_Q(X))^* \to \mu(X^*)$ is a complete contraction. Also, the embedding $X \subset (\mu(X^*))^*$ gives a Q-space structure on X. Hence the identity map $id : \mathrm{Max}_Q(X) \to (\mu(X^*))^*$ is a complete contraction. Taking the duals, we see that $id : \mu(X^*) \to (\mathrm{Max}_Q(X))^*$ is a complete contraction. Thus $(\mathrm{Max}_Q(X))^* = \mu(X^*)$. The other part follows by duality. □

Corollary 3.15. *An operator space X is a μ-space if and only if its bidual X^{**} is a μ-space.*

Proof. Let $X = \mu(X)$. Then $X^* = \mu(X)^* = \mathrm{Max}_Q(X^*)$, so that

$$X^{**} = (\mathrm{Max}_Q(X^*))^* = \mu(X^{**}).$$

Thus X^{**} is a μ-space. The converse part follows from the fact that $X \subset X^{**}$ and from Corollary 3.8 □

Acknowledgement

The authors extend thanks to the referee for his valuable suggestions. The first author acknowledges the financial support by University Grants Commission of India, under Faculty Development Programme. The authors are grateful to Prof. Gilles Pisier and Prof. Eric Ricard for having many discussions on this subject.

References

[1] D. Blecher, 'The standard dual of an operator space', *Pacific J. Math.* 153 (1992), 15–30.

[2] D. Blecher and V. Paulsen, 'Tensor products of operator spaces', *J. Funct. Anal.* 99 (1991), 262–292.

[3] M.-D. Choi and E. Effros, 'Injectivity and operator spaces', *J. Funct. Anal.* 24 (1977), 156–209.

[4] E. Effros and Z.-J. Ruan, *Operator spaces* (London Math. Soc. Monographs. New Series, Oxford University Press, 2000).

[5] T. Oikhberg, 'Subspaces of maximal operator spaces', *Integr. Equ. Oper. Theory* 48 (2004), 81–102.

[6] V. Paulsen, 'Representation of function algebras, abstract operator spaces and Banach space geometry', *J. Funct. Anal.* 109 (1992), 113–129.

[7] V. Paulsen, 'The maximal operator space of a normed space', *Proc. Edinburgh Math. Soc.* 39 (1996), 309–323.

[8] V. Paulsen, *Completely bounded maps and operator algebras* (Cambridge University Press, 2002).

[9] G. Pisier, 'The operator Hilbert space OH, complex interpolation and tensor norms', *Mem. Amer. Math. Soc.* 122, 585 (1996), 1–103.

[10] G. Pisier, *Introduction to operator space theory* (Cambridge University Press, 2003).

[11] E. Ricard, *Décomposition de H^1, multiplicateurs de Schur et espaces d'opérateurs* (Ph.D. thesis, Université Paris VI., 2001).

[12] E. Ricard, 'A tensor norm for **Q**-spaces', *J. Operator theory* 48 (2002), 431–445.

[13] Vinod Kumar P. and M.S. Balasubramani, 'Submaximal operator space structures on Banach spaces', *Oper. Matrices*, 7(3), (2013), 723-732.

[14] Z.-J. Ruan, 'Subspaces of $\mathbf{C}^*$-algebras', *J. Funct. Anal.* 76 (1988), 217–230.

Vinod Kumar P.
Department of Mathematics, Thunchan Memorial Govt College
Tirur, Kerala, India
e-mail: `vinodunical@gmail.com`

M.S. Balasubramani
Department of Mathematics, University of Calicut
Calicut University. P. O., Kerala, India
e-mail: `msbalaa@rediffmail.com`

Appendix

IWOTA 2013 Schedule of Plenary and Semi-plenary Talks

The inauguration is at 9 AM on 16 December.

Schedule of talks for Monday 16 December

		Speaker	Chair
9:15–10:15	Plenary Talk	Dinesh Singh	G. Misra
10:15–10:45	Coffee		
10:45–11:45	Plenary Talk	James A. Jamison	T.S.S.R.K. Rao
12:00–13:00	Semi Plenary Talks	Wolfgang Arendt	T.S.S.R.K. Rao
		Michael Dritschel	G. Misra
13:00–14:15	Lunch		
14:15–15:45	Thematic sessions		
16:00–17:00	Public lecture	Persi Diaconis	
17:00–17:30	Coffee		
17:30–18:30	Plenary Talk	Harald Upmeier	N.J. Young

Schedule of talks for Tuesday 17 December

		Speaker	Chair
9:15–10:15	Plenary Talk	Rajendra Bhatia	K.B. Sinha
10:15–10:45	Coffee		
10:45–11:45	Semi Plenary Talks	Joseph A. Ball	Sanne ter Horst
		Denis Potapov	Ajit Iqbal Singh
12:00–13:00	Semi Plenary Talks	Victor Vinnikov	Sanne ter Horst
		Paul Binding	Ajit Iqbal Singh
13:00–14:15	Lunch		
14:15–16:30	Thematic sessions		
16:30–17:00	Coffee		
17:30–18:00	Plenary Talk	Giles Pisier	I. Chalendar
18:00–18:30	Video on IISc		

Schedule of talks for Wednesday 18 December

		Speaker	Chair
9:15–10:15	Plenary Talk	K.B. Sinha	D. Goswami
10:15–10:45	Coffee		
10:45–11:45	Semi Plenary Talks	Debasish Goswami	B.V.R. Bhat
		Nicholas Young	Zinaida Lykova
12:00–13:00	Semi Plenary Talks	Ken Dykema	B.V.R. Bhat
		Tom ter Elst	Zinaida Lykova
13:00–14:15	Lunch		
14:15–16:30	Thematic sessions		
16:30–17:00	Coffee		
17:30–18:00	Plenary Talk	Raul E. Curto	R. Bhatia
18:00–18:30	Video on ISI		

Schedule of talks for Thursday 19 December

		Speaker	Chair
9:15–10:15	Plenary Talk	Jean Pierre Antoine	S.T. Ali
10:15–10:45	Coffee		
10:45–11:45	Semi Plenary Talks	S. Twareque Ali	A. Odzijewicz
		Fumio Hiai	Takashi Sano
12:00–13:00	Semi Plenary Talks	M.A. Kaashoek	A. Odzijewicz
		Il Bong Jung	Takashi Sano
13:00–14:15	Lunch		
14:15–16:30	Thematic sessions		
16:30–17:00	Coffee		
17:30–18:00	Plenary Talk	Isabelle Chalendar	J.A. Ball

Schedule of talks for Friday 20 December

		Speaker	Chair
9:15–10:15	Plenary Talk	Birgit Jacob	V. Vinnikov
10:15–10:45	Coffee		
10:45–11:45	Semi Plenary Talks	Ajit Iqbal Singh	Kaushal Verma
		Pierre Portal	E.K. Narayanan
12:00–13:00	Semi Plenary Talks	Dan Timotin	Kaushal Verma
		Daniel Alpay	E.K. Narayanan
13:00–14:15	Lunch		
14:15–16:30	Thematic sessions		
16:30–17:00	Coffee		
17:30–18:00	Plenary Talk	V.S. Sunder	M.A. Kaashoek

Talks at the Faculty Hall are in red, talks at the UG lecture Hall at the old Physics building are in blue.

Plenary Speakers and Their Titles

Antoine, **M. Jean-Pierre**, Louvain, Belgium.
Coherent states and wavelets, a contemporary panorama.

Bhatia, **Rajendra**, Delhi, India.
Inertias of some special matrices.

Chalendar, **Isabelle**, Lyon, France.
Inner functions in Operator Theory.

Curto, **Raúl E.**, Iowa, USA.
Berger measures for transformations of subnormal weighted shifts.

Jacob, **Birgit**, Wuppertal, Germany.
Linear port-Hamiltonian systems on Infinite-dimensional spaces.

Jamison, **James E.**, Memphis, USA.
Hermitian operators on vector valued function spaces.

Pisier, **Gilles**, College Station, USA.
Quantum Expanders and Growth of Operator Spaces.

Singh, **Dinesh**, Delhi, India.
Operators on Function Spaces and Applications of the H^1*-BMOA Duality.*

Sinha, **Kalyan B.**, Bangalore, India.
An Approximation Theorem and Two-variable Trace Formula.

Sunder, **V.S.**, Chennai, India.
Minimax theorems in Non-commutative Probability Spaces.

Upmeier, **Harald**, Marburg, Germany.
Hilbert spaces of cohomology and Radon transform.

Semi-plenary Speakers and Their Titles

Ali, **S. Twareque**, Montreal, Canada.
Some families of complex Hermite polynomials and their applications to physics.

Alpay, **Daniel**, Be'er-Sheva, Israel.
Non commutative stochastic distributions, free processes with stationary increments and stochastic integration.

Arendt, **Wolfgang**, Ulm, Germany.
Regularity of Semigroups: Asymptotic Behaviour at 0 *and Multipliers.*

Ball, **Joseph A.**, Blacksburg, Virginia, USA.
Transfer-function realization and zero/pole structure for multivariable rational matrix functions: the direct analysis.

Binding, **Paul**, Calgary, Canada.
Some two parameter eigenvalue problems.

Dritschel, **Michael A.**, Newcastle upon Tyne, UK.
Dilations and constrained algebras.

Dykema, **Ken**, College Station, USA.
Hyperinvariant subspaces and upper triangular decompositions in finite von Neumann algebras.

ter Elst, **Tom**, Auckland, New Zealand.
The Dirichlet-to-Neumann operator via hidden compactness.

Goswami, **Debashish**, Kolkata, India.
Quantum Isometry Groups.

Hiai, **Fumio**, Tohoku, Japan.
Higher order extension of Löwner's theory: Operator k-tone functions.

Jung, **Il Bong**, Taegu, Korea.
Unbounded quasinormal operators and related properties.

Kaashoek, **M.A.**, Amsterdam, The Netherlands.
The inverse problem for Ellis–Gohberg orthogonal matrix functions.

Klep, **Igor**, Auckland, New Zealand.
Linear Matrix Inequalities and Positive Polynomials.

Portal, **Pierre**, Canberra, Australia.
Holomorphic functional calculus and square functions.

Potapov, **Denis**, Sydney, Australia.
Recent successes in perturbation theory.

Timotin, **Dan**, Bucharest, Romania.
An extremal problem for characteristic functions.

Vinnikov, **Victor**, Be'er Sheva, Israrel.
Transfer-function realization and zero/pole structure for multivariable rational matrix functions: the converse analysis.

Young, **Nicholas**, Leeds and Newcastle, UK.
Operator monotone functions and Löwner functions of several variables.

Singh, **Ajit Iqbal**, Delhi, India.
Quantum channels and entanglement.

The detailed abstracts of all talks are available from IWOTA 2013 website
http://math.iisc.ernet.in/∼iwota2013

Thematic Session I: Operators, functions and linear systems

Organizers: Rien Kaashoek and Sanne ter Horst

- Denis Potapov, Frechet differentiability of Schatten–von Neumann p-norms.
- Sourav Pal, Spectral sets and distinguished varieties in the symmetrized bidisc.
- Roland Duduchava, Calculus of Gunter's derivatives and a shell theory.
- Alexander Sakhnovich, Explicit solutions of linear and nonlinear evolution equations depending on several variables.
- Jacob Jaftha, Dissipative linear relations in Banach spaces and a multivalued version of the Lumer–Phillips Theorem.
- Alon Bulbil, Continuous stochastic linear systems.
- Santanu Dey, Functional Model for multi-analytic operators.
- Snehlatha Ballamoole, A class of integral operators on spaces of analytic functions.
- Joseph A. Ball, Convexity analysis and integral representations for generalized Schur/Herglotz function classes.
- Nicholas Young, Realization of symmetric analytic functions of noncommuting variables.
- Sanne ter Horst, Equivalence after extension and Schur coupling coincide, on separable Hilbert spaces.
- Daniel Alpay, Schur analysis in the setting of slice hyper-holomorphic functions.
- Salma Kuhlmann, Application of Jacobi's Representation Theorem to locally multiplicatively convex topological real algebras.
- Mehdi Ghasemi, Moment problem for continuous linear functionals.
- Victor Vinnikov, Determinantal representations of stable and hyperbolic polynomials.
- Zinaida Lykova, 3-extremal holomorphic maps and the symmetrised bidisc.
- Eli Shamovich, Determinantal Representations and Hyperbolicity on the Grassmannian.

The detailed abstracts of all talks are available from IWOTA 2013 website http://math.iisc.ernet.in/~iwota2013

Thematic Session II: Geometry of Banach spaces in operator theory

Organizer: T.S.S.R.K. Rao

- Lajos Molnar, Isometries of certain nonlinear spaces of matrices and operators
- V. Indumathi, Polyhedral conditions and Best Approximation Problems
- Vrej Zarikian, Bimodules over Cartan Subalgebras and Mercer's Extension Theorem
- Jiri Spurny, Baire classes of Banach Spaces and C*-Algebras
- A.K. Roy, On Silov boundary for function spaces
- S. Dutta, Predual of completely bounded multipliers
- K. Paul, Jayanarayanan C.R., A. Bhar, T. Paul, Short talks

The detailed abstracts of talks other than short talks are available from IWOTA 2013 websiteb http://math.iisc.ernet.in/∼iwota2013

Thematic Session III: Concrete operators

Organizers: Isabelle Chalendar, Alfonso Montes Rodriguez and Ilya Spitkovsky

- Hervé Queffelec, Approximation numbers of composition operators on the Dirichlet space
- Romesh Kumar, Composition Operators and Multiplication Operators on Banach Function Spaces
- Ilya Spitkovsky, On the kernel and cokernel of some Toeplitz and Wiener–Hopf operators
- George Exner, A weak subnormality condition bridging Agler–Embry and Bram–Halmos
- Gerardo Chacón, Composition Operators and derivative-free characterizations of Dirichlet-type Spaces
- György Pál Gehér, Tree-shift Operators and their Cyclic Properties
- Hocine Guediri, The Bergman Space as a Banach Algebra
- Mubariz T. Garayev, Some Concrete Operators and their Properties
- Patryk Pagacz, On wandering vectors for isometries and Szegő measure properties
- Gopal Datt, Hankel to weighted Hankel operators
- Aneesh M, Supercyclicity and frequent hypercyclicity in the space of self-adjoint operators
- Frantisek Stampach, The characteristic function for infinite Jacobi matrices, the spectral zeta function, and solvable examples

The detailed abstracts of all talks are available from IWOTA 2013 website http://math.iisc.ernet.in/∼iwota2013

Thematic Session IV: Functional and harmonic analytic aspects of wavelets and coherent states

Organizers: S. Twareque Ali, M. Jean-Pierre Antoine and Jean-Pierre Gazeau

- Anatol Odzijewicz, Positive kernels and quantization
- P.K. Das, Generation of a superposition of coherent states in a resonant cavity and its nonclassicality and decoherence
- J.-P. Antoine, Wavelets and multiresolution: from NMR spectroscopy to the analysis of video sequences
- S.T. Ali, Quaternionic Wavelets and Coherent States

The detailed abstracts of all talks are available from IWOTA 2013 website http://math.iisc.ernet.in/~iwota2013

Thematic Session V: General session

Organizer: Kaushal Verma

- Surjit Kumar, Spherically Balanced Hilbert Spaces of Formal Power Series in Several Variables
- Piotr Budzy'nski, On subnormality of unbounded weighted shifts on directed trees
- J.J. Grobler, Stochastic processes in Riesz spaces: The Kolmogorov–Centsov theorem and Brownian motion
- Janusz Wysoczanski, On generalized anyon statistics
- Peter Semrl, Adjacency preserving maps
- Martin Weigt, Unbounded derivations of commutative generalized B*-algebras
- Lucijan Plevnik, Maps preserving complementarity of closed subspaces of a Hilbert space
- M.N.N. Namboodiri, Korovkin-type theorems via completely positive/bounded maps on operator algebras – recent developments
- Prahlad Vaidyanathan, E-theory for Continuous Fields of C*-Algebras
- P. Vinod Kumar, Minimal and Maximal Operator Space Structures on Banach Spaces
- Vladimir Peller, Estimates for Lipschitz functions of perturbed self-adjoint operators based on finite-dimensional estimates
- Sachin Bedre, Fixed point theorems for M-contraction type maps in partially ordered metric spaces and applications to fractional differential equations
- Sangeeta Jhanjhee, Joint spectral theory using Clifford algebras

The detailed abstracts of all talks are available from IWOTA 2013 website http://math.iisc.ernet.in/~iwota2013

Thematic Session VI: Multivariable operator theory

Organizers: Gadadhar Misra and Jaydeb Sarkar

- Santanu Dey, Characteristic function of liftings
- Sameer Chavan, Conditional completely hypercontractive tuples
- Eli Shamovich, Lie Algebra Operator Vessels and General Taylor Joint Spectrum
- Sasmita Patnaik, Subideals of Operators
- Il Bong Jung, On quadratically hyponormal weighted shifts
- Gregory Knese, Canonical Agler decompositions
- Bata Krishna Das, Tensor product of quotient Hilbert modules
- Vinayak Sholapurkar, Rigidity theorems for spherical hyperexpansions
- Victor Vinnikov, Vessels of commuting selfadjoint operators
- Sanne ter Horst, Stability of noncommutative multidimensional systems and structured Stein inequalities
- Kalpesh Haria, Outgoing Cuntz scattering system for a coisometric lifting and transfer function
- Santanu Sarkar, The defect sequence for contractive tuples

The detailed abstracts of all talks are available from IWOTA 2013 website http://math.iisc.ernet.in/~iwota2013

Thematic Session VII: Spectral theory and differential operators

Organizers: Paul Binding, Tom ter Elst and Carsten Trunk

- Carsten Trunk, On a class of Sturm–Liouville operators which are connected to PT quantum mechanics
- Rostyslav Hryniv, Reconstruction of Sturm–Liouville operators with energy-dependent potentials
- Vadim Kostrykin, The div A grad without ellipticity
- Pierre Portal, Non-autonomous parabolic systems with rough coefficients
- Jonathan Rohleder, Titchmarsh–Weyl theory for elliptic differential operators
- Alexei Rybkin, On the Hankel operator approach to completely integrable systems
- Andreas Ioannidis, The eigenvalue problem for the Cavity Maxwell operator
- Petr Siegl, Root system of perturbations of harmonic and anharmonic oscillators
- Maria Kovaleva, Stokes graph and non-oscillating solutions
- Kiran Kumar, Truncation method for random bounded self-adjoint operators
- Anton Popov, Spectral analysis for three coupled strips quantum graph

The detailed abstracts of all talks are available from IWOTA 2013 website http://math.iisc.ernet.in/~iwota2013

The manufacturer's authorised representative in the EU is Springer Nature Customer Service Centre GmbH, Europaplatz 3, 69115 Heidelberg, Germany. If you have any concerns regarding our products, please contact ProductSafety@springernature.com

Printed and bound by CPI Group (UK) Ltd, Croydon, CR0 4YY

29/07/2026

02180627-0003